建筑模式图则

——营造良好社区的工具

[美] 雷·金德罗兹　罗布·罗宾森
URBAN DESIGN ASSOCIATES 主编
焦怡雪 译

中国建筑工业出版社

著作权合同登记图字：01-2005-5899号

图书在版编目（CIP）数据

建筑模式图则——营造良好社区的工具/[美] 金德罗兹，罗宾森主编；
焦怡雪译.—北京：中国建筑工业出版社，2008
ISBN 978-7-112-10262-4

Ⅰ.建… Ⅱ.①金…②罗…③焦… Ⅲ.城市规划－建筑设计 Ⅳ.TU984

中国版本图书馆CIP数据核字（2008）第118152号

责任编辑：董苏华
责任设计：郑秋菊
责任校对：兰曼利 王 爽

建筑模式图则
——营造良好社区的工具
[美] 雷·金德罗兹 罗布·罗宾森 URBAN DESIGN ASSOCIATES 主编
焦怡雪 译

*
中国建筑工业出版社出版、发行（北京西郊百万庄）
各地新华书店、建筑书店经销
北京嘉泰利德公司制版
北京方嘉彩色印刷有限责任公司印刷
*
开本：889×1194毫米 1/20 印张：11⅗ 字数：350千字
2008年11月第一版 2008年11月第一次印刷
定价：72.00元
ISBN 978-7-112-10262-4
(17065)

THE ARCHITECTURAL PATTERN BOOK

URBAN DESIGN ASSOCIATES

AUTHORS

Ray Gindroz and Rob Robinson
principal authors

Donald K. Carter
Barry J. Long, Jr.
Paul Ostergaard

with contributions by
David R. Csont
Donald Kaliszewski
James H. Morgan
Donald G. Zeilman

PREFACE BY
David Lewis

EDITOR AND
CONTRIBUTING WRITER
Karen Levine

回复：城市设计事务所组织编著的《建筑模式图则》

我心情愉快地把《建筑模式图则》的终稿收存起来，这部书刚刚由诺顿 专业建筑师出版社出版。

创作完成了《城市设计手册》的这家事务所，又给我们带来了在今天制定和使用模式图则——一种可以回溯到维特鲁威和帕拉第奥的传统，并且是许多美丽房屋的资料来源——来进行邻里设计的实践指导。《建筑模式图则》回顾了21世纪指南手册的前身，从文艺复兴时期到乔治时期和巴黎风格时期的组合部件图集，以及殖民时期的美国模式图则。本书表明了通过使用制图技巧能够为当前的设计和建设过程给予指导的技术与工作方法，传统的建筑模式图则，能够作为一种在大尺度的开发建设中实施城市设计的手段而得以复兴。同时，本书还展示了能为当代设计师提供范例的样本。

城市设计事务所，是一家以匹兹堡为基地在全美执业的事务所，它创造性地与市民共同投身于复兴城市邻里，改造公共住房，为城市中心区和滨水区带来新的活力，并建设新的传统城镇。

致以最美好的祝愿，

伊丽莎白

目 录

致谢 6

序言 7

前言 10

第一部分 模式图则——过去与现在

第一章 现今模式图则的前身 13
第二章 模式图则的衰退与复兴 35
第三章 模式图则的用途与结构 47
第四章 模式图则对当前实践活动的价值 59
第五章 编制模式图则的过程 73

第二部分 UDA 模式图则范例

第六章 总则册页 83
第七章 社区模式册页 113
第八章 建筑模式册页 147

后记 模式图则的复兴 224

参考文献 226

索引 227

致 谢

城市设计事务所（Urban Design Associates）对本书所引用工程项目的客户致以诚挚的感谢：

BAXTER,FORT MILL,
SOUTH CAROLINA
Baxter Clear Springs, Inc, and
Celebration Associates

CRAWFORD SQUARE
PITTSBURGH, PENNSYLVANIA
McCormack Baron Salazar

DIGGS TOWN
Norfolk Redevelopment and Housing
Authority and the City of Norfolk, Virginia

DUCKER MOUNTAIN
BILTMORE, NORTH CAROLINA
Biltmore Farm, Inc.

EAGLE PARK
BELMONT, NORTH CAROLINA
Graham Development

EAST BEACH
Norfolk Redevelopment and Housing
Authority in association with
East Beach Company, LLC

EAST GARRISON
MONTEREY COUNTY, CALIFORNIA
East Garrison Partners I, LLC
Urban Community Partners
Woodman Development Company, Inc.
William Lyon Homes

GÉNITOY EST
BUSSY SAINT GEORGES, FRANCE
Establissements Publics d'Amenagement
Marne-La-Vallee and
E.R.A.S.M.E.Etudes Urbaines

LIBERTY PATTERN BOOK
LAKE ELSINORE, CALIFORNIA
The Town Group

MASON RUN
Crosswinds Community and
the City of Monroe, Michigan

MIDDLETOWN ARCH
Norfolk Redevelopment and Housing
Authority and the City of Norfolk, Virginia

NORFOLK PATTERN BOOK
City of Norfolk, Virginia

PARK DUVALLE
The City of Louisville,
Housing Authority of Louisville and
The Community Builders

PINEWELL-BY-THE-BAY
Norfolk Redevelopment and Housing
Authority and the City of Norfolk, Virginia

THE LEDGES OF
HUNTSVILLE MOUNTAIN
AMERECO Real Estate Services, Inc.
The Ledges Village Builders
John Blue

WATERCOLOR
WALTON COUNTY, FLORIDA
ARVIDA,a St.Joe Company

序　言

作为城市设计师，我们所关注的是公共领域的设计。我们的目标是营建良好而适宜的街道、公园和广场，在这些地方，个体建筑的特征是与周边的其他建筑、场地平面的处理以及空间尺度相协调的。

城市设计师通常会对某个特定空间绘制透视图，在其中我们希望能体现出对城市居民来说具有重要意义的特性。那些图纸所说明的并不仅仅是限定了空间的建筑物，而且还包括景观美化、地面铺设和其他存在于建筑物之间的要素。遗憾的是，在当前的设计实践中，建设结果常常未能实现这些图纸所设定的目标，因为建成的建筑品质并不符合空间视觉的要求。

在邻里的设计中，许多不同的建造者或开发者可能会参与设计过程，他们中的每一个都对住房建造和交易的最佳途径有不同的观点。最通常的是，他们仅仅重视住房的室内部分，而很少关注会影响到公共空间的室外部分。自从出现了占据部分住宅沿街立面的前置式车库之后，这一问题变得尤为突出。结果，传统邻里街道最重要的特征——临街建筑的多元复合方式——已经遭到了损害。

另外，建筑构件的标准化和工业化已趋向于削弱地方传统的影响。在建筑学专业（及其对现代主义的坚定主张）和建筑工业（了解顾客们需要“传统”建筑）之间所出现的分歧，导致了一种淡化传统建筑形式的变体。

因此，从 1970 年代中期开始，城市设计事务所开始探求设计住宅工业的工作方法，以便找到进行城市设计的有效途径。

找到解决办法的答案并不需要走得太远。在我们的周围，在每一个美国的城镇和城市中，我们都能找到一些传统的邻里街坊，在那里有趣味而富于变化的建筑和众多的小店铺共同形成了良好的街道和公共空间。通常，这些房屋作为工业革命后大城市扩张所引发的建设浪潮的一部分，大多是在 1880 年至 1930 年之间迅速建成的。

比这些邻里街坊丰富的视觉效果更为引人关注的是如此众多的不同建造者完成作品的方式，不同的房屋共同构成了一系列的公共空间，而同时并没有妨碍房屋建造者或拥有者的个人创造力。

在我们看来，这些传统邻里能够在三个方面给当前的实践活动带来启示：

第一，每一个邻里都有其自身独一无二的特性，这将它与同一城市中的其他邻里区别开来——这也是与其他城市和地区的邻里所不同的特性。

第二，每一栋单体住宅都与街道或公共空间具有明确的关系，从而使其表现出对公共领域友好宽容的态度。

第三，这里没有不和谐或比例失调的房屋。尽管它们所展示的是包括混合风格在内的多种多样的建筑风格，这些住宅和小店铺是与其独特风格和街道尺度的特点相符的。

所以我们要问："这是如何形成的？还有为什么今天我们无法做到？"

显然，第一个问题的答案在于这些房屋是良好的地方建筑传统和当时可采用的建造技术制约条件下的共同产物。第二，这些房屋是在关于街道特性的文化共识中形成的。还有第三点，我们认为这要部分地归因于由建筑师完成，而由建造者使用的建筑模式图则在引导良好设计方面所发挥的作用。

我们确信模式图则的传统能够再次被用来在建筑师和建造者之间提供有效的沟通方式，然而，为了实用，模式图则的传统形式需要进行修改以适应当今的需要。这些需要包括：

- 在邻里建设过程中所有参与者之间形成可以共享的图景；
- 明确界定房屋所处公共空间的特征和形式；
- 提供比例均衡、风格适宜的建筑的启蒙读本。

基于这一认识，城市设计事务所的模式图则由三个部分构成：

- 总则部分，描绘了特定开发项目的图景，并包括公共空间和建筑的地方传统的文献资料；
- 社区模式部分，对房屋与公共空间之间的关系处理提供指导；
- 建筑模式部分，针对邻里街坊内房屋的不同建筑风格分别给出适用的建筑要素。

为了使现今模式图则的应用前后连贯，本书的第一部分首先综述了模式图则的发展和使用历程，从公元 1 世纪罗马维特鲁威时代模式图则的最初应用直到现今它们的使用情况。其后，我们对城市设计事务所用以发展能够适应 21 世纪需求的模式图则的方法进行了讨论。本书的第二部分从城市设计事务所的模式图则中选择了具有代表性的册页范例，用以表明这种方法广泛的适用性，无论项目领域或位置如何，都有助于表达和塑造城市社区及其街道景观、房屋的理想形式与特征。本书将模式图则册页范例组织为三章，与城市设计事务所的典型模式图则的三个组成部分协调一致：总则、社区模式和建筑模式。在每一章，你都能够找到从不同的城市设计事务所模式图则中精选出的册页复制品，说明了我们所采用的方法是如何转化为供客户使用的最终模式图则的。

我们希望本书能够有助于推动建筑师、开发商、规划师和建造者在建设工作中的密切合作的进程，从而重新恢复曾经存在于那些参与我们城镇建设和更新的人们之间的共识。

雷·金德罗兹（Ray Gindroz）和罗布·罗宾森（Rob Robinson），主编

城市设计事务所

前 言

30年前，城市设计事务所在宾夕法尼亚州的约克市开展业务。在与普林斯顿的朱尔斯·格雷戈里和历史约克组织的约翰·沙因的合作项目中，我们的任务是为约克市国家登录历史地区进行分类编目。当时，这是美国最大的历史地区之一。

一个鲜为人知的（也许是无关紧要的）事实是，在1777年一个短暂的时期内，约克市曾经在大陆会议迁至费城之前作为13个州的首府。随后较为重要的影响是，一条连接巴尔的摩市和约克市的铁路在1840年代建成了；此后，约克市快速并高密度地发展起来。

实际上，我们很快认识到我们在约克市中心拥有一处国家登录历史地区，它是在大约30年的时间跨度内建成的，全部是“如出一辙”。

最初，我们赞叹于房屋立面的各种细部装饰——砖砌、楣石、凸窗、托架、天窗、门、台阶和铁艺。当我们更仔细地观察时，我们意识到这些街区是一种重复的网格。城市住宅和主要街道商铺的空间序列在街区内是以重复性的地块尺度布置的。并且，重复性的楼层平面和构件中的潜在数学规律共同形成了这些街道的丰富的视觉效果。

在这些城市住宅的内部，我们再次发现了丰富多变的细部装饰——硬木地板、壁炉架、镶板台阶、室内门和刻花玻璃窗。在沿街立面上，每一栋住宅的细部都是各不相同的，但又始终在空间尺度上保持协调一致。因而我们发现，建造者和他们的委托客户都从依照标准模数生产产品的供应商的目录册里选择他们的细部装饰。因此造就了这美丽城镇中一条又一条街道的丰富而统一的景观。

在这一地区旅行游历的时候，我们在雷丁、哈里斯堡、巴尔的摩、费城和其他城市中发现了相似的建筑物和细部装饰，这表明，在伴随工业革命出现的建设热潮中，在一个广阔的市场地域内建造者们使用了模式图则和供应商目录册。但最令我们惊异的是，这些建筑的基本语言和词汇是如何深入大众头脑的。

作为建筑师，我们学习过詹姆斯·马斯顿·费奇、文森特·斯卡利和约翰·萨姆森的作品，并亲身体验过乔治时期伦敦建筑的范例。但我们开始认识到，这些建筑范例也不知不觉地渗透于这些街道建造者的建筑语汇中，并因而形成了城市社区建设的基础。

与我们在约克市的工作相似，在其他城镇和城市中，我们也在社区和邻里中作为城市设计师开展工作。我们常召开公共会议和讨论会，以判断市民们认为对于他们和他们的未来而言什么才是重要的。在这些会议中，我们不断地为古老社区里的非专业建筑师的人们如何处理他们自己的城市街道的基本语言而留下深刻印象。他们谈论的不单纯是房屋建筑，而且也涉及父母、孩子常常如何使用门廊、人行道、街道宽度和他们的邻近设施——会议上，其他的人都能完全理解他们的邻居极力描述的东西对他们也很重要。

在那些日子里，我们很少被委托在环境良好的城市社区中开展工作。我们被邀请进入的城市社区是老旧的内城街坊，它们受到民权运动骚乱、工业经济低迷和郊区化带来的人口外流的影响。毫无例外，这些街坊看起来就像是破旧的、拼凑缝补而成的百纳被。

由于我们的部分工作是不得不对其进行重新修补，我们有时在公共会议中要求人们绘制"记忆地图"。正是在这时，我们才会了解这些残缺的街道在三、四代人之前的完整面貌。我们会看到和听到建筑和城市设计中所反映出的迁入居民如何将古老而遥远的文化，无论是从欧洲还是从美国南方腹地，带入到模式图则的房屋和街道中，而且带入到教堂和商店、习俗和食品中——完全因为这些原生模式的语汇是今天还生活在这些社区里的人们所仍然普遍接受并沉浸其中的。

我们能够理解到邻里的建筑语言既是人们共同的记忆又是共同的渴求，并且正如历史联系着过去，习俗也是联系传统与未来的门槛。作为城市设计者，这种关于传统、城市演进发展、创造和再现城市整体的概念，已成为我们工作的首要关注点。

我们常常注意到，当我们在公共会议上要求人们描述一条主要的商业街或居住性街道时，我们所听到的答案并不是购物中心或占地广阔的郊区住宅。而是由店铺主人居住的、带有帆布篷和二层公寓的小商铺构成的连续商业街，以及两侧排列着住宅和门廊的树荫浓密的邻里街道。

如同我们的《城市设计手册》一样，本书致力于恢复这种建筑语言传统，以使规划师、建筑师和市民能够延续美国的良好街道和宜居城市邻里的传统。

戴维 · 刘易斯 (David Lewis)，事务所创办者

城市设计事务所

第一部分

模式图则——过去与现在

这一部分对始于希腊和罗马时期传统的建筑论文丛集、模式图则和建造者指南的发展历程进行了概述。论述集中于预期用途、信息结构、材料的处理技巧、组织和展现。这些已有的范例，成为涉及当今建筑类型、技巧和方法的新的模式图则发展的范例，其相关内容在本部分中也进行了说明和论述。

第一章

现今模式图则的前身

概述

作为城镇和城市建筑的一种辅助工具，模式图则在现今的复兴，仅仅是建筑师努力与建造者们进行沟通的漫长历史中的最近的一个实例。从历史上来看，每一个历史时期都会产生参考指南、手册或模式图则，从而通过提供既特别符合时代技术方法，又适应过程目标的设计图样来协调建设过程。考虑到那些促成历史上模式图则的广泛认同、使用和成功——以及它们在当今的适用性——的因素，这些图则的三个方面特别引起我们的关注：

1. 这些图则的版式、结构以及带有素材的系列资料所提供的范例。
2. 这些图则分别详细地描述了建造过程的一些方面，但在其他关键因素方面却未加叙述。这为理解那些作者所熟悉和认同的建造实践的相关部分提供了线索。图则所给予极大关注的题目，正是那些在当时新兴的东西，或是认为需要讲授的内容。
3. 在图则表达的观点和当时建造实践之间的关系也是启发性的，就这些观点来说，作者认为正是需要交流的关键概念。

通过研究历史上的模式图则，我们识别出五种独立的类型：

1. *论文丛集*（Treatises）：介绍建筑或建筑设计的理论。其中很多用来描述和说明古典建筑元素，就所有处理建筑风格的方法来说，似乎受目标和数学描述的影响最深。
2. *范例图则*（Precedent Books）：提供历史建筑的平面、立面和细部，旨在作为建设工程的参考范例。

3. *平面图则*（Plan Books）：提供范例建筑的平面、立面和三维表现图，这是作者鼓励读者仿效或直接使用的。
4. *建造指南*（Construction Manuals）：提供关于如何去设计和建造房屋的实用信息。
5. *目录图册*（Catalogs）：描述标准部件并说明如何在大量设计中应用它们。一般来说，目录图册是用于销售建筑材料的，但在某些情况下，一些公司（如西尔斯和罗巴克公司）也成套供应整栋住宅。

本书中讨论的大部分历史先例是在一本书中将一种或几种类型结合在一起的。例如，帕拉第奥的图则以论文丛集作为开始，并提供了包括施工建议的范例和很多建筑平面。

无论是在历史上还是在当代，模式图则的读者都是非常多样化的。模式图则已经证实它对于建筑师、政府机构和其他管理者、政客、开发商、建造者和消费者来说都具有价值，通过随后的讨论，其中的各种原因将得到清晰的解释。

模式图则是从什么时候开始使用的呢？为了回答这个问题，我们必须追溯到古罗马时期。

维特鲁威

维特鲁威，一般被认为是公元1世纪奥古斯都统治时期的一名建筑师，他记录下了其所在时代的罗马建筑实践。在他的《建筑十书》第一卷的序言中，维特鲁威直接面对皇帝介绍了该论文丛集的写作目的。这本书为罗马帝国提供了良好服务，规范了整个罗马世界的军事营地和城镇的规划与建筑形式。

在《建筑十书》中，维特鲁威对如何进行正确建设提出了实用的建议，包括对正确使用建筑柱式——这是指可供选择的立柱及柱头的风格——的规定。《十书》的排列次序是特别关注思考过程的，它提出在设计城市的过程中要重视先后的步骤以及在该过程中建筑的系统化作用。

在对建筑师的教育作为建筑总体的、技能的和最基本的原则进行论述之后，在第一卷书中，维特鲁威阐述了设计城市的过程。他劝告读者要去寻找具有温和气候，并与食物产出地联系密切的“健康的场地”。在对墙体进行讨论之后，维特鲁威将如何确定道路和街巷的模式作为建设健康城市的一种途径进行了说明，要使城市免受强劲冷风的侵袭，并能在炎热季节中保证空气流通。一旦道路和街巷的骨架建立起来之后，他写道，下一步就要确定城市中主要的纪念性建筑和公共建筑的位置。维特鲁威没有提到建筑肌理，但很明显他想要让读者明白建筑肌理是由作为整个罗马帝国规范的标准城市住宅和商业建筑类型构成的。

在第二卷书中讨论了居住场所的起源后，维特鲁威记述了建筑材料和建造技术，这是所有建筑赖以存在的基础。直到涉及神庙建筑的第三卷和第四卷书中，维特鲁威才介绍了三种柱式——多立克柱式、爱奥尼柱式和科林斯柱式——和正确使用它们的规则。这些规则包括每一种柱式的比例体系和其构造细节。正是这种正确建筑元素的设计指南，与随后的论文丛集一样，影响了一代又一代的设计者。

在对柱式进行描述之后，第五卷书继续介绍城市中其他公共建筑的设计原则。在第六卷书中，维特鲁威讨论了住宅设计的问题，强调建筑设计应与地域气候和场地几何平面相适应。随后他说明了房间的设计——首先是中庭，然后是住宅中各种类型的其他房间。

除了在第一卷书中作者在对“适宜”一词的含义进行解释时指出住宅的入口应与室内装饰的格调和尺度保持一致之外，书中没有论及住宅的外部装饰问题。这是合乎情理的，因为住宅通常是处于城镇街道的框架之内的，与纪念性建筑相比，住宅更属于“肌理性的”建筑。建筑的沿街立面通常是商店或铺面，仅通过一个入口与门廊相连。

从第七卷到第十卷的其余四卷书，讨论了建筑材料和建造方法，例如像供水控制这样的技术问题和各种机械装置的设计。

《建筑十书》对建筑问题提供了一个综合的视野。单个的建筑作品，从整个城镇的文脉环境来看，则是其中的组成部分。尽管仍有很多问题没有论及，我们还是能够通过该书认识到整体的建造过程和方式，并从而使单个建筑的细部与城镇，实际上是社会的总体形象相协调匹配。维特鲁威完整地说明了在罗马帝国境内建造新城的清晰计划，并且通过这种方法建立起我们今日所熟知的许多城市的基本框架。我们可以假定，他未涉及的那些东西，都是当时根深蒂固的习俗，而维特鲁威则把关注重点放在可能是被忽略或他认为是没有处理得好的东西上。

阿尔伯蒂和皮恩扎

维特鲁威的《建筑十书》在罗马帝国末期就失传了，直到15世纪早期才又被重新发现。佛罗伦萨的建筑师和理论家莱昂·巴蒂斯塔·阿尔伯蒂以此作为基础完成了他自己题为《建筑十书里的建筑艺术》（*On the Art of Building in Ten Books*）的十卷著作。阿尔伯蒂受过的教育和培训定会令维特鲁威满意的，因为他们都将广泛的文化培训与详尽的技术知识相结合。的确，阿尔伯蒂常常被称为“文艺复兴时期的多面手”。

市政厅

尽管维特鲁威的文献资料状况不尽如人意，甚至其语法也不例外，但阿尔伯蒂仍然把它视作文艺复兴早期的最有影响的论文基础。在他的著作中，阿尔伯蒂极大地拓展了和谐理论，并进一步完善了建筑柱式的规则。此外，该书的结构组织直到今天也是非常有价值的。

主教宫邸

与维特鲁威的做法一样，阿尔伯蒂也是从建筑与其所在场地之间的关系开始进行论述的。接下来，他讨论了每一栋房屋的建筑部件彼此之间及其与建筑整体之间相联系的方式，然后是每一个单元内部的关系，随后他以对经典柱式的论述作为结束。阿尔伯蒂比维特鲁威更清晰地论证了在局部及其所形成的整体之中存在的和谐关系。这也进一步发展成为在建筑的部件之间存在的“和声共鸣”关系，既体现在它们的比例关系方面，也体现在它们的构图组织方面。他相信在音乐中存在的和声体系，是能够直接适用并支配房间以及建筑部件的比例关系的。

主教堂

通过在建筑的沿街立面中应用和谐原理，阿尔伯蒂改变了城市空间的属性。例如，佛罗伦萨的鲁切拉伊宫的立面雅致而尺度近人，而它与首层用沉重石材砌筑的典型佛罗伦萨风格宫殿的纪念性形式存在着显著差异。在这些建筑中，沿街立面使街道成为一个使人产生防范之心、多少有些令人生畏的场所。街道仅仅是两排建筑之间的通道。而借助阿尔伯蒂的和谐立面，把街道变成了城市的会客厅，成为一个文明而舒适的场所。

皮科洛米尼宫

这一点在皮恩扎的皮奥二世广场，我们能够看得格外清晰，这要归

皮奥广场（朝南）

功于阿尔伯蒂的门徒罗塞利诺，但其实际上可能正是老师所提出的概念。四栋新建的建筑限定出广场空间。每一栋建筑的立面都与其自身条件协调一致：大教堂最复杂精美，然后是教皇的宫邸，再次是主教的宫邸，最后是市政厅。这些建筑立面所采用的柱式各不相同，但在每一个立面中比例关系和柱式的使用都是恰到好处的。广场的横向水平表面被用来组织所有的建筑要素——在明显遵循了维特鲁威街道组织原则的城市主要街道之侧，它真正形成了城市的会客厅。这些街道挡住了寒冷的北风而向南面敞开。各种建筑要素可以被看作是一种成套的部件，其中很多能够供其他建造者在城镇各处此后建成的建筑中使用，从而使整座城市的肌理具有和谐感。

阿尔伯蒂的著作极具影响力。从其中，我们能够追寻到文艺复兴早期的其他类似著作和非常丰富的建筑作品实例。对阿尔伯蒂和其他作者来说，一个关键性的问题是如何在 15 世纪当时的建筑技术条件下应用“古代人的知识”。建筑所产生的美感，部分来自对古典准则进行诠释并使其与当地审美与传统的方式相适宜。例如，在皮恩扎，建筑的细部设计明显少于罗马和托斯卡纳地区，而多纳托·伯拉孟特在伦巴第所完成的作品则从当地建筑传统中汲取了建筑语汇。

《塞巴斯蒂亚诺·塞利奥论建筑》
（上图）范例：竞技场
（中图）建筑模式
（下图和右图）建筑的可能形式

塞利奥和图鉴

在一百年之后，塞巴斯蒂亚诺·塞利奥开始出版他的五卷建筑著作。在塞利奥的著作中（参见由沃恩·哈特和彼得·希克斯所翻译的《塞巴斯蒂亚诺·塞利奥论建筑》的英文版本），我们看到了模式图则出现的最初版本形式。这五卷著作以设计和构图的总体原则作为开始。塞利奥将阿尔伯蒂关于和谐和比例的理论从哲学语汇转化为一系列的图解和图表。

第二卷书对透视图（以及三维投影图的几种不同类型）的画法进行了介绍，并以三幅著名的舞台布景作为结束，其中两幅表现了城市街道的三维透视图，而另一幅表现了乡村树林场景。在街道布景中包括各种建筑，大都适合喜剧使用的朴实无华（即对人们日常生活的表现），而适合悲剧使用的则具有纪念性气氛（一般表达更多的戏剧性和更夸张的人性冲突）。建筑与营造城市空间意象结合起来，使我们能够认识到建筑细部是如何对空间塑造作出贡献的。

塞利奥的第三卷和第五卷书通过将建筑细部与完整或局部的建筑立面图相结合的方式，对由“古人”建造的公共建筑形式的范例进行了评述。建筑被视为能够以不同方式进行装配的成套部件。

影响最大的是我们在第四卷书中所看到的，作者通过绘制建筑立面图的方法对柱式进行了详细论述，说明了这些建筑元素能够怎样被用来营造各种不同的建筑类型。这种方法使建造者，无论是否识字，都能够学会如何正确地去建造古典风格的建筑。当然，建造者会根据当地的材料和方法来进行诠释，但建筑形式、比例关系和细部仍然是正确的。而这也正是现代模式图则得以发展的根基所在。

Of Perspectiue

HOuses for Tragedies, must bee made for great personages, for that actions of loue, strange aduentures, and cruell murthers, (as you reade in ancient and moderne Tragedies) happen alwayes in the houses of great Lords, Dukes, Princes, and Kings. Therefore in such cases you must make none but stately houses, as you see it here in this Figure; wherein (for that it is so smal) I could make no Princely Pallaces: but it is sufficient for the workeman to see the manner thereof, whereby he may helpe himselfe as time and place serueth: and (as I sayde in the Comicall) hee must alwayes study to please the eyes of the beholders, and forget not himselfe so much as to set a small building in stead of a great, for the reasons aforesayd. And for that I haue made all my Scenes of laths, couered with linnen, yet sometime it is necessary to make some things rising or bossing out; which are to bee made of wood, like the houses on the left side, whereof the Pillars, although they shorten, stand all vpon one Base, with some stayres, all couered ouer with cloth, the Cornices bearing out, which you must obserue to the middle part: But to giue place to the Galleries, you must set the other shortening Cloth somewhat backwards, and make a cornice aboue it, as you see: and that which I speake of these Buildings, you must vnderstand of all the rest, but in the Buildings which stand far backward the Painting worke, must supplie the place by shadowes without any bearing out: touching the artificiall lights, I haue spoken thereof in the Comicall worke. All that you make aboue the Roofe sticking out, as Chimneyes, Towers, Piramides, Obliskes, and other such like things or Images; you must make them all of thin bords, cut out round, and well coloured: But if you make any flat Buildings, they must stand somewhat farre inward, that you may not see them on the sides. In these Scenes, although some haue painted personages therein like supporters, as in a Gallery, or doore, as a Dog, Cat, or any other beasts: I am not of that opinion, for that standeth to long without stirring or mouing; but if you make such a thing to lie sleeping, that I hold withall. You may also make Images, Histories, or Fables of Marble, or other matter against a wall; but to represent the life, they ought to stirre. In the latter end of this Booke I will shew you how to make them.

塞巴斯蒂亚诺 · 塞利奥为一出悲剧设计的舞台布景

《建筑四书》
（上图）建筑模式：第一卷书中的科林斯柱式的设计；
（中图和下图）建筑可行性：瓦尔马拉纳宫邸

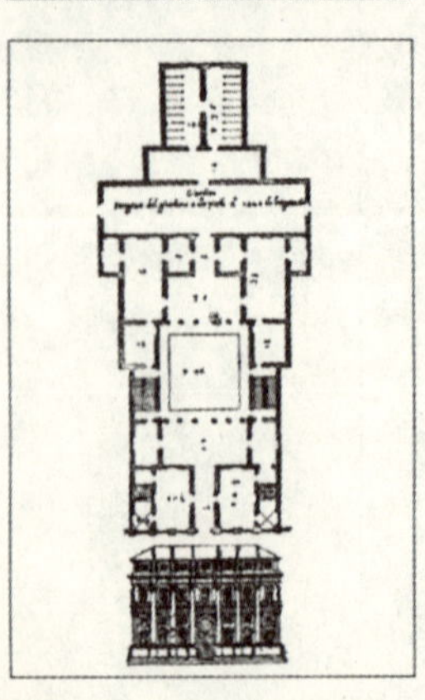

帕拉第奥和帕拉第奥主义

16世纪，另一位享誉世界的意大利建筑师，安德烈亚·帕拉第奥出版了他的《建筑四书》，这可能称得上是建筑史上最具影响力的著作。在其中，帕拉第奥以木版画方式刊印了他毕生设计的郊区住宅和豪华别墅的平面图与立面图。他所选择的是将房屋设计与建筑细部相结合的版式，这很明显是其获得成功并具有持久影响力的关键所在。帕拉第奥的基本版式成为模式图则产生以来就一直采用的版式。

同样，帕拉第奥著作内容的组织方法，帮助我们设计出了现代模式图则的可行结构。帕拉第奥的第一卷著作首先论述了建筑材料和施工技术，包括结构体系，然后介绍了五种柱式（三种古典柱式和罗马人在其基础上发展形成的塔司干柱式与组合柱式）的准确细部和它们的正确应用准则。帕拉第奥以楼梯等“机械”部件的图样作为第一卷书的结束。实际上，这本书提供了正确建造房屋的成套部件。

第二卷和第三卷书包括了帕拉第奥的代表作选集，用以说明第一卷书所介绍的成套部件被用来有效营造各种建筑物的方法。第二卷书主要讨论了住宅，包括城市豪华宅邸和乡村别墅，而第三卷书则主要涉及公共建筑。

在第四卷书中，帕拉第奥介绍了古罗马建筑的范例和详图，他从其中得到启发，并以其作为自己建筑设计的基础。

帕拉第奥介绍了两种主要的新方法。在他的作品中，我们第一次看到古典建筑元素被应用于住宅建筑的尺度中。另外，该书的版式为之后出现的所有模式图则打下了良好的基础。帕拉第奥所创造的版式，脱离了其他早期先例所采用的将平面图、立面图、细部图与建筑理论相结合的方式，这使他的论述更为清晰，设计也更容易模仿采用。由于图样既具有独特性又具有普适性，使人们能以各种不同的方式对其进行诠释，并因而适用于不同建筑传统和气候条件。

帕拉第奥的作品在1570年出版后就广受欢迎并风靡欧洲。到1616年由伊尼戈·琼斯按照帕拉第奥的原理，在英格兰格林威治建造的女王官邸就产生了轰动效应。出于对帕拉第奥的深深仰慕，琼斯在他的有生之年收集了帕拉第奥的全部240幅图纸，并作为他自己从未出版的论文丛集的插图。所有这些甚至先于帕拉第奥《建筑四书》在英格兰的实际出版；这些卷册先在法国出版，最终才复制成第一个英文版本。在随后的150年之中，又出版了几个版本，最具权威性的是由伯灵顿勋爵资助的莱奥尼出版社版本，而勋爵本人就是一位建筑师。

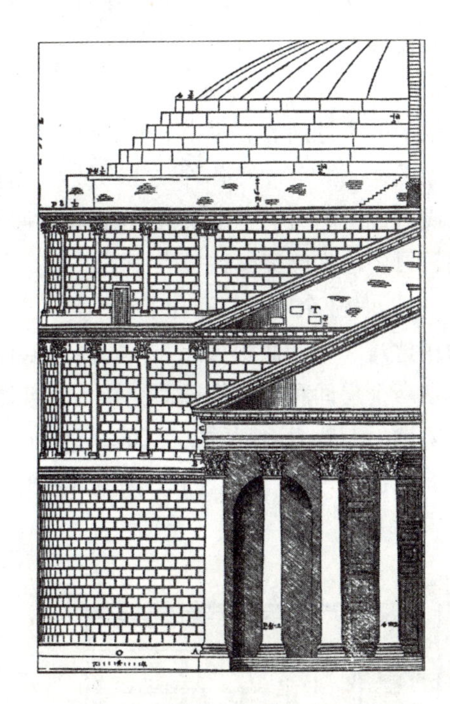

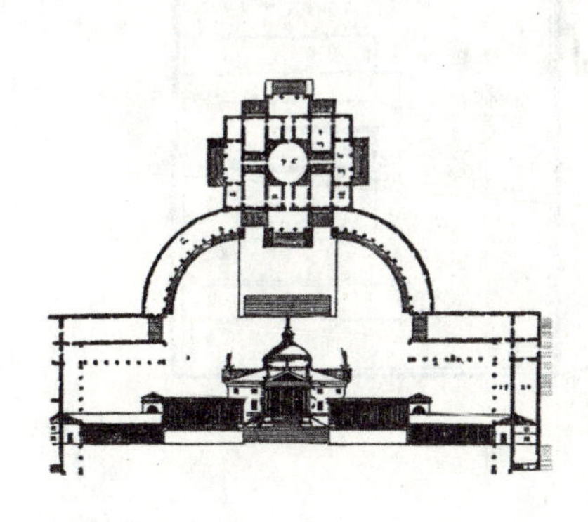

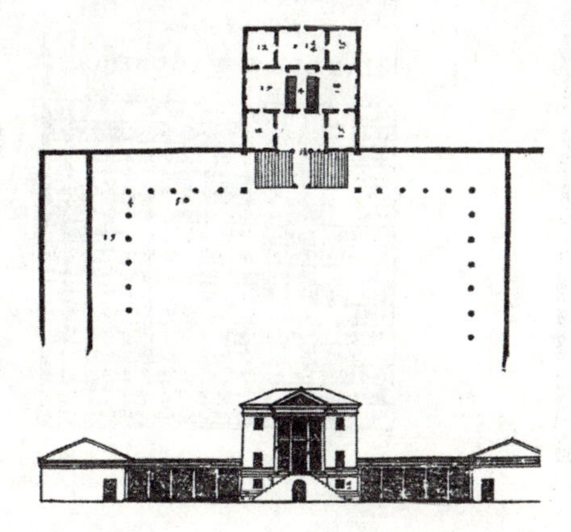

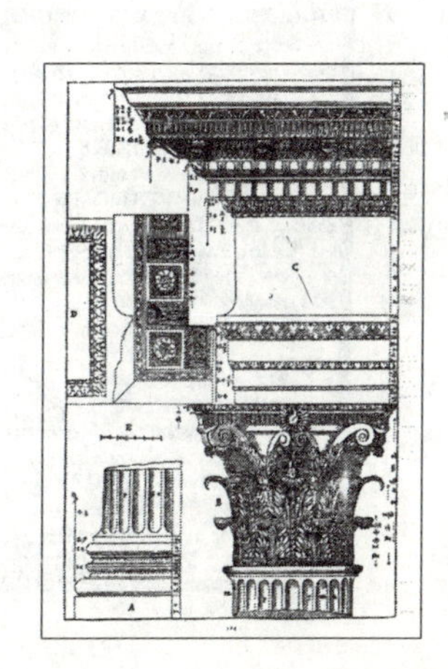

《建筑四书》
（上图）范例：万神庙的立面和剖面研究；
（下左图）特里西诺别墅，梅莱多设计；
（下中图）拉戈纳别墅，吉佐莱设计；
（下右图）范例：朱庇特神庙的建筑细部

早期城市组合部件图集：伦敦乔治时期

其他关于古典设计原理的书籍，例如科伦·坎贝尔（Colen Campbell）的《英国的维特鲁威》（Vitruvius Britannicus），在伦敦和其他英国城市的一连串建设热潮和城市拓展过程中得到了广泛的应用。这些书籍在当时的城市发展过程中——这是一个与今天的发展形成有趣对比的过程——发挥了重要的作用。

在伦敦，大片的土地掌握在少数一些土地所有者手中，他们也充当了主要开发者的角色。他们制定总平面图，一般会涉及各种住宅的尺寸，通常包括一个或更多的街区。以单独或者成批的方式，地块被出售给各种开发者，既包括大公司也包括手艺人，只要他们能攒够钱成立一家公司。这些不同团体的技术水平和设计鉴赏力，也从业余水平到经验丰富不等。

伦敦的城市住宅是简洁的方盒子建筑。尽管面积可能各不相同，但其布置方式基本上都是每层带楼梯包括两个房间和相邻的辅助空间。面向公共空间紧密排列的建筑立面优雅而协调，而背立面则常常因加建部分显得参差不齐。建筑部件被应用于这种标准化、重复性、批量生产的住宅类型中。

威廉·哈夫彭尼所著《实用建筑》的扉页和插图

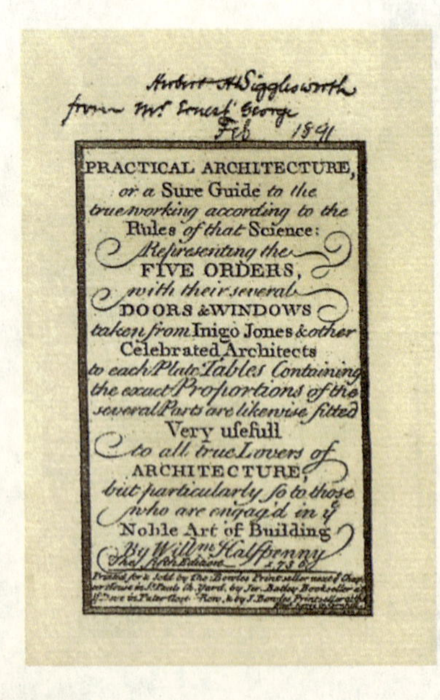

PRACTICAL ARCHITECTURE,
or a Sure Guide to the
true working according to the
Rules of that Science:
Representing the
FIVE ORDERS,
with their several
DOORS & WINDOWS
taken from Inigo Jones & other
Celebrated Architects
to each Plate Tables Containing
the exact Proportions of the
several Parts are likewise fitted
Very usefull
to all true Lovers of
ARCHITECTURE,
but particularly so to those
who are engag'd in y^e
Noble Art of Building
By Willm Halfpenny

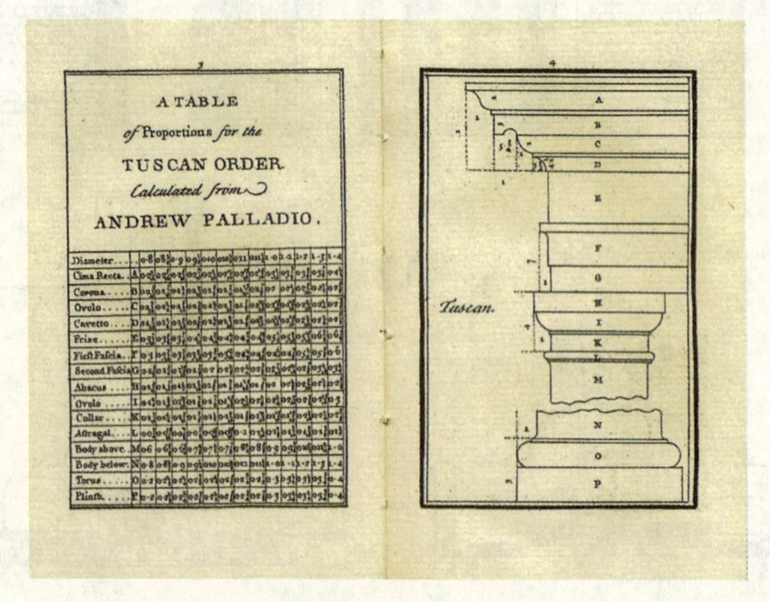

A TABLE
of Proportions for the
TUSCAN ORDER
Calculated from
ANDREW PALLADIO.

Diameter		0·8	0·8½	0·9	0·9½	0·10	0·10½	0·11	0·11½	1·0	1·1	1·2	1·3	1·4
Cima Recta	A	[illegible]	[illegible]	[illegible]	[illegible]	[illegible]	[illegible]	[illegible]	[illegible]	[illegible]	[illegible]	[illegible]	[illegible]	[illegible]
Corona	B	[illegible]	[illegible]	[illegible]	[illegible]	[illegible]	[illegible]	[illegible]	[illegible]	[illegible]	[illegible]	[illegible]	[illegible]	[illegible]
Ovolo	C	[illegible]	[illegible]	[illegible]	[illegible]	[illegible]	[illegible]	[illegible]	[illegible]	[illegible]	[illegible]	[illegible]	[illegible]	[illegible]
Cavetto	D	[illegible]	[illegible]	[illegible]	[illegible]	[illegible]	[illegible]	[illegible]	[illegible]	[illegible]	[illegible]	[illegible]	[illegible]	[illegible]
Frize	E	[illegible]	[illegible]	[illegible]	[illegible]	[illegible]	[illegible]	[illegible]	[illegible]	[illegible]	[illegible]	[illegible]	[illegible]	[illegible]
First Fascia	F	0·3	[illegible]	[illegible]	[illegible]	[illegible]	[illegible]	[illegible]	[illegible]	[illegible]	[illegible]	[illegible]	[illegible]	0·6
Second Fascia	G	[illegible]	[illegible]	[illegible]	[illegible]	[illegible]	[illegible]	[illegible]	[illegible]	[illegible]	[illegible]	[illegible]	[illegible]	[illegible]
Abacus	H	[illegible]	[illegible]	[illegible]	[illegible]	[illegible]	[illegible]	[illegible]	[illegible]	[illegible]	[illegible]	[illegible]	[illegible]	[illegible]
Ovolo	I	[illegible]	[illegible]	[illegible]	[illegible]	[illegible]	[illegible]	[illegible]	[illegible]	[illegible]	[illegible]	[illegible]	[illegible]	[illegible]
Collar	K	[illegible]	[illegible]	[illegible]	[illegible]	[illegible]	[illegible]	[illegible]	[illegible]	[illegible]	[illegible]	[illegible]	[illegible]	[illegible]
Astragal	L	[illegible]	[illegible]	[illegible]	[illegible]	[illegible]	[illegible]	[illegible]	[illegible]	[illegible]	[illegible]	[illegible]	[illegible]	[illegible]
Body above	M	0·6	[illegible]	[illegible]	[illegible]	[illegible]	[illegible]	[illegible]	0·8	[illegible]	[illegible]	[illegible]	[illegible]	1·0
Body below	N	0·8	0·8½	0·9	0·9½	0·10	0·10½	0·11	0·11½	1·0	1·1	1·2	1·3	1·4
Torus	O	[illegible]	[illegible]	[illegible]	[illegible]	[illegible]	[illegible]	[illegible]	[illegible]	0·3	[illegible]	[illegible]	[illegible]	0·4
Plinth	P	[illegible]	[illegible]	[illegible]	[illegible]	[illegible]	[illegible]	[illegible]	[illegible]	0·3	[illegible]	[illegible]	[illegible]	0·4

通过提供各种正确的帕拉第奥风格建筑细部的样本，乔治时期模式图则确保了这一时期的住宅建筑能够具有始终如一的特征品质。在这些模式图则中介绍的建筑部件，包括能够应用于标准方盒子建筑上的檐口、窗套和门套、特殊的窗子（例如凸窗、帕拉第奥式窗和保温窗）和门廊。这些模式图则也是与建筑部件批量生产的发展趋势相吻合的。其中介绍的建筑细部也成为商铺加工窗子和装饰构件的样本。新的工业产品，例如科德斯石材，能为建造者提供价格更为低廉的装饰部件。当然，装饰部件的设计也是与模式图则中所能找到的图样相一致的。

正如我们在伦敦和其他英国城市中看到的那样，这一过程所带来的成效是在大量美丽、比例匀称的街道两侧所分布的优雅住宅建筑。尽管在形式上（在有些情况下是建筑特征）非常相似，围合这些空间的住宅建筑展示出了丰富多样的细部，从而营造出令人难忘的空间。同时，这种多样性得到了良好控制，并常常被用来营造大尺度、宫殿式的建筑形式。例如，伦敦的贝德福德广场拥有相似的住宅，在广场每条边的中部是三角山墙的住宅而两端是转角住宅。其带来的效果是每栋建筑个性的表达均处于可控制的范围之内。从而形成了英国与美国在建筑历程上的极大差异。

其他的模式图则，例如威廉·哈夫彭尼的那些著作，涉及了建筑和结构的实践方面，以及与帕拉第奥风格不同的其他建筑风格。

早期城市组合部件图集：J·N·L· 迪朗与巴黎

在拿破仑时期，让·尼古拉斯·路易·迪朗（Jean Nicolas Louis Durand）是在培训工程师的工艺学校中任教的一名工程师。迪朗与克洛德·尼古拉斯·勒杜（Claude Nicholas Ledoux）和艾蒂安-路易·部雷（Étienne-Louis Boullée）共同工作，并建造了一些建筑。迪朗也出版了著作《古代及现代各类大型建筑物的精选及比较》(Recueil et parallèle des édifices de tous genres anciens et modernes)，对柱式及其应用进行了对比性的分析。当时，拿破仑要求建筑师必须接受系统性、实用性建造方法的培训，从而执行他对城市建设的想法。迪朗被任命负责在工程学校中培训建筑师。

迪朗的讲义被以各种形式出版，包括他的《建筑讲义概要》(Précis of

the Lectures on Architecture）汇总。在这些著作中，他介绍了一种直接、简单的建筑方法。在第一卷的导言中，迪朗说明存在着很多不同的建筑类型和环境，不可能教会人们如何去建造全部这些类型，但是，通过确定建筑要素和将其进行组合的方法，就有可能提供使建筑师能够设计任何风格建筑的技巧。

迪朗的培训体系以精确表现的建筑部件和局部，特别是柱式，作为开始。然后以简洁的线描插图说明建筑类型以及代表了各种建筑要素的符号，例如拱廊、三角山墙等等。然后用示意图的方式来说明这些部件和局部的各种组合方式。借助这套方法，第一部真正的城市组合部件图集诞生了。

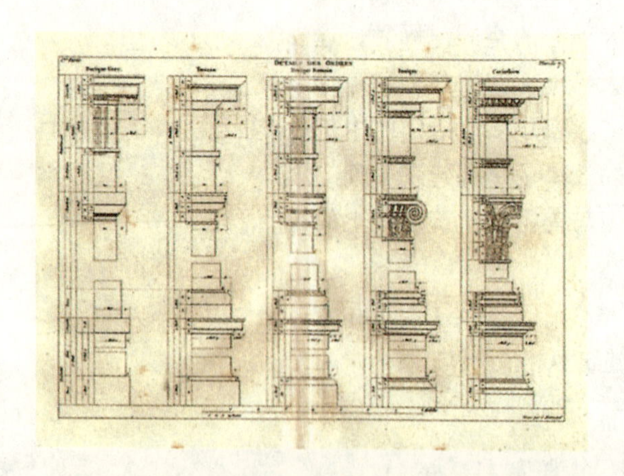
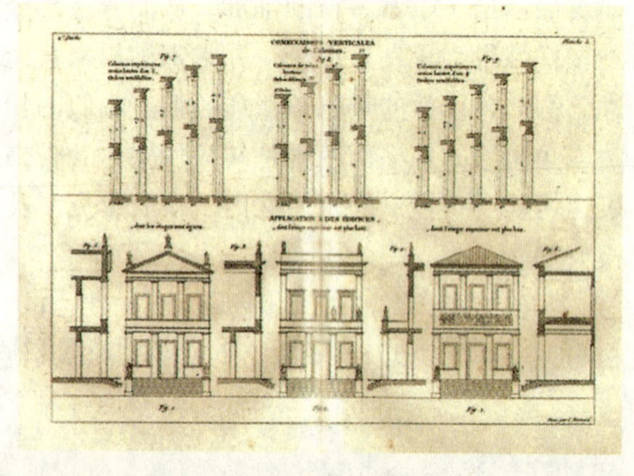
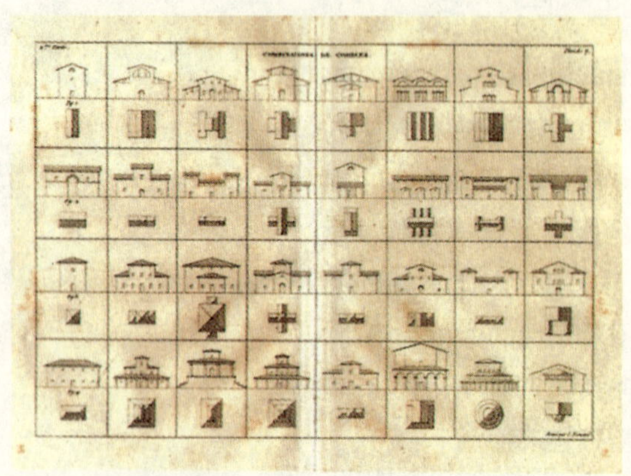
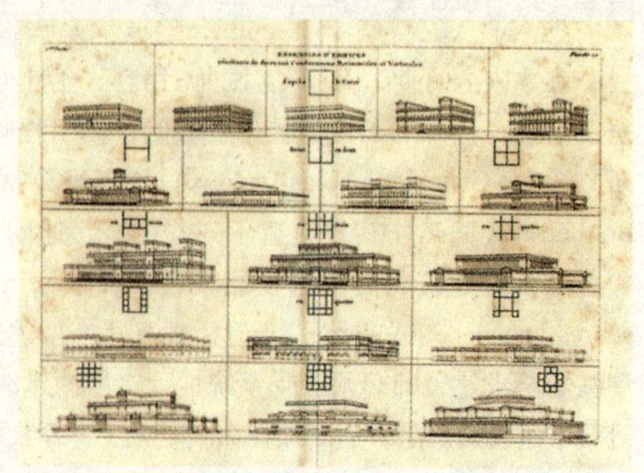

J·N·L· 迪朗所著《建筑讲义概要》的插图

早期的美国模式图则

在殖民地边缘地带和早期联邦制美国中远离城市的地方，并不难找到比例合宜、细部精美的住宅建筑。在如此偏远的地点，它们是如何被建成的呢？答案非常简单。当地的工匠使用了模式图则，这教给他们适宜住宅的基本建筑细部。

美国模式图则包括很多不同的类型。其中一些源自英国的模式图则，包括古典论文丛集和实践性的“建造者指南”或木工手册。经典论文丛集继续得到提炼改善，19 世纪末期威廉 · R · 威尔（William R.Ware）所著的《美国的维尼奥拉——古典建筑建造指南》（The American Vignola：A Guide to the Making of Classical Architecture）达到了其顶峰。此后模式图则开始表现为新的形式：能够参考复制的蓝图指南；可供选用的建筑产品的目录；甚至还包括前机械制造时代住宅的平面图。

鉴于所有这些类型的存在，模式图则的版式与其内容同等重要，因为在当时，模式图则既是一种教育工具，帮助建造者建设在建筑学上来说是“正确”的住宅，又是一种用来推销住宅的市场工具。通过这种方式，在建造技术和开发者的目标之间形成了一致性。对某种特定窗子或门廊细部的详细说明，有赖于对市场中住宅需求的了解。这并不是强加于建造者的规定，倒不如说是一种能够帮助他们实现成功开发的手段。

大多数的模式图则，无论是使用了古典风格的元素还是介绍了一种新的风格，都包含了高水平的建筑设计。训练有素的建筑师贡献出他们的设计，并且因为他们在工作中使用了传统的建筑语汇，因此用以评估设计效果的客观基准得以建立起来。

建成于 17 世纪和 18 世纪初的早期美国住宅，遵循的是移民工匠所掌握的传统技术和方法。通常，施工和设计都非常简单。例如，在这一时期的新英格兰和弗吉尼亚，住宅都始于带有烟囱的一个房间的方盒子形式；然后发展成带有楼梯和阁楼的两个房间的房子；再后来，形成了带有厨房和食品贮藏功能的单坡顶住宅。这种殖民时期的住宅类型，通过建造者 / 工匠所获得的知识而被大量制造出来，一般来说并没有借助任何形式的参考手册。在早期定居点的环境情况和有限资源条件下，住宅和房屋建筑都是用木材建造的——笨重的木材梁柱框架结构连同木制墙板，取代了移民工匠在欧洲旧大陆所习惯使用的砖石结构技术。这种改变产生了一种源于中世纪住宅建筑的独特的美国住宅类型。

随着定居点成熟，出现了更多的工匠，并开始使用新增加的资源。殖民地的建造工艺开始从基本的遮蔽风雨的需要，转向在某种程度上更为精致和完善的商业活动。随着财富和资本的逐渐增长，出现了更为复杂的建筑物。富裕的土地所有者开始倾向于进口已雕刻加工好的室内和室外装饰部件，例如门楣和门道。这些部件被添加到当地建造的住宅和房屋建筑上。一些来自英格兰的商人，也开始参与建造伦敦及其周边地区的古典风格细部装饰的建筑。这些细部装饰最初应用于作为城市公共建筑的砖石建筑之中，例如威廉和玛丽学院的初期建筑，或者波士顿 1701 年建成的国会大厦。在 1700 年代早期，这些细部装饰也开始使用于住宅建筑之中。

遵循了在由帕拉第奥和塞利奥发展起来的较早期建筑论文丛集译本中所记录的构图原则后建成的最早的住宅实例，可以追溯到 18 世纪中期。在这一时期，工匠们开始使用数量越来越多的英格兰出版的早期模式图则（美国有据可查的建筑图则可以在一些重要的文献资料中找到，例如贾尼斯 · G · 席梅尔曼（Janice G.Schimmelman）所著的《早期美国的建筑图则》（Architectural Books in Early America）和海伦 · 帕克（Helen Park）所著的《美国革命时期以前的建筑图则目录》（A List of Architectural Books Available in America Before the Revolution）。由于图则实用性的增强，早期的乡土建筑随之转变为一种美国古典主义风格。有趣的是，这些图则给不同地区带来的影响也各不相同。南加利福尼亚州的德雷顿市政厅（建成于 1738 年）源于帕拉第奥的设计风格，其特征是深深的可以遮蔽炽热骄阳的门廊；而在新英格兰地区，同一时期建成的建筑却没有门廊而代之以更为精致的门套和细部装饰。

这些早期的资料图书可以分为两类。大型论文丛集的卷数较多，价格也较贵；而图则相对较少。除了帕拉第奥的《建筑四书》之外，还包括詹姆斯 · 吉布斯（James Gibb）的《建筑一书》（A Book of Architecture，1728 年）和《若干建筑部件的制图标准》（Rules for Drawing the Several Parts of Architecture, 1732 年）；威廉 · 肯特（William Kent）的《伊尼戈 · 琼斯的设计》（Designs of Inigo Jones，1727 年）；亚伯拉罕 · 斯沃恩（Abraham Swan）的

《英国建筑师：又名，建造者宝典》(The British Architect: Or, The Builder's Treasury,1745 年)；巴蒂·兰利（Batty Langley）的《设计宝典》(Treasury of Designs,1745 年)；罗伯特·莫里斯（Robert Morris）的《建筑精选：城乡皆宜的平、立面设计规则》(Select Architecture:Being Regular Designs Of Plans and Elevations Well Suited to Both Towns and Country,1755 年)；威廉·哈夫彭尼 (William Halfpenny)、约翰·哈夫彭尼 (John Halfpenny) 和罗伯特·莫里斯的《现代建造者辅助参考：又名，建筑完整体系的简明摘要》(The Modern Builder's Assistant: Or a Concise Epitome of the Whole System of Architecture, 1742 年)；威廉·萨尔蒙（William Salmon）的《伦敦的帕拉第奥风格》(Palladio Londinensis,1734 年)。这些图则和其他书籍一起为 18 世纪中后期的美国建造者们提供了参考资料。

很多价格比较便宜的袖珍版本，例如威廉·佩恩（William Pain）所著的《实用住宅木工技术》(The Practical House Carpenter,1796 年）和带有精选插图的小册子，都可以从零售商那里买到。

这些书籍最初是为工匠和市场交易设计的，主要由详细说明的插图构成，并供建造者作为细部设计的参考书使用。这些书籍也提供组合好的立面图，并成为住宅整体形式和构图的一种来源。所有这些书籍都是以罗马风格和帕拉第奥古典主义风格为基础的，并成为伊尼戈·琼斯、罗伯特·亚当和詹姆斯·亚当等英国建筑师和建造者用来进行建筑设计的准确参考资料。杰斐逊因在建筑论著中总结了帕拉第奥、塞利奥、莫里斯、吉布斯、

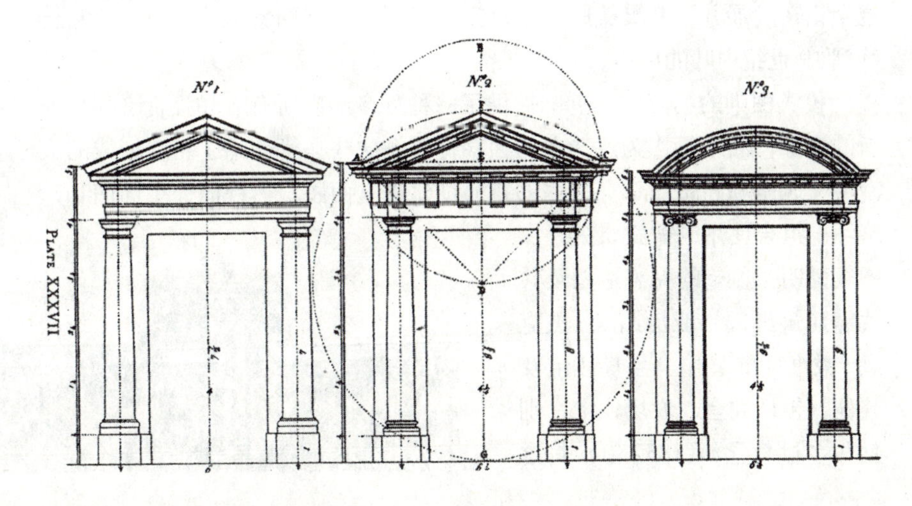

詹姆斯·吉布斯所著《若干建筑部件的制图标准》的插图

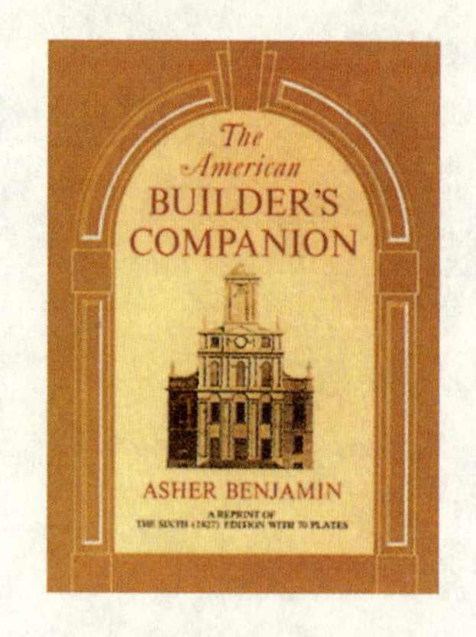

坎贝尔、弗雷亚尔·德·钱布雷等大师的朱庇特神殿、蒙蒂塞洛宅邸和其他工程，而声名鹊起。早期的美国建筑师——例如查尔斯·布尔芬奇、约翰·麦考康伯和本杰明·亨利·拉特罗布——都借鉴了这些同样的范例。

这种“时髦样式”的参考资料已渗透到住宅、教堂和公共建筑的日常应用中，并遍及整个殖民地和独立战争后新成立的各州。建造者指南/木工手册的趋势一直以美国建造者/建筑师所出版书籍的方式稳定地持续到19世纪。其中包括阿舍·本杰明所著的《乡村建造者辅助参考》(1797年)和《美国建造者手册》(1806年)，以及欧文·比德尔所著的《青年木工辅助参考》(1805年)。这些书籍所参考的仍然是修正了罗马模式的英式建筑细部。这些建筑细部通常会被转化为适合在美国环境应用的木构建筑细部设计。它们常常被应用于对标准的本地住宅形式进行装饰和细部设计，而不是作为有机的整体设计。

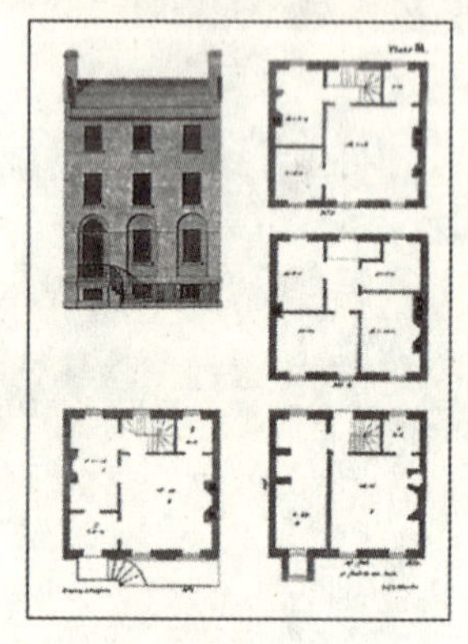

建造者指南的第二波热潮，源自聚焦希腊范例并将其作为建筑细部和住宅形式新的参考资料的书籍。约翰·哈维兰在1818年至1822年之间出版的三卷本《建造者辅助参考》，阿舍·本杰明在1827年出版的《美国建造者手册》和米纳尔·拉费沃尔所著的《现代建造者指南》(1833年)，都是向美国建造者阐释希腊建筑的构成要素、比例和细部设计的重要出版物。这种仿照希腊细部设计和构图比例的潮流迅速流传开来——纽约州、俄亥俄州、宾夕法尼亚州、威斯康星州、佛蒙特州、北卡罗来纳州和弗吉尼亚州，都通过这些书籍在其技术和风格的迅速的传播中有所受益。这一阶段，这个正在发展并不断繁荣的国家真正形成了一种“国家风格”——一种持续到19世纪中叶的风格。

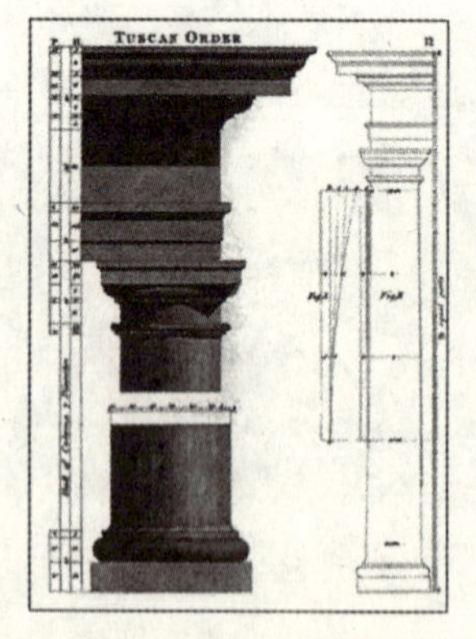

(上图)阿舍·本杰明所著《美国建造者手册》的封面和插图；(下右图)建筑的可能形式：米纳尔·拉费沃尔所著《现代建造者指南》的插图

模式图则的第三波热潮则提供了一种更多地供所有者和投资者使用的新的资源，而不再作为建造者的独享资料。这些书籍介绍了正在形成的浪漫主义风格和建筑设计。这些书籍通过建筑立面图、透视图、平面图和建筑细部组合使用的方式，第一次图解说明了住宅的完整造型及其关键建筑细部。亚历山大·杰克逊·戴维斯的著作《乡村住宅及其他，包括农舍、农场住宅、别墅和乡村教堂》(1837年)是这

类书籍中最早出版的。这一作品之后又出现了很多其他著作，其中也包括了安德鲁·杰克逊·唐宁的一些著作，他在浪漫主义风格的乡村别墅和住宅设计市场中占据了主导地位。作为接受过专门训练的园艺家，唐宁出版了几部著作，特别是其中的第一部，《农舍住宅；或，乡村住宅和乡村别墅，及其花园与庭院的系列设计》（1842 年）是广受欢迎的一部。

这些新的书籍仍然是以英国建筑范例和英国著作为基础的，但已经过改进使其能够适应美国对乡村住宅和乡村别墅日益增长的需求。很多美国建筑师所作的原创设计成为这些书籍的特色所在；戴维斯、唐宁和其他建筑师发扬了他们的原创设计。其来源是哥特风格和意大利风格，并以在 18 世纪和 19 世纪在英格兰发展起来的如画原则为基础，该原则鼓励在景观和建筑之间建立联系，从而使人产生诗情画意般的联想。

这些设计是与住宅设计和建造方面的另一项全国性发展趋势相吻合的。锯木场和轻型框架技术的发展，使如画风格的住宅得到繁荣发展，它成为一种主流的木构建筑类型，具有精致的装饰部件、丰富而轻巧的形式以及复杂的集成类型。这些成品和建造技术作为新的风格，包括安妮女王风格和乡间风格，得以继续发展直至 19 世纪末期。

塞缪尔·斯隆在 1852 年出版的著作《典范建筑师》，涵盖了范围很广的建筑类型，包括住宅、学校、教堂的设计，以及为建造者提供的良好实践要点和技术数据。这大概是到那时为止所出版的最为完整的模式图则了。

除了哥特风格的设计灵感之外，其他来自异国的建筑影响在 19 世纪后

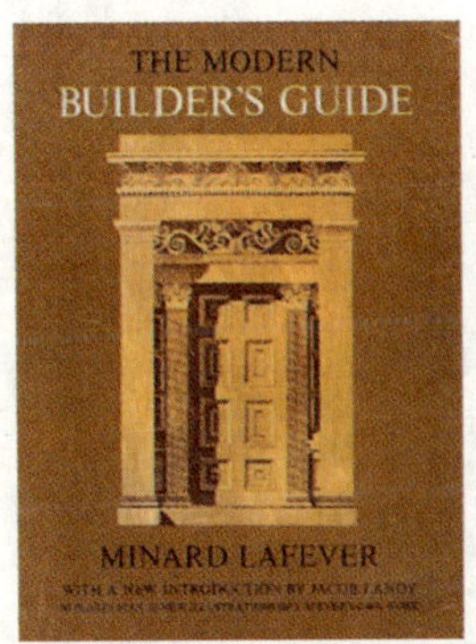

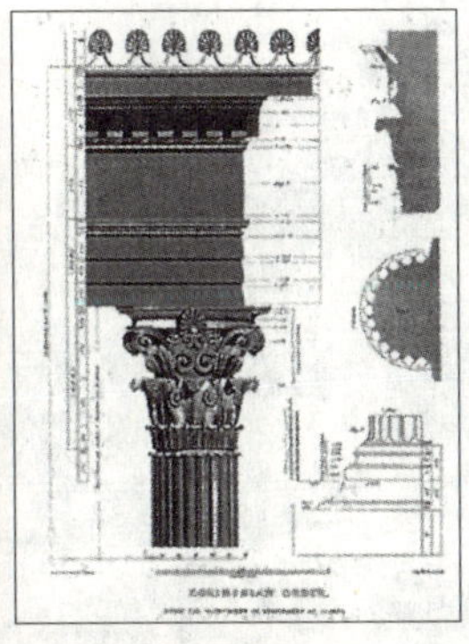

米纳尔·拉费沃尔所著的《现代建造者指南》

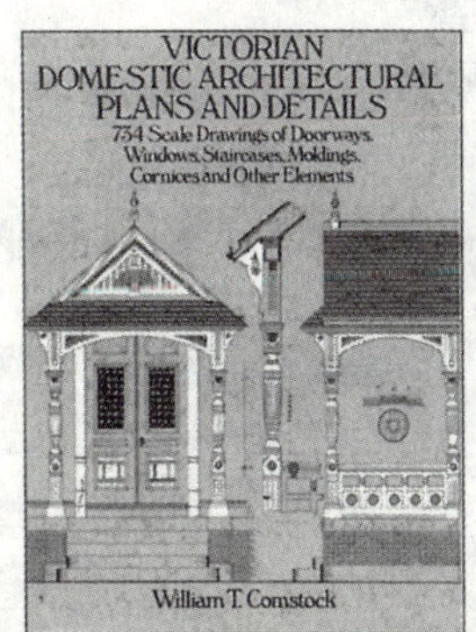

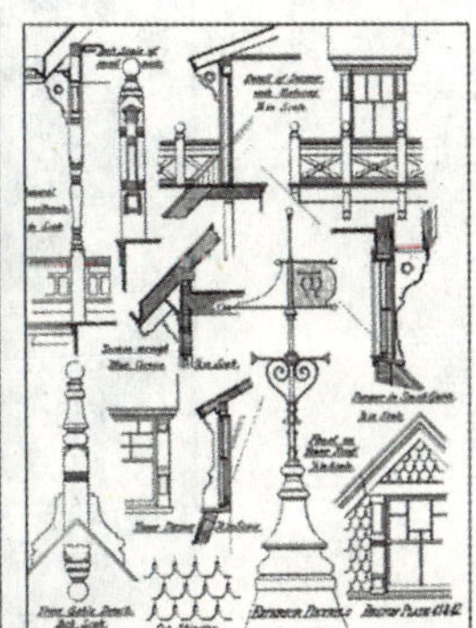

威廉·T·康斯托克所著的《维多利亚式家居建筑平面图和细部设计》

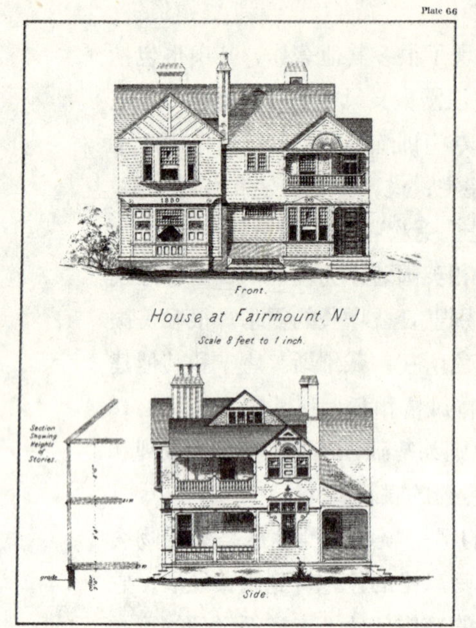

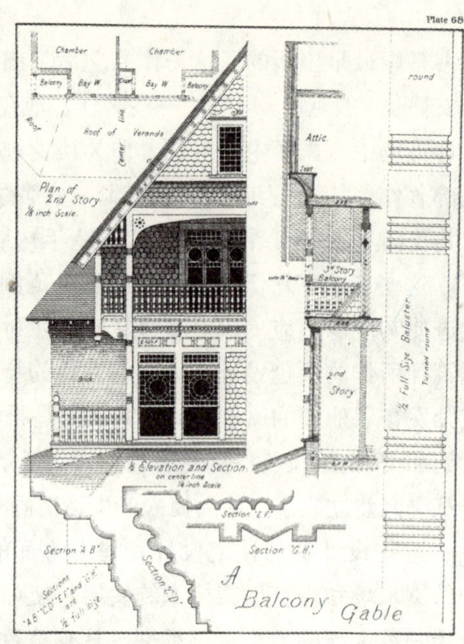

威廉·T·康斯托克所著的《现代建筑设计及细部》插图

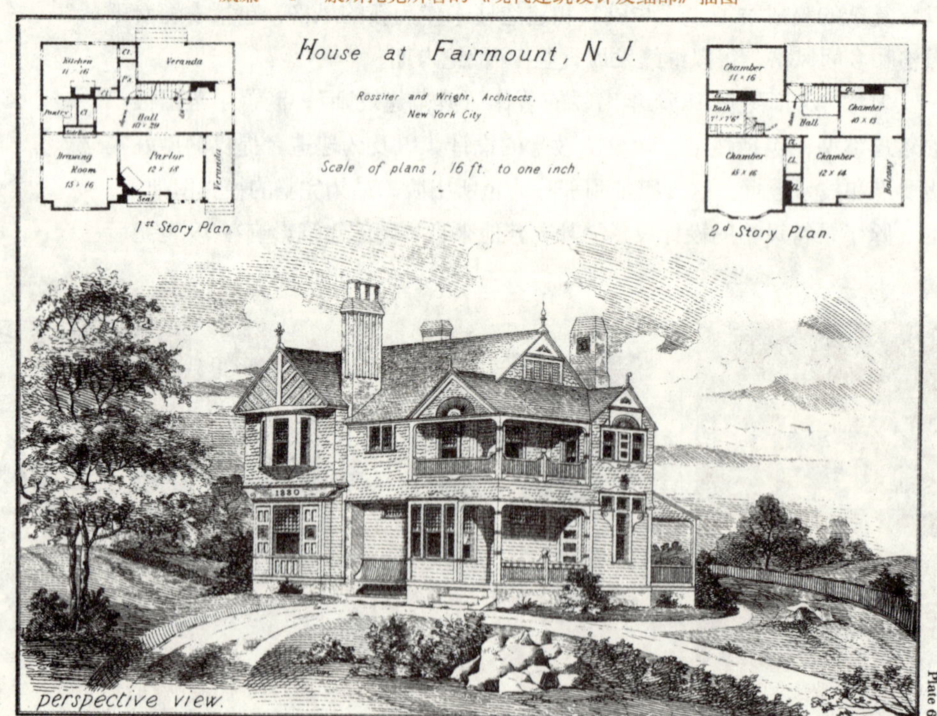

期模式图则的出版热潮中也开始被引进了。例如，卡尔弗特·沃克斯1857年出版的著作《乡村别墅和村舍》中所包括的法兰西风格的复折式屋顶、建筑设计和建筑细部。截至此时，模式图则已成为建筑师用来推动住宅设计服务的一种有效营销工具，卡尔弗特·沃克斯则使用他自己的模式图则来广泛地推销其设计服务。

对先前模式图则的改进出现于1860年代。这一时期的模式图则开始以大比例、清晰绘制的建筑细部为特征，与较早版本中的小插图相比较来说更容易进行测量和复制。阿莫斯·J·比克内尔和M·F·卡明斯等发行人开创性地对模式图则重新进行排版，使其更便于使用和清晰易读。也有很多出版物开始包括了详细说明和合同范本等内容。一直有一些公司在认真推行这种售房计划与设计服务的融合，例如帕利泽、帕利泽联合公司，这是一家以康涅狄格州为基地的建筑制造公司。他们的产品目录包括透视图和平面图，并带有很大部分的住宅部件和产品的制造商的广告宣传。

（左侧插图）西尔斯、罗巴克及其公司出版的《荣誉建成的现代住宅》；
（下图）R·W·肖佩尔所著的《跨世纪住宅、乡村别墅和郊区别墅》

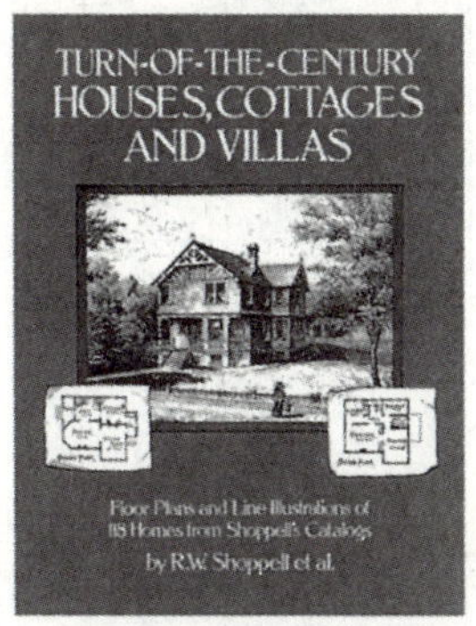

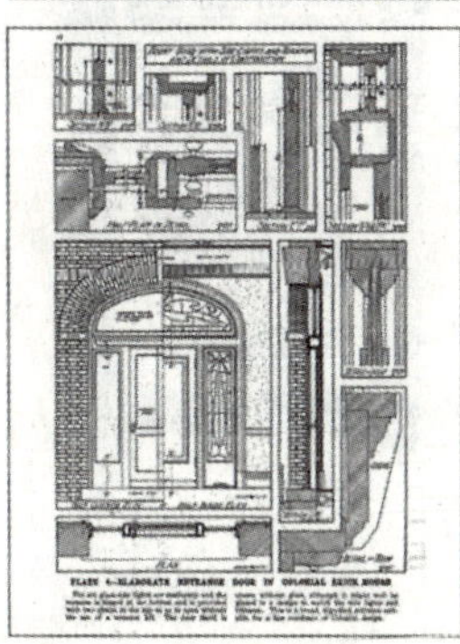

在19世纪晚期，“平面图则”成为模式图则发展的下一个趋势。例如，R·W·肖佩尔的合作建筑协会，在1881年开始出版住宅平面目录，并最终开始以“季刊”的形式出版住宅平面图则，而且一致持续到1900年代。尽管很多实际上并没有进行订购的当地建造者也模仿了其中的很多平面，这些书仍然非常畅销。在威斯康星州拥有木材生意的威廉·拉德福德，在1900年代早期创造了得到广泛应用的一系列平面图则。该系列包括范围宽广的风格变体和集成类型，它们追随着不断变化的特定风格而不断发展。其中住宅的类型包括从村舍和平房直到更大、更精致的住宅。

芝加哥住宅拆除公司，是通过住宅平面图则的销售而将建筑材料*直接地*卖给购买者的第一家公司。随后，这种形式被西尔斯和罗巴克、阿拉丁以及标准住宅等公司所改进，并成为遍及美国的重要商业活动。到1920年代，传统的模式图则已经被能够进行订购并运送到业主场地的住宅产品目录，或出售整套图纸和详细说明书的平面图则所取代。

惠好公司和白松系列

为了促进木材的使用，惠好公司在 20 世纪 20 年代资助了很多平面图则。“白松”（White Pine）系列是一种定期刊物，其中的一些对美国的历史性建筑进行了纪实评述，并达到了其前代对古罗马建筑进行评述的同等详细程度。这激起了对美国历史遗产的新的尊重，它是一种在始终将自身视为新生的文化环境中被低估了的遗产。

其他的“白松”特刊出版了设计竞赛的成果。很多在乡村的优秀建筑师参加了在预先确定的资金数额范围内设计最好的住宅或最好的度假住房的竞赛。最佳的几个方案会在定期刊物中登出。图纸所遵循的版式与历史性住宅所采用过的版式相似：通过场地总平面图来说明建筑与其文脉环境的关系，随后是平面图和一些典型的建筑细部。而封面则常常采用三维的表现图，例如透视图或照片。平面图则被认为是一种市场销售的工具和教材，并如前面所介绍过得那样，包括了价格相对适中的详细建造文件的订购表。

在 1923 年，在竞赛中获得优胜的住宅在华盛顿特区的林荫大道建成，成为体现美国文化的住宅设计的重要范例。这些图则和项目的影响是引人注目的。它们激发并规范了低造价、大众市场的住宅建造设计。这些图则优美而简明的图示设计，是它们取得成功所不可或缺的部分，并清楚地表现了本卷所说明的当前模式图则的版式。

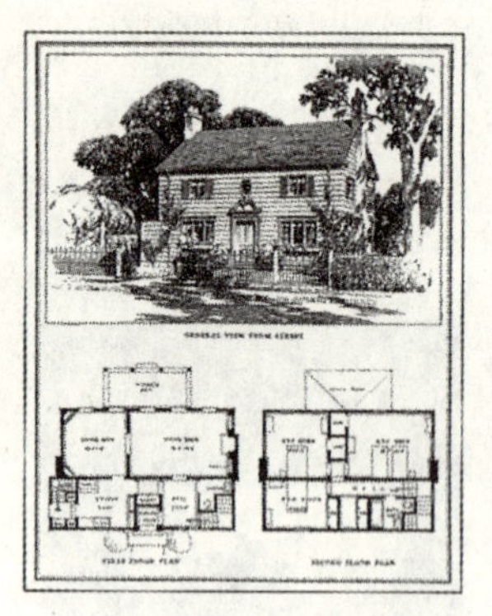

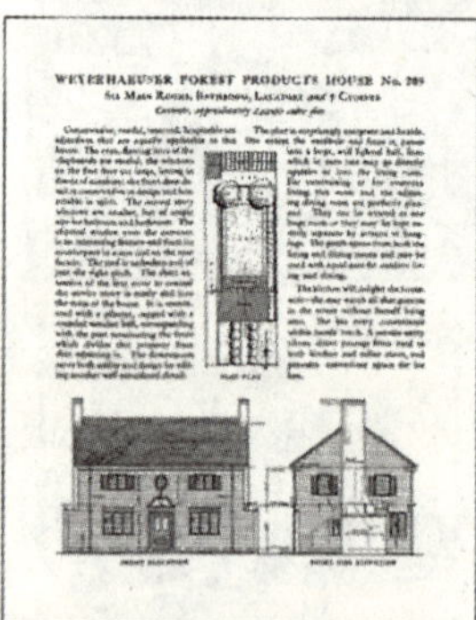

（顶图、中图和右侧插图）惠好公司的《十二个现代小住宅》；
（底图）1922 年 2 月号《建筑专题的白松系列》的封面

模式图则和美国的城市主义

尽管英国和法国的模式图则被用来营造统一和谐的城市整体效果，而美国精神则一直是鼓励个性发展。在美国城市化和欧洲城市化之间存在的显著差异，不仅影响了城市的建成效果，也对模式图则产生了影响。

在殖民地时期情况并非如此，但在美国独立战争之后就逐渐变得明显起来，对个性的表达——每栋住宅的不同建筑形式——变成了重要的目的和主题。因此，随着图则的发展，它们带来了更为显著的多样性结果，在很多邻里社区中，再没有两所住宅建造得相同的了。住宅也许具有同样的楼层平面和同样的基本方盒子结构，但它们以不同的建筑风格进行装饰，采用不同的独特元素，例如门廊和各种各样的建筑细部与色彩。

因此，自相矛盾的是，通过在美国提供标准化的模式，模式图则却成为建造格外多样化的城市建筑的手段。在我们邻里社区中限定了街道空间的建筑和多样化的建筑细部设计之间存在着和谐一致的联系，这形成了在我们所知世界中最为美丽的一些城市空间。

一条传统的邻里街道

（本页图和下页图）20 世纪晚期的住宅和邻里街道：传统丧失

第二章

模式图则的衰退与复兴

衰退

到 20 世纪末期，在美国的住宅开发实践已造成了世界上已知的最差的一些城市空间。这到底是如何发生的呢？我们又是如何从营造最佳邻里跌落到建造最差空间的呢？

已有很多论述涉及在第二次世界大战后造成不断蔓延的郊区环境的力量，在那里细分地块的概念已取代了邻里的概念。在战前和战后开发实践之间存在的种种差异，导致了邻里开发方式的恶化。

第一，出现的是建筑行业的标准化，以应对人们对新建住宅急切而贪婪的欲望。

第二，是将开发过程分割成一系列的专业，而在他们之间缺乏有效的沟通，这打破了以前在开发过程中的各种参与者之间存在的联系。在逐渐变得“高效”的过程中，街道的设计、公共设施的配置、土地的整合以及

建筑的设计，都被视为生产线程序中相互分离的步骤，而不是营造整体场所的完整过程的组成部分。

第三，曾经存在的关于邻里和邻里街道形式的意见共识也开始瓦解了。增加了的街道宽度和断头路模式的扩展，取代了由既作为社会场所又作为交通通道的街道所组成的相互连通的路网，形成了密度过低以至降低了社交品质的孤立飞地。在这一过程中的下一阶段，就是像门廊这样的能够促进社会交往的住宅要素的丧失，而取代它们的是能停放 2 或 3 辆小汽车的车库，这使街道变成格外宽阔、景观优美的道路。

第四，标准化的区划政策和法规的强制性要求，并不是在场所物质环境发展构想的基础上，而是在预防不良情况发生的基础上制定的，这导致了在美国消耗大量土地的低密度、小块分割的环境的形成。

第五，建筑学专业和建造行业的疏离，导致了模式图则和房屋建设的设计质量的降低。拒绝传统建筑语汇的现代主义运动，试图改变公众的喜好——“教育”和说服美国公众去购买现代派的住宅并拒绝传统建筑。没有人赞成这一观点。对大多数家庭来说，一套住宅是他们最大的单笔投资，也在他们的净资产中代表了最大的一项。由于这一原因，大多数住宅购买者所寻求的是那些让他们感到可靠和稳定的东西，是那些作为不断发展的传统的组成部分的东西，而不是那些能够很容易变得时髦的东西。因而，在 20 世纪 50 年代、60 年代和 70 年代的时期中，越来越少有具有才华的建筑师参加到住宅建造产业中。建筑学院也不再为大量生产的住宅设计提供培训。在第二次世界大战之前，最有才华的建筑师都为模式图则提供设计，包括弗兰克·劳埃德·赖特这样的革新者，也包括传统的建筑师。

这样，我们在现今所面对的问题，是由以下方面共同形成的：构思贫乏的城市形式、不对等的立法和实施程序，以及在设计专业和建造行业之间缺乏沟通。

新城市主义和模式图则的复兴

起始于1980年代的新城市主义运动，开始借助对传统的城市主义的复兴来处理这些问题。佛罗里达州的锡赛德被作为典范城镇，来说明有如此广泛的建筑风格能够被用来创造一座传统城镇。由于真正的社区所具有的魅力，在马里兰州的肯特兰和其他城市应用传统的建筑语汇建设了已获得成功的邻里和城镇。这些不断增加的应用实践，已在全美国各地完成了很多成功的开发项目；反过来，这又对建筑工业产生了非常积极的影响。很多新城市主义的实践者已复兴了制定标准化住宅平面的经验，将其作为恢复一贯高质量的成品住宅设计的途径。

不同的公司采用不同的方式来复兴城市主义和设计邻里，在这些邻里中每栋住宅都有利于提高公共空间品质，而不是孤立分割。本书介绍了城市设计事务所（以下简称为UDA）所使用的方法，这在当前的实践活动中促成了模式图则的复兴。UDA的方法经过了很长时间的发展，比新城市主义运动的开始还要早15年。

事务所最为关注的始终是对有助于形成社会资本的公共领域和环境的设计。因此，我们对住宅的关注是由外而内的。住宅建筑如何能够塑造出宜人的街道和广场空间呢？

我们全部的工作，特别是在已建成的城市和城镇中的，是在广泛的公众参与程序背景下完成的，因此努力去满足社区居民的利益。无论是在物质环境还是在社会环境方面，他们都毫无例外地对现今的开发活动感到不满意，而对传统邻里充满热情。我们早期项目中很多是邻里复兴计划，在其中现代住宅和开发建设在提出后遭到了拒绝。在这一时期，我们目睹了在现在已很普遍的社区实践的产生，在这些实践中居民被组织起来保护他们的邻里免遭现代主义建筑和开发行为的破坏。在下面的几节里记述了UDA将模式图则作为解决这些问题的方式的发展过程。

1970 年代的保存运动和历史性填充建设

1970 年代，就在保存运动蓬勃发展的时候，UDA 参与了很多社区建设。最初集中于历史建筑的保存运动，开始涉及作为历史建筑的总体集合而具有价值的整个地区。UDA 接受委托为宾夕法尼亚州的约克市准备历史地段国家登录导则，那是当时美国最大的该类地区。在这项工作的过程中，我们用照片记录了每一栋住宅，并提出改建和改善的建议，以使其与最初的建筑保持一致。为了完成这一任务，我们研究了在这个邻里中所采用的传统建筑语汇。然后我们开始分析很多居住街区的街道立面，这一分析揭示了建筑的异乎寻常的复杂性，在联排住宅的简单体系中，通过添加凸窗和门廊，或者改变屋顶轮廓线，或者使用不同的窗子，或者通过色彩变化，都能够产生多样性。这次的经历完全改变了我们的习惯。UDA 认识到住宅生产有可能被视为建筑部件的组合体。建筑的平面和基本体量都是相同的，但一系列可互换的建筑要素则可能形成各种各样不同的建筑，并随之产生具有不同特征的邻里街道。

在 1970 年代后期，我们开始在匹兹堡的谢迪赛德邻里开展一个填充式开发项目。开发商很希望建筑是传统形式的，并在一开始就建议我们使用最近在附近建成的乔治风格联排住宅开发项目的建筑语汇。我们提出一个建议，是采用相邻地区住宅的建筑风格，在那里带有更多的维多利亚风格。这些住宅结合使用了各种山墙处理方式、屋顶天窗、凸窗和门廊，从而既具有一致性又具有多样性。我们开始用基本的城市住宅平面做试验，包括两种形式——三层的和三层半的住宅。在这些基本的方盒子上，我们开始应用不同的建筑元素。我们把这种使用部件组合工具包的方法称为装配住宅。

历史性联排住宅的立面分析，宾夕法尼亚州约克市

1978年,UDA开始进行弗吉尼亚州里士满市伦道夫邻里的再开发规划。再开发规划是针对这个传统邻里的拆除改造部分的。在以前的一项规划中提出过一系列郊区风格的“方案”,不同方案是针对不同收入群体的。在公众参与的程序中,社区成员表达了他们对这种方法的关注和对重建一个传统邻里社区的期望。这导致了一个被认为是“激进的”,但实际上是回归传统城市主义的规划方案。该方案由两侧排列着带有前后院落的住宅的街道构成。街道被设计成为相互连接的网络,其中包括三个公园和很多社区服务设施。尽管邻里的不同部分所获得的资金不同、居民的收入不同,但是在它们之间并不存在实质差别——至少没有城市形式方面的差别。

所有的住宅都将按照最低预算进行建设,因此其建筑形式就必然是简洁且标准化的。同时,另一个基本要求是要能够表现在至今仍繁荣发展的邻里部分中所存在(并将继续存在)的属性感和持久性。这些正面采用砖墙的住宅,有些是独户的,有些是联排的,都是在1920年代按照模式图则建造的。尽管住宅平面是标准化的,但其立面处理的方式却具有丰富的多样性。主要特征都是红色砖墙和白色装饰,但通过凸窗、屋顶细部、门廊细部和窗子的位置,增加了住宅的多样性。

方案推荐持续了这种传统,并包括一系列草图来说明将标准的住宅平面处理为不同街道立面的方法。在这个方案中,这些意在成为开发商用于住宅设计的设计导则没有奏效。设计质量与社区发展的目标不相符,因此UDA被邀请设计每一栋住宅。我们拒绝了这一要求,因为我们感到这太像同一个方案。而城市景观应该是很多声音共同作用的结果,而不是一个。作为替代,我们提出制定一部模式图则,它将为住宅的沿街立面提供正确的建筑细部设计。为了更为有效,必须提供立面、门廊和外部细节的建设文件层面的信息,就这样,UDA模式图则诞生了。

模式图则住宅立面,弗吉尼亚州里士满市

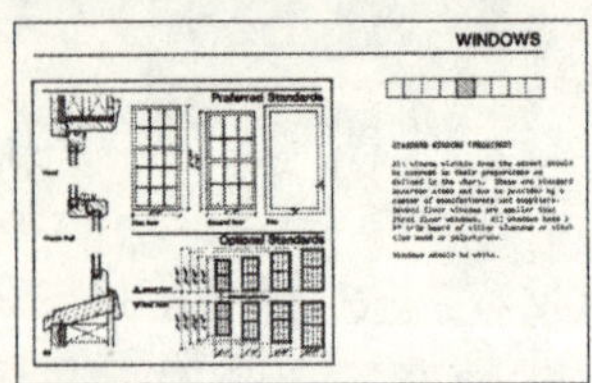

模式图则和建成效果，弗吉尼亚州诺福克市的米德尔敦阿尔赫

诺福克的新建邻里：1980 年代

1986 年，UDA 为弗吉尼亚州诺福克市的一个邻里，它现在被称为米德尔敦阿尔赫，制定了总体设计。它位于一处失败的公共住房项目的场地上，名声很差。城市原来计划将这块用地作为工业区进行再开发，但相邻的邻里对此提出反对，并说服城市领导将再其开发为私有产权的邻里。很多人对于这一处于这样位置的计划是否能够行得通持怀疑态度。为了取得成功，UDA 提出的设计必须要为这一地区塑造全新的形象。

我们采用了在诺福克最受欢迎的一些邻里中所找到的半圆形总平面形式。这一方案将由很多小的住宅建造商依照我们被要求提供的设计导则来共同建设完成。房地产营销的专业人士坚定地认为，在这样一个初创的场地位置，建筑应该是传统且舒适的。在该地区，在建造商和住宅购买者中最流行的风格是威廉斯堡式风格。UDA 对被归类为这种风格的常规新建住宅所进行的分析，清晰地表明它们与这种风格的真正代表还存在着很大的差异。为了纠正这种状况，UDA 制定了一份供住宅建造商使用的、简单的模式图则。在提出模式图则的过程中，我们发现我们只能指定相对较少的条目来要求他们一定遵守。因此我们问自己，“如果我们可以要求正确地做 5 件事情，那么应该是哪些呢？”对这个问题的答案引导了我们所设计模式图则的内容——详细内容如下：

1. 体块，包括屋顶坡度和檐口细部；
2. 立面构图，包括窗子和门的位置；
3. 窗子与门的类型和比例；
4. 沿街立面所使用的材料；
5. 在场地上的位置、前院和前门的位置。

建造商接受了这份模式图则，因为它增强了他们的市场目标（即，“威廉斯堡式”风格），保证了其他建造商都将保持同样的品质，因而使他们能够在住宅内部进行自由发挥。

一年以后，我们被邀请与乔纳森·巴奈特和经济学家菲利普·海默进行合作，共同完成诺福克市欧申维尤社区的总体设计。再一次地，我们的工作对象是一个名声较差的邻里，尽管它位于切萨皮克海湾并具有滨水特色。规划包括一项被称为滨湾派恩维尔的开发项目。这座城市没有在这片滨水场地上开发高密度的共同产权的公寓，而是选择开发3个街区的小型、独户住宅地块，以希望吸引人们来建造豪华的大型别墅。这是被称为提升欧申维尤市场形象的再开发策略的一部分。

规划确定50英尺×75英尺的地块，并对其进行布置以最大程度地提供眺望海湾的视野。并要求建筑延续在这一地区存在的传统海滨住宅的某些形式，例如安排在二楼的起居空间和围绕式的游廊。狭长的场地导致住宅类型与带有长长的、多层侧廊的查尔斯顿独立住宅相似。尽管一些场地被出售给生产商，大多数场地还是销售给个人住宅购买者的。因此，人们希望这些住宅遵守能够形成和谐邻里特征的设计标准。UDA被要求制定一份模式图则来提供体块、窗子、门廊、材料和色彩方面的模式。在这一项目中，模式图则的理念得到了更为充分的发展，而较少强调建筑细部。

（上左图）滨湾派恩维尔，弗吉尼亚州诺福克市；（下图）模式图则页面实例

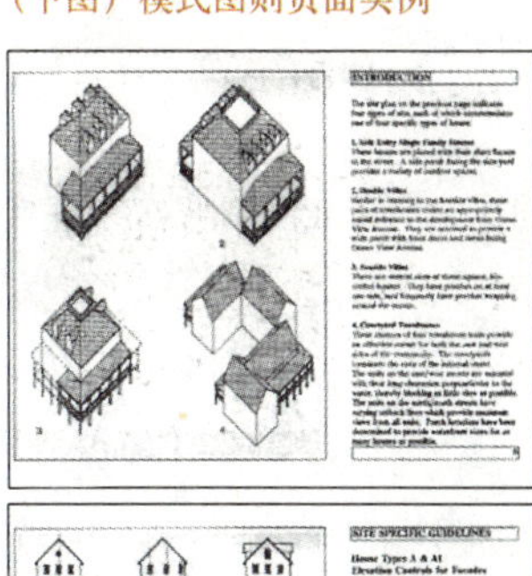

INTRODUCTION

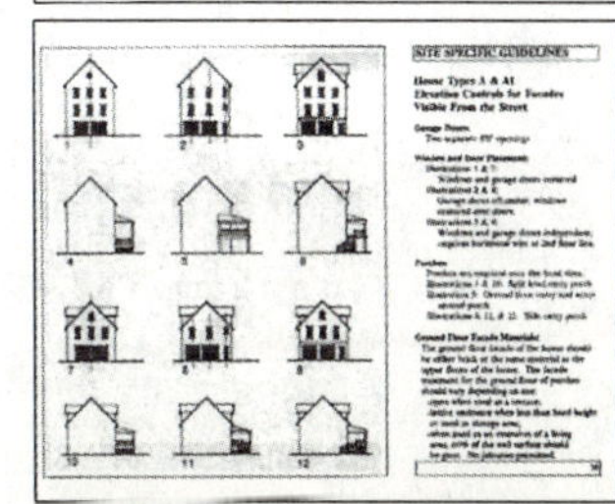

SITE SPECIFIC GUIDELINES

Elevation Controls for Facades Visible From the Street

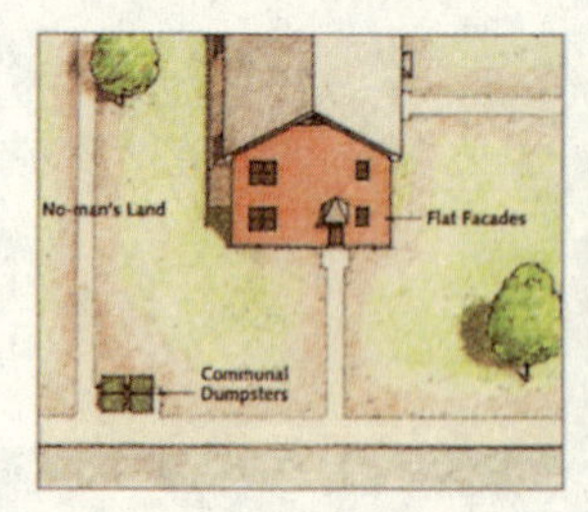

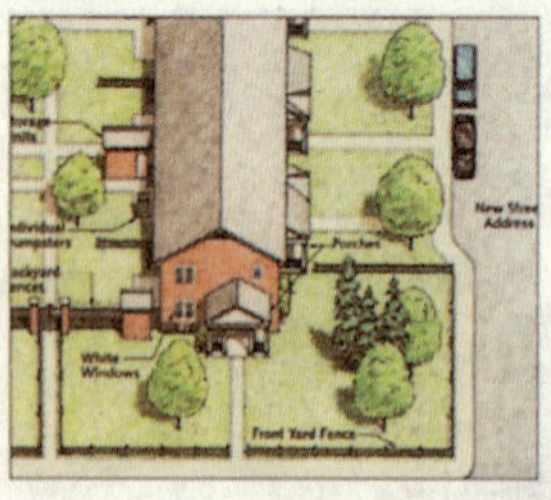

（上左图）已存在的没有社区模式的场地；
（上右图）按照社区模式改建过的场地；
（下图）弗吉尼亚州诺福克市的迪格斯镇，外部改造之前的景观和外部改造之后的景观

邻里公共场地的社会影响

在1990年代早期，有两个项目促进了UDA模式图则的发展——弗吉尼亚州诺福克市的迪格斯镇和宾夕法尼亚州匹兹堡市的克劳福德广场。

迪格斯镇是一个公共住宅“社区”。到1990年代早期，它因其高犯罪率而臭名昭著。大部分居民生活在对发生在公共场地里的犯罪团伙打斗的恐惧之中。孩子和家养植物都要关在房子里，因为户外空间是非常危险的。戴维·赖斯，诺福克再开发和住房局的执行理事，要求UDA提出一些有助于弥补这种令人绝望的社会环境的物质空间改善方案。他问与简单地更换新窗户相比，一般的现代化投入是否能够给人们的生活质量带来更为实质性的改变。很明显，如果没有这种改变，那么改善工作就仅仅相当于重新布置泰坦尼克甲板上的椅子。

设计过程以鼓励居民参与作为开始，这样我们就能够尝试去了解他们所面对的问题，以及造成该项目机能不良的物质环境形式所在。我们得知居民们感到无法控制在其住宅单元周围的开放空间，这些空间是被犯罪团伙所掌控的。住宅“单元”既没有前院也没有后院，也没有门廊或栅栏。很多单元面朝“绿地”（通常是棕色的）而不是街道。

居民梦想获得带有门廊、有栅栏的前院、有栅栏的后院和能够让他们

克劳福德广场，宾夕法尼亚州匹兹堡市

在住宅前停车的街道。借助这些传统邻里的元素，居民们认为他们就能够从歹徒手中夺回邻里并使其成为他们自己的邻里。设计方案提供了这些元素，并在1993年至1995年之间进行建设，到1994年中期居民们就获得了对邻里的控制。犯罪率急遽下降，学校状况得到改善，就业也增加了。

这种变化完全改善了在住宅、街道和后院之间的地带。通过在公共空间建立传统的邻里形式，提供了一种使居民在其中获得成功的环境。

这一经验增强了UDA的决心，提高对邻里建设最重要（也是最被忽视）的公共领域的认识。

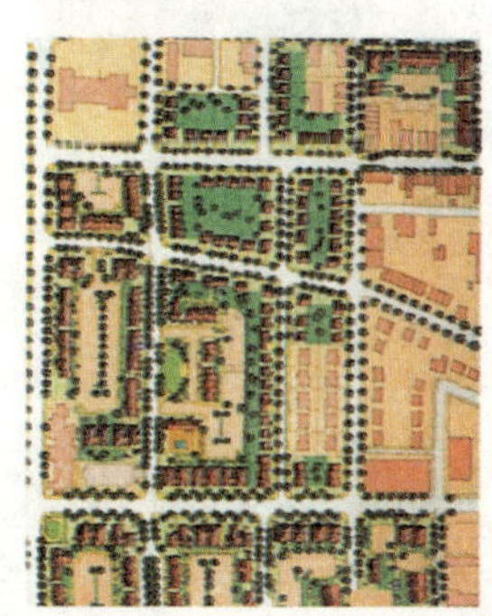

克劳福德广场是紧邻匹兹堡市中心的一个新建邻里。它是1990年至1998年期间在一处失败的城市更新项目的20英亩空地上建成的。位于匹兹堡希尔地区的边缘，克劳福德广场所在区域多年来被认为是不安全的地带。当开发商，麦克马克·巴伦事务所的理查德·巴伦提议建设一处混合收入的邻里时，遭到了地产所有者的置疑，他们认为中等收入或高收入者在

这一地区进行投资或租赁是难以想像的。该设计方案是在要求建设具有林荫道、前门廊和后院的传统匹兹堡邻里的公共程序中产生的。模式图则确定了建筑类型、体量、门窗类型和门廊与凸窗等特殊元素。它被用来设计出租房屋和出售住宅。其成果是住宅的多样化形式，尽管它们是新建的，但与那些评价很高的匹兹堡邻里很相似。

通过这种意象和公共空间品质所形成的稳定感，促使克劳福德广场在吸引各种收入群体的租住者和购买者方面获得了成功。从这一经验中，UDA认识到模式图则在混合收入邻里，特别是在被认为存在风险的地段的开发中所具有的作用。

新城

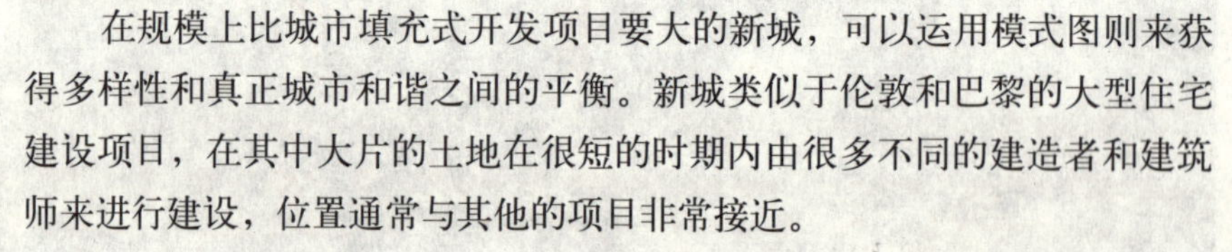

巴克斯特，南卡罗来纳州福特米尔市

在规模上比城市填充式开发项目要大的新城，可以运用模式图则来获得多样性和真正城市和谐之间的平衡。新城类似于伦敦和巴黎的大型住宅建设项目，在其中大片的土地在很短的时期内由很多不同的建造者和建筑师来进行建设，位置通常与其他的项目非常接近。

在UDA的经验中，通过一开始用图像描绘社区的预期意象——反过来这又是新建社区获得市场成功的关键，模式图则能够促进所有这些不同参与者的共同协作。为了实现预期意象，每一栋住宅和建筑的单体都必须为公共空间的总体品质作出贡献。确定的模式是直接与预期意象相关的，而且不仅可以被视为对市场销售过程的支持，更是其获得成功所不可或缺的。

这适用于较小的新建邻里和城镇，例如密歇根州门罗市的梅森兰和亚拉巴马州的莱奇斯，并同样适用于较大的城镇，例如佛罗里达州的塞里布瑞恩和沃特卡勒，以及南卡罗来纳州的巴克斯特。它也被证明能够同样

（上图和下页上图）梅森兰，密歇根州门罗市

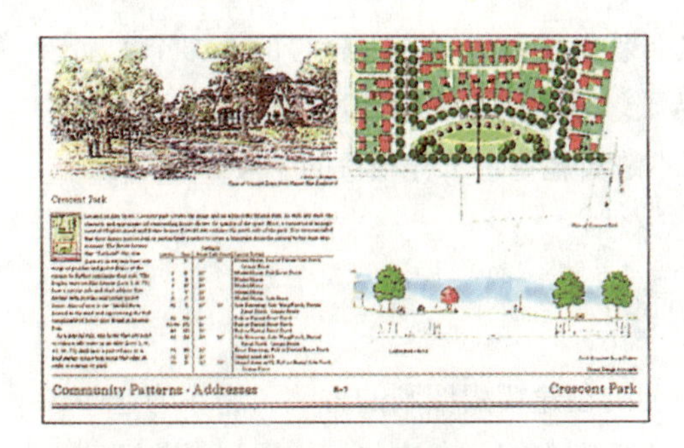

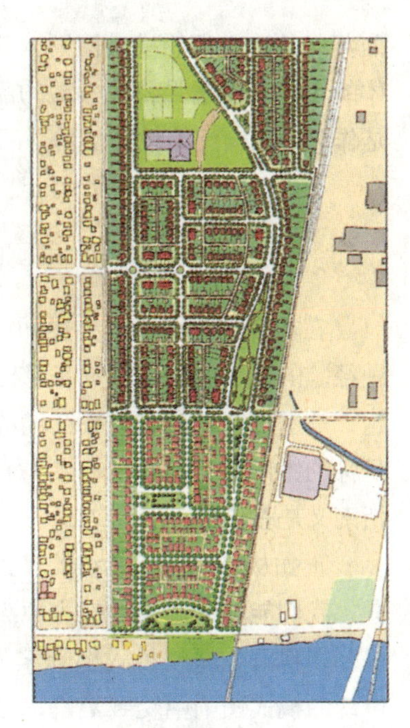

的应用于大规模城市邻里的重建中，例如肯塔基州路易斯维尔市的帕克杜瓦拉。

UDA 模式图则的现有版式和内容，在为几个新城制定模式图则的过程中得到了重要的发展。由于正确的市场销售意象对于一座新城的成功而言是至关重要的，我们所创建模式图则的第一部分——总则部分——在我们的方法中变得日益重要。在总则部分，照片、图纸和文字描述了开发项目的概念和预期品质。如果开发商和建造者响应并支持这一部分，他们将更可能遵从模式图则的建议。

城市和区域

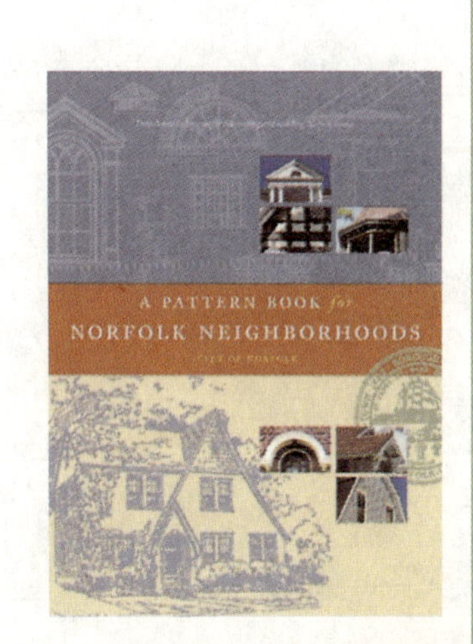

模式图则也能够应用于城市和区域。例如，诺福克市希望提升在已有历史性邻里中的新建项目的品质，从而强化而不是削弱这些地区的已有特征。但由于新建邻里缺乏建筑特征，城市将发展目标确定为鼓励能够营造更为丰富、更具“诺福克风格”特征的建设项目。UDA 也最先准备了一系

列英格兰约克郡的模式图则，其目标是协调参与开发建设的众多个人和团体的工作，以增强公共领域的品质。城市范围和区域性的模式图则的目的是作为应用广泛的教育工具，而不是实施特定工程项目的手段。

为城市化服务的模式图则

在本章后面部分所描述和说明的创建模式图则的方法，是作为这些各种经验的成果并随着时间推移而不断完善的。在这些模式图则中共同的主题，是需要在向参与建设活动的不同群体传达开发的目标，并就实现这些目标的手段和方法在所有参与者之间建立共识。因此，模式图则既是操作指南又是教科书。

正如从第一章所总结的模式图则的发展历史中所使人清晰认识到的那样，我们能够从过去的努力中汲取灵感。近年来，在美国和其他国家的一些公司也开始采用模式图则作为一种工具。例如在英国，罗伯特·亚当从传统的模式图则中受到启发，已为康沃尔郡的王室直辖领地（庞德贝里也包括在其中）的各种开发活动制定了模式图则。

UDA 模式图则的版式包括了这些历史前例的要素。例如：

- 每一部模式图则都始于对场地位置和街道形式的关注，正如维特鲁威和塞利奥所做的那样。
- 社区模式和建筑模式的范例都是建立模式图则的重要参考。古典论文从集借助大量的论述发展了这一方法。
- UDA 模式图则将建筑细部设计与二维或三维的图像联系起来，正如维特鲁威的大部分著作和 20 世纪模式图则，特别是“白松”系列所做的那样。

UDA 模式图则不同于它们的前身，因为我们的关注点聚集在城市空间并将营造有凝聚力的公共领域作为其首要目标。前期阶段致力于在公共领域品质方面形成较大范围的共识，这可能是因为需要花费较长时间来建设城市和城镇。因此，创建模式图则的过程和其版式，都能够成为克服我们目前缺乏意见共识问题的一种途径。通过在一起共同工作，正如中世纪建造者所做的那样，所有群体都能够在对公共领域的共同预期中进行工作。

第三章

模式图则的用途与结构

现代模式图则的核心内容是城市空间的特征和属性，这是通过对三方面内容的认真关注而确定的：开发的总体规划、规划中典型城市空间的发展意象，以及具有自身建筑细部的单个建筑。我们如此经常地看到在规划中的可能已很优美的空间，但却难以在视觉上产生愉悦，其原因是比例失调的建筑形式，或缺少足够的窗户，或者是阴暗沉闷的建筑材料和色彩！在所有的三个方面，模式图则都提供了标准和模式来确保各要素在一定程度上保持协调关系。

以上三个方面所需要应对的城市化建设的挑战来自：

- 使不同领域的专家能够参与城镇和城市的建设，并在对未来发展的物质环境形态一致认识的条件下共同开展工作，而这种物质环境形态是来源于项目所在特定区域的特性的。
- 建立有形而直观的、每栋单体建筑都有助于形成的公共空间和社区模式形态。
- 提供关于适合于项目所在地的独特特征和传统的建筑风格模式与要素的入门读本。

现代模式图则所介绍和说明的内容是为完成这一任务而准备的。很多模式图则是为特定开发项目制定的，其委托任务来自私营部门的主要开发商或行使职权的公共机构。无论客户来自公共部门还是私营部门，模式图则都为其提供服务，来协调由很多不同的独立建造者在较长时期内进行建设的大型而复杂的开发项目。

在这里所介绍的模式图则，是在前面章节中所描述的发展历程的产物，也是应对在后面章节中所讨论的当前开发活动所面临挑战的结果。这些图则的版式和内容是通过与每一开发项目的参与者合作制定出来的，这将在第五章进行介绍。

不同于它们的前身，这些模式图则关注于公共领域，而不是某种特定的建筑语汇或风格的构成要素（尽管很多模式图则确实包括了各种风格的建筑模式)。与设计导则形成对比的是，现代模式图则在场地建筑布局和使用特定建筑要素方面提供了系统化的方法。

设计导则常常被开发企业视为严格的、常规之外的处罚，因为它们强制要求建造者遵循他们认为是专断的规则。然而，由于具有前面所讨论过的建设者参与构思的传统，这些模式图则的设计旨在帮助建造者完成开发的市场目标，获得支持审批程序的共同意见并加快开发进程。

通常，这些模式图则具有“服务工具”的功能，换言之，它们作为所涉及群体之间达成的契约性安排的一部分来指导建筑和公共空间的建设。由于模式图则的制定获得了所参与群体的认同和共同契约形式的支持，通常对不同群体之间土地转让的行为形成了约束。

通过这种方式，模式图则不同于能够被纳入公共立法并从而依法提供开发控制的区划条例或法规，甚至不同于新城市主义章程。模式图则所依赖的，是其在提供开发的总体图景和建设这一图景所需的详尽要求方面所具有的说服力和吸引力。在某些案例中，由于正确运用了模式图则所确定的模式，常常能够在更短的程序中获得市政当局的认可，模式图则已帮助开发商更为顺利地完成审批和授权程序。

城市集成工具包的概念

可以将模式图则视为一种建筑部件的工具包，它使建筑师和建造者在继续保持特定地区固有的传统邻里设计的独特特征的同时，获得创造各种类型住宅的灵活性。

通过经典的儿童玩具“土豆头先生”的比喻，能够很好地说明这种方法。这种玩具为眼睛、鼻子、嘴巴和耳朵提供了几种不同的选择，这让儿童能够通过在可利用的面部器官中进行选择来形成各种不同的人物形象。基本的形状是已知的——一只土豆——同时不同部件的大致位置也是已知的：成对的眼睛和耳朵具有正确的位置，鼻子、嘴巴、胡子和领结也一样。经过认真设计的模式图则的建筑模式部分所提供的一系列内容，所做的正是同样的事情。集成册页建立了基本的形状（土豆）。门窗的样式选择，就像眼睛和耳朵一样，在立面上占有正确的位置。门廊和其他的特定部件，就像鼻子和胡子，在正面占据显著的位置。而主要的大门（嘴巴）则在它们下面找到自己的位置。

这种建筑形式的简化方法，正如在后面的章节中我们将看到的那样，是为满足建造工业的标准化生产技术的需求所设计的，同时提供了建筑的一致性和形式。

然而，建筑模式仅仅是大范围城市化途径中的一个要素。邻里和城市都是由各种要素集合而成的复杂系统。这些要素包括街道格局、公共的开放空间、街区、地块和建筑物。位于肯塔基州路易斯维尔市的帕克杜瓦拉邻里的分解透视图（右图所示）说明了这些各种各样的要素。于是，同样的技术也被应用于在模式图则中的社区模式，作为确定公共空间特征及建筑物在其中所发挥作用的一种方法。而建筑模式则将这一方法拓展到建筑的细部设计。

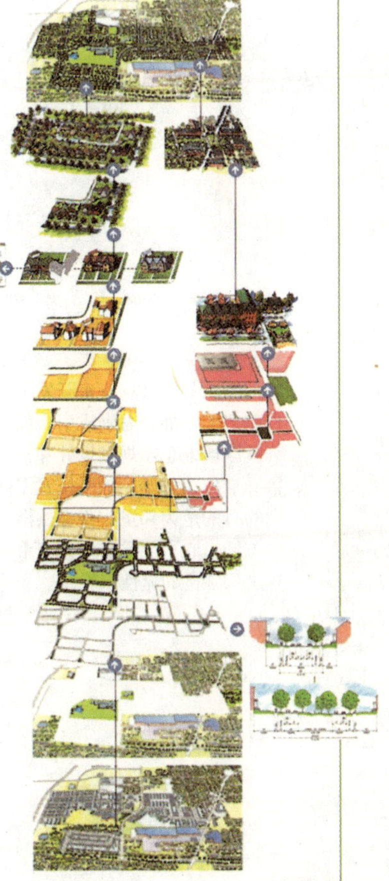

尽管这种方法提供了创造城市生活的体系，它是被设计用来在每个场所营建符合当地传统、文化和喜好的独特环境。这是应对在开发过程中所存在风险的一种途径，同时使开发项目具有当地特征。由于不同区域对建筑风格的当地解释是各不相同的，这种方法为所有涉及者提供了认识城市空间的一种综合手段。

同样的，尽管大多数现代模式图则都具有共同的普适属性和模式，它仍然是一种使建筑要素与每一处地方环境相适应的特殊方法，我们认为这是复兴和延续独特而丰富的美国城市环境的最重要的方法。

结构

在通常情况下，现代模式图则包括三个部分：

Ⅰ. 总则

Ⅱ. 社区模式

Ⅲ. 建筑模式

在本书的第二部分提供了每一部分的范例。在某些案例中，也包括了其他附加的部分——例如，在景观设计、附件或术语词汇表方面。

Ⅰ. 总则

总则部分向所有参与者介绍了开发项目的基本特点。这是一种在项目过程中所有参与者之间建立共识的方法，也作为模式图则所要求的特定条件的现实基础。

为了完成这些任务，总则部分包括描绘了计划开发项目的预期特征和意象的视觉图像。这一部分的详细内容在各个模式图则中是各不相同的，但总体而言都包括下列内容：

对开发项目总体的描述。如果模式图则是为某一特定开发项目制定的，那么总则或引言部分就以场地总体规划设计和销售标准作为开始。如果模式图则的目的是为邻里改善提供指导方针，它则侧重那些社区希望进行强化的属性。

The Townscape of East Beach

The Neighborhood Parks

Towns and villages along the East Coast have both formal parks and court-house squares as well as wonderfully landscaped local streets with a surprising variety in character. East Beach neighborhoods are designed around a series of organic parks along the streets that take advantage of existing landscape and mature trees. These relaxed neighborhood parks are complemented by more formal civic spaces such as the square on Pleasant Avenue and form a network of open spaces that link to both shorelines.

The Pretty Lake Marina

Many coastal towns have a main street that leads to the water as in Edenton, North Carolina or Annapolis, Maryland. East Beach plans to have a marina precinct alive with a mix of residential and commercial uses in mixed-use buildings that add contrast and character to the inland residential neighborhoods. This precinct will have a continuous series of public spaces and thoroughfares that provide access to the waterfront for residents, slip owners and visitors.

The Bay Front

The neighborhood streets connect Pretty Lake to the Chesapeake Bay. Residents are never more than two blocks from the water living in East Beach. The Bay front will have a continuous public beach with access points at the ends of streets or through public greens and paths that open up to the spectacular views.

Shore Drive

The principal gateway into East Beach is along Shore Drive. The waterfront village meets Shore Drive across a continuous park. This park address connects the Pretty Lake mixed-use district overlooking the marinas with the stately housing designed in the tradition of an 'Admiral's Row' and a neighborhood shopping precinct at the entrance with Pleasant Avenue.

The Townscape of East Beach

INTRODUCTION 4

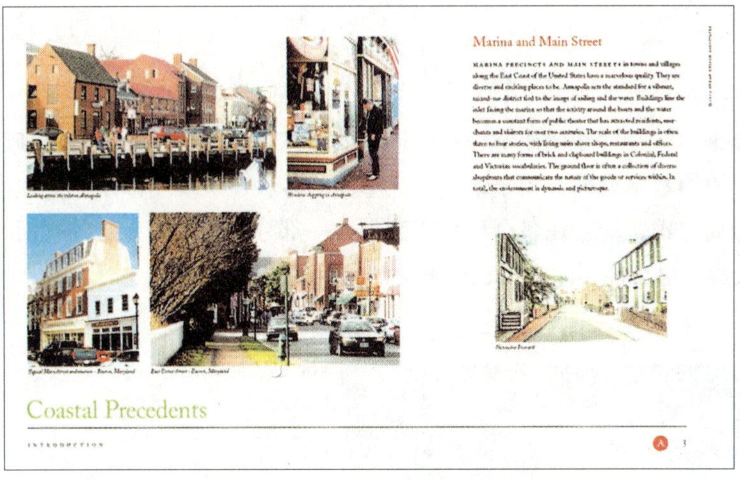
Marina and Main Street

Coastal Precedents

INTRODUCTION

对文脉环境和范例的描述。为了启发和鼓舞模式图则的使用者，这些描述包括：场地自身的属性；与场地相关的自然系统；相邻的邻里、城镇和开发情况；场地所在社区和区域的城市与建筑传统；街道和人行道的尺度、建筑后退、建筑风格、主要的材料和色彩以及景观特征。

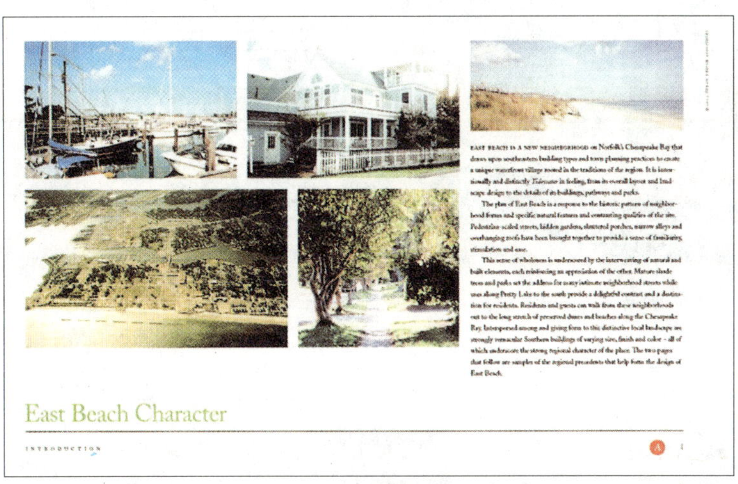
East Beach Character

INTRODUCTION

对基本属性和特征的描述。接下来，对详细条款的总结可能包括典型场所、建筑原型和重要建筑细部的图像。它明确了最适合于开发项目的模式和既能够满足地方传统又符合当前市场需求的基本属性。

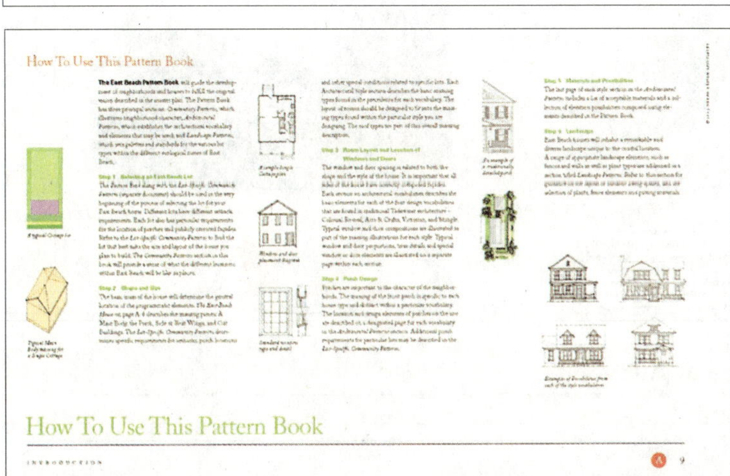
How To Use This Pattern Book

How To Use This Pattern Book

INTRODUCTION

如何使用模式图则。分步骤的操作指南，应采用图表方式来介绍使用图则的各种方法。

II. 社区模式

基于对地方建筑传统的研究，模式图则的社区模式部分描述了开发项目中公共空间的基本属性、用于营造这些空间的各种建筑类型以及它们在地块上的布局方式。尽管在各个模式图则中这一部分都是独一无二的，而且在篇幅长短上也可能不同，但总体来说都包括以下内容：

布局平面及要求。总平面图用图表的形式说明对城市的要求条件，并明确平面中的关键场所和公共空间。建议的街道与公共空间的形式是通过平面图和剖面图来表现的，这些图纸标明了公共空间和私人领域的建筑后退、立面比例、建筑高度和景观特征的处理。

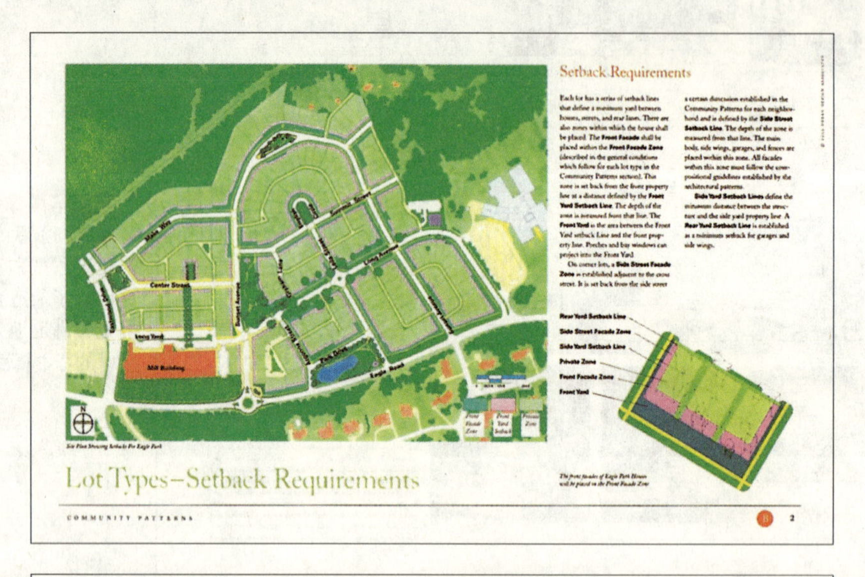

Setback Requirements

Each lot has a series of setback lines that define a minimum yard between houses, streets, and rear lanes. There are also zones within which the house shall be placed. The **Front Facade** shall be placed within the **Front Facade Zone** (described in the general conditions which follow for each lot type in the Community Patterns section). This zone is set back from the front property line at a distance defined by the **Front Yard Setback Line**. The depth of the zone is measured from that line. The **Front Yard** is the area between the Front Yard setback Line and the front property line. Porches and bay windows can project into the Front Yard.

On corner lots, a **Side Street Facade Zone** is established adjacent to the cross street. It is set back from the side street a certain dimension established in the Community Patterns for each neighborhood and is defined by the **Side Street Setback Line**. The depth of the zone is measured from that line. The main body, side wings, garages, and fences are placed within this zone. All facades within this zone must follow the compositional guidelines established by the architectural patterns.

Side Yard Setback Lines define the minimum distance between the structure and the side yard property line. A **Rear Yard Setback Line** is established as a minimum setback for garages and side wings.

Lot Types–Setback Requirements

COMMUNITY PATTERNS　B　2

场所。剖面图也对社区中的不同场所进行了描述和说明。可以用一页附有对特定邻里中住宅布局标准说明的地块平面图的册页，来对每一处独特场所进行描述。通常配合平面图配有一张正常视点的透视图和一张街道剖面图，来表达发展目标的完整图景。

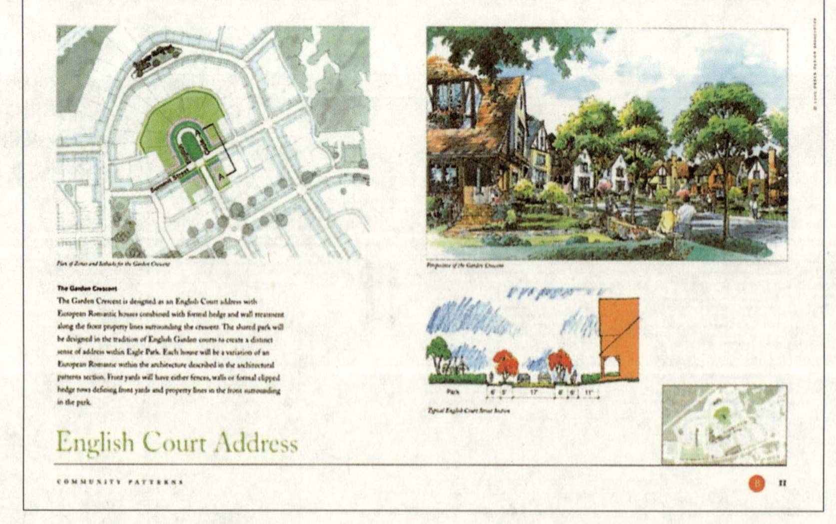

The Garden Crescent

The Garden Crescent is designed as an English Court address with European Romantic houses combined with formal hedge and wall treatment along the front property lines surrounding the crescent. The shared park will be designed in the tradition of English Garden courts to create a distinct sense of address within Eagle Park. Each house will be a variation of an European Romantic within the architecture described in the architectural patterns section. Front yards will have either fences, walls or formal clipped hedge rows defining front yards and property lines in the front surrounding in the park.

English Court Address

COMMUNITY PATTERNS　B　11

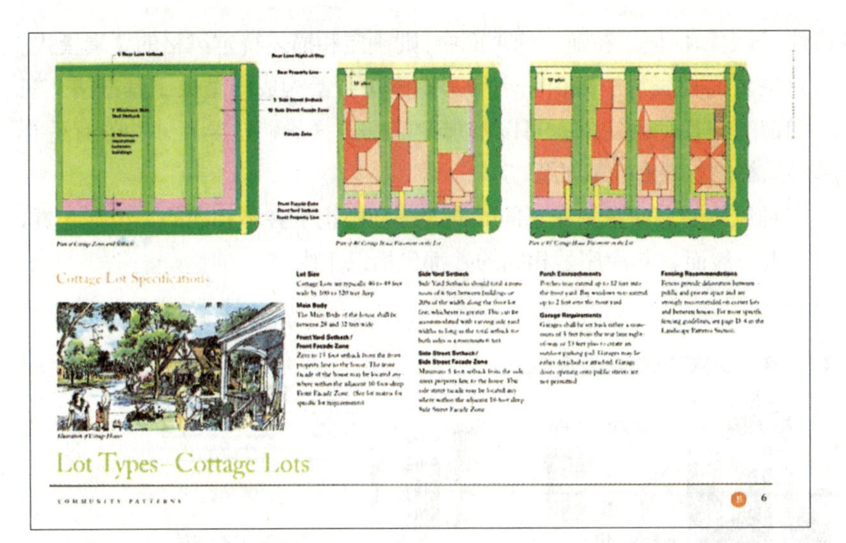
Cottage Lot Specifications

Lot Types–Cottage Lots

COMMUNITY PATTERNS

地块类型。对开发项目中的各种地块类型分别进行描述，同时在这些册页中确定单体建筑的位置、体量和建议在设计中给予特别关注的区域。

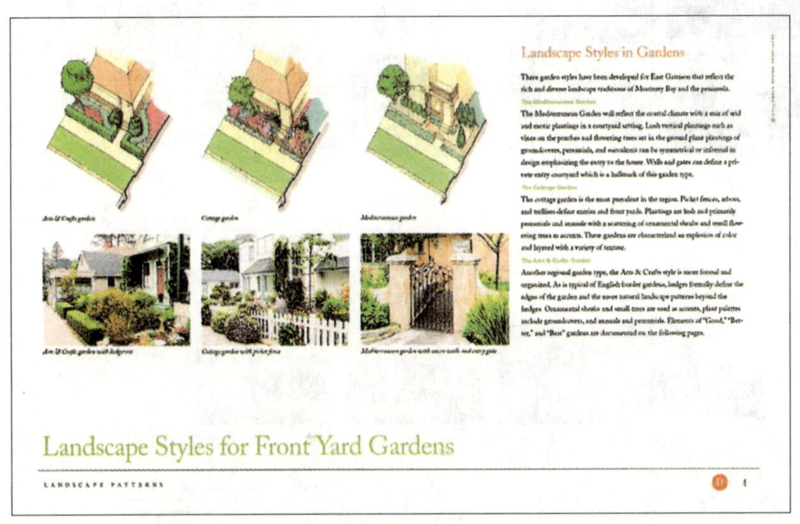
Landscape Styles in Gardens

Landscape Styles for Front Yard Gardens

LANDSCAPE PATTERNS

景观和场地设计要素。如果没有安排单独的景观模式部分，这些页面要确定栅栏、巷道结构和景观要素。如果合适的话，这一部分可能还会包括对灯光照明的说明，在商业性的模式图则中还常常包括对公共空间设施（例如：长凳、垃圾箱、自行车停放支架等）的风格导则或说明。

Ⅲ. 建筑模式

建筑模式部分阐明了为项目选定的建筑风格。风格类型的数量在不同模式图则中各不相同。风格的选择则是范例研究和所有群体合作过程的结果，在合作中人们共同决定哪些建筑风格最适宜用来营造城镇或邻里的可取品质。

每种建筑风格都用一系列页面来进行介绍，其中最先描述该风格的基本属性、历史和特征。在随后的页面中，则确定和描述特定风格的要素模式。

这些页面代表了一种由住宅的基本体块要素构成的集成工具方法，在其中门窗的位置是确定的，可以增加像门廊这样的特定要素，并附有适合于该种风格的材料和色彩的图卡。

每种风格的页面数量在不同的模式图则中也是各不相同的，少则 3 页，多则 8 页。然而，其次序是相同的并都包括以下内容：

历史和特征。对风格的历史和关键特征进行描述的目的，是为了确定应给予尊重并正确建造的基本要素。在任何建设过程中，对能够进行控制的各个方面的数量都是有限的，因而确定最为重要的东西就是十分关键的，4–6 图就列出、勾画和描述了这些特性。

伊斯特比奇模式图则中的泰德沃特木瓦风格是典型的历史和特征页面的实例。在该地区，这种风格常常是各个时期建设的综合复合体，而所造成的变化已改变了住宅的最初形式。这种风格的基础是能够被细分为殖民风格或维多利亚风格图卡的简单形式。其特征包括简单的体量，在基本形体上窗子的对称构图，简化的殖民风格或维多利亚风格的建筑细部，人字型或棚式的老虎窗，简单的一层或二层门廊，切割木瓦壁板和白色的装饰。

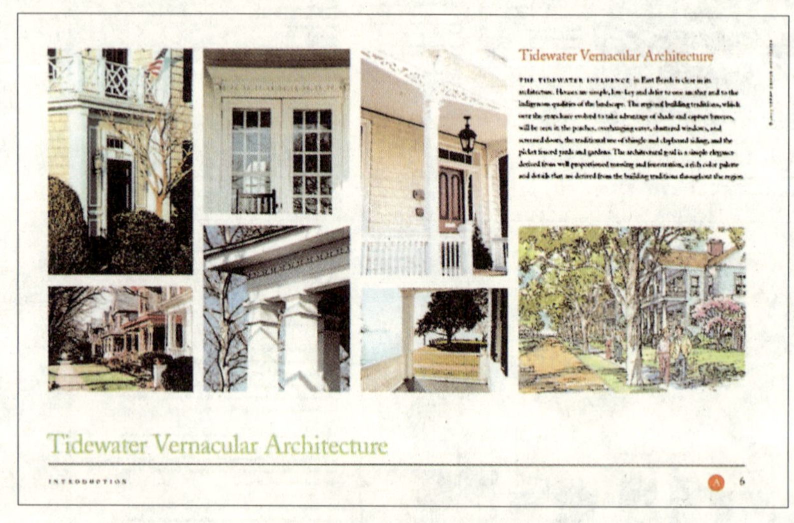

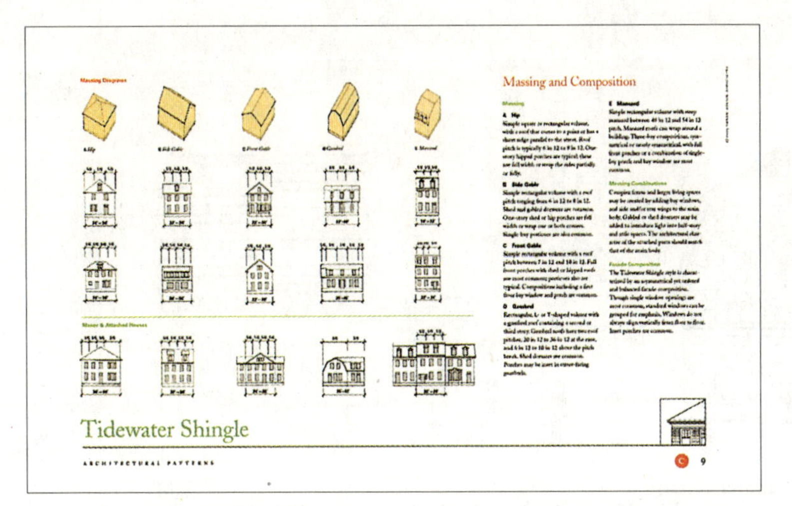

体块和构图。住宅的体块包括已经确定前门位置的建筑主体部分和用来构成较复杂住宅的侧翼部分。体块模式确定了屋顶坡度、高度和建筑的总体形式。立面构图，特别是窗子的位置，是与建筑体块密切相关的。例如，位于住宅主体部分的窗子通常被布置为3开间或5开间的构图。

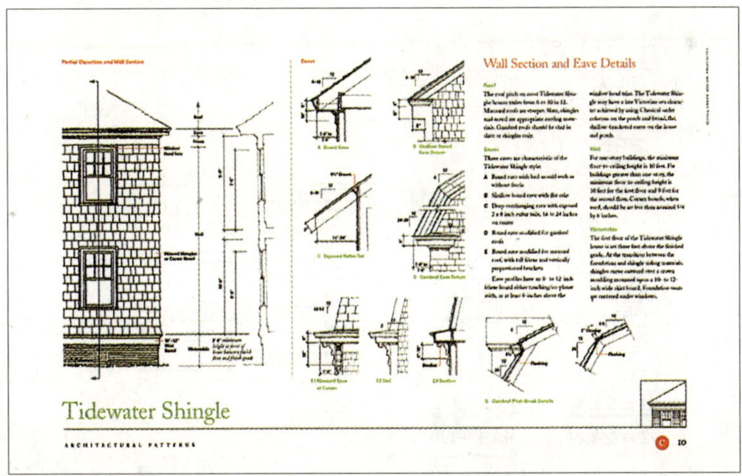

体块和屋檐细部。住宅的体块是与因不同风格而异的屋檐和拱腹结构相连接的。在泰德沃特木瓦风格的实例中，屋顶坡度通常是6∶12或8∶12。复合坡度的复斜屋顶类型和复折式屋顶则采用了不同的比例关系。屋顶的悬挑部分是各不相同的，在维多利亚风格细部中悬挑多达20英寸，而在殖民风格细部的拱腹结构中则只有12英寸。

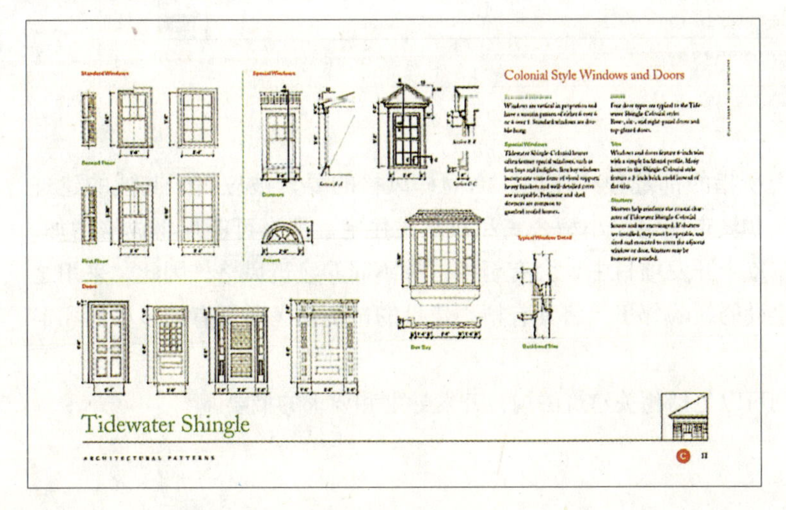

窗和门。模式图则确定了窗的长宽比例和在住宅体块中进行布置的原则。泰德沃特木瓦风格采用了两种基本的窗子类型——殖民风格的窗子类型(6∶9或6∶1的窗格模式)或维多利亚风格的窗子类型（2∶1的窗格模式）。通常，首层的窗子会比第二层的窗子高一些。

门廊。门廊是与泰德沃特木瓦风格联系最紧密的特殊要素。它可以是一层或两层高，并可以有各种形式的屋顶和立柱组合。这一页的图纸指出了在各种风格的体块类型中门廊的布置方法，并包括栏杆和立柱类型等正确的细部和构成。

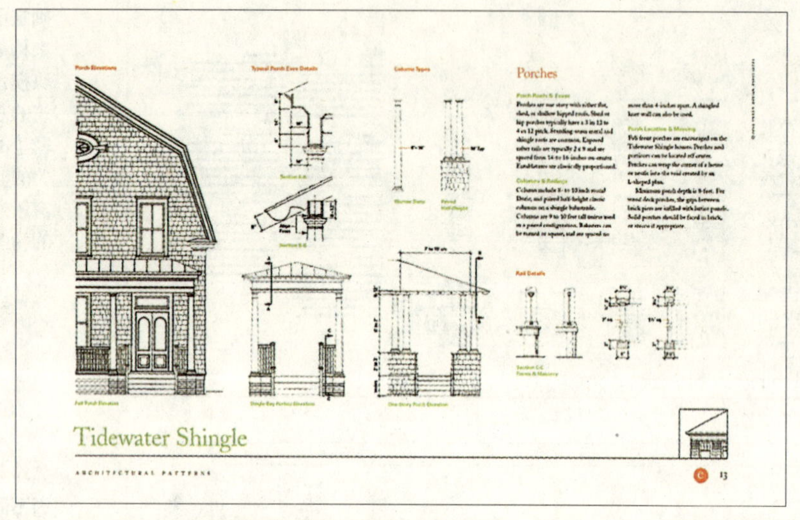

材料、色彩及其可能性。要提供一份材料清单。适宜的色彩图卡通常是在独立的色彩图卡手册中的。对大量住宅立面的可能形式进行图解说明，是为了展示通过联合使用某种建筑风格的体量模式、立面构图、窗、门、特殊要素、材料和色彩所能够创造的多种多样的住宅可能形式。

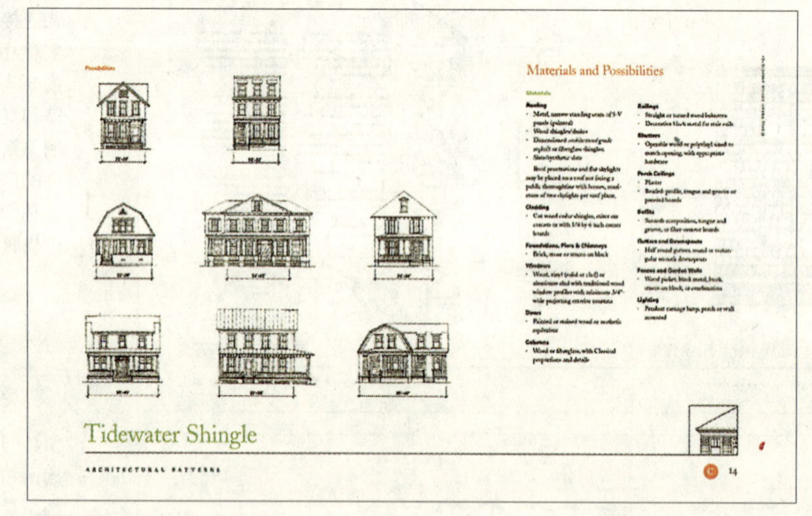

以上所指的都是独户住宅。当同样风格的要素被应用于其他的建筑类型，例如城市住宅、小型公寓和双联式住宅，那么可能性将成倍增加。通常，在每个开发项目中，会有至少三种不同的建筑风格。因此，采用这种方法建设的新的邻里，将具有精心建设的传统社区所拥有的多样性和丰富性。

附则可以包括相关建筑法规、开发导则和要求的摘要。

模式图则封面

完成模式图则的最后一部分工作是它的封面。与任何书籍相同，封面必须能够吸引读者。因此，应该认真地将注意力放在强烈的、有吸引力的图像选择上。UDA 模式图则的封面包括三种比例的图像。

封面图像强调了开发项目的总体意象、有助于感受这一意象的公共空间的设计和能够使这一意象变成现实的单体建筑细部设计之间所存在的相互联系。

Design Guidelines for a Waterfront Neighborhood
East Beach Pattern Book
URBAN DESIGN ASSOCIATES
单体建筑和建筑细部的意象
作为背景的整个开发项目的总平面图或鸟瞰图
典型城市空间的图象

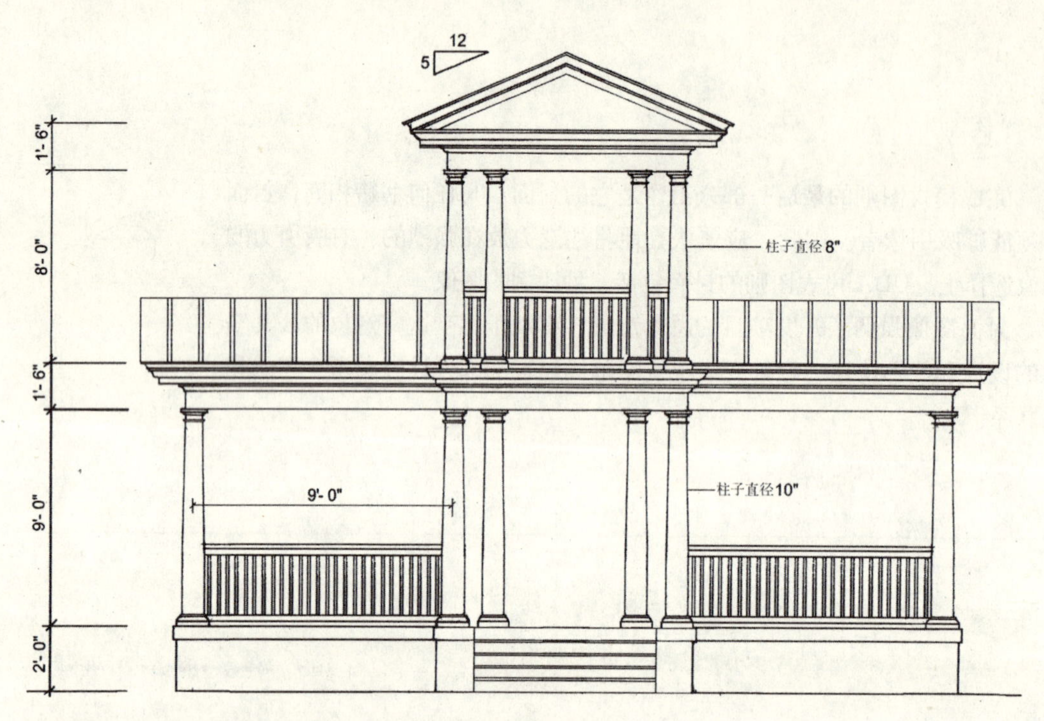

南卡罗来纳州福特米尔市的巴克斯特，经典建筑模式和建成效果

第四章

模式图则对当前实践活动的价值

模式图则的方法，为我们提供了一种处理关键问题的手段，而这些问题正是开发团队在社区和新城设计中必须解决的。在本章中，我们将首先总结当前城市化进程中开发实践所遇到的问题，然后介绍模式图则通过提供有益指引和鼓励各方合作来解决这些问题的方法，最后以使用这种方法建成的邻里和新城的实例来作为本章的结束。

当前的开发实践

开发是一种有风险的商业活动。开发商在工程项目的规划和前期工作中努力将这种风险最小化，并喜欢在意外情况和偶然性相对较少的可预期的环境下开展工作。为了取得成功，开发商必须应对一系列的挑战，而其中最重要的则是下文所列出的那些。

*市场。*首要的也是最基本的挑战就是要去应对市场。开发项目的总体规划必须是在对市场潜在可能性的清晰认识基础上做出的。公共空间的属性，户型单元类型的分配，商业、居住和其他用途的混合——所有这些都必须要满足开发商所力图吸引的市场的需求。因此，总体规划和其所采用的方法，应该成为市场计划中的一个关键性部分。当然，为了营造城市生活环境，开发行为必须实现总体设计所描绘的目标。

这种要在市场中取得成功的要求，对开发团队来说常常是一种可预知的风险。当前的开发实践通常在销售价格和权属类型——例如出租、共同产权或独立销售——的基础上，将多种建筑类型和相似规模的住宅搭配进

范例

模式图则的可能效果

住宅设计

建成效果

行开发。这在新开发的社区中产生了明显的居住分异，因为它向购买者或承租者传递了关于价格和品质的特定信息。这种模式与许多传统邻里和城镇的基本原则与特征是背道而驰的，在那些地方，建筑类型、权属、规模、用途和价格等方面的混合搭配都要自由得多。从开发商的观点来看，这种对产品类型的明确界定能够形成针对购买者的直接销售定位。一种模糊这些标准分类之间差别的更为有机的模式，对销售和市场行销团队提出了不同的挑战。

虽然设计者可能理解总体设计中不同场地的关系和固有品质，但大多数人并不能很好地看懂总平面图，他们也无法对场地建成后的情况产生一致的认识。在开发策划阶段，与可能的购买者和建造者就场地的基本属性进行讨论交流以达成共识就显得很重要。很多新开发的场地，通过使用二维的总平面图、住宅立面图和能够带来积极联想的照片来进行销售。然而，让未来的购买者或租用者仅仅通过这些参考材料就预先了解该场地在建成后的实际效果几乎是不可能的。

*适应性。*在总体设计中所存在的内在风险，是超越可预见的时间段而在建筑的功能混合和类型选择方面进行的刚性配置。在新城市主义的开发活动中这点尤为如此，一旦第一阶段的工作完成后，它们能够改变市场定位。通过营造具有强烈场所感的、有吸引力的环境，这些开发活动与常规开发相比能够吸引更为广泛和更为多元化的人群。很多依赖于特定市场定位的规划，在开发过程中形成了良好的场所感并拓展了原有的总体规划。在许多混合用途的开发项目中一个常见的问题，就是它们在总体计划中期待的、能吸引众多建筑类型和设计众多用途的建筑面积的能力。通常，如果开始的设想不能按照预期落实，那么就需要彻底重做方案。因此，很重要的一点就是要具备应对市场变化的手段，例如改变单元类型和价格的灵活性，而无需损害总体设计的完整性。

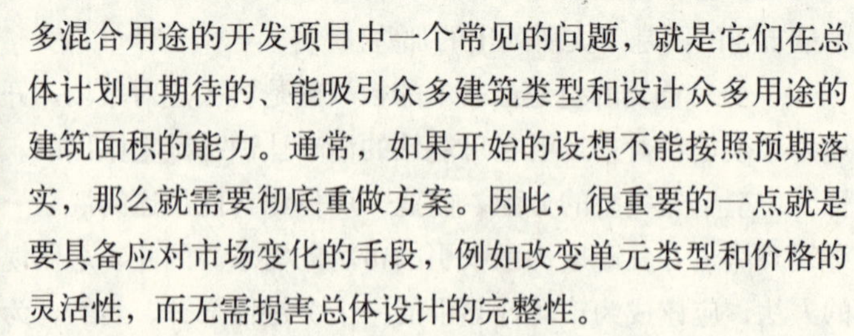

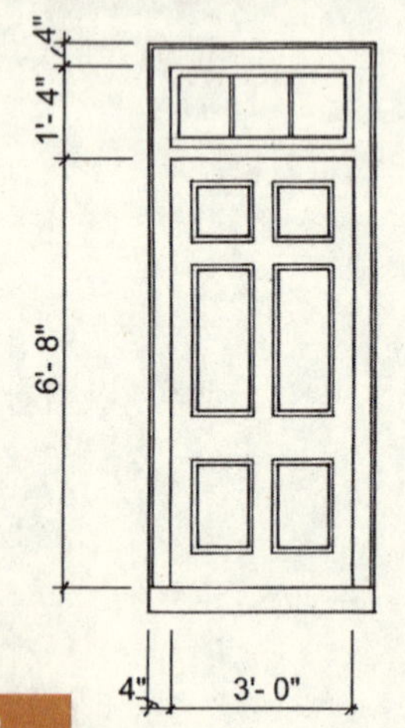

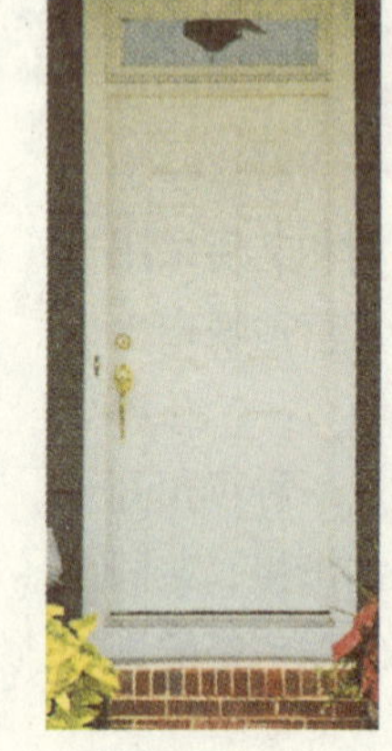

*标准化生产。*为了最大限度降低风险、成本和时间，在大多数的开发实践中存在着标准化生产的压力。建造商和开发商普遍喜欢使用他们过去曾用过、并知道能够取得

成功的住宅设计图。在成品住宅供应中，预算非常有限，同时很少有时间为单个客户的要求去修改已经通过的符合总平面图的住宅设计图纸。

在过去几十年中，大规模的居住区开发以美国全国性或地区性的生产建设为主，建造商能够通过高效生产来大量建造住宅并获得利润。住宅建筑主要源自建造者所得到的建筑平面图，通过不断的试验和失误，最终成功运作于根据目标购买者或租用者而细分的各种市场中。室内装修、厨房和浴室布局以对大型室内空间的感受乃是这些平面设计的重点。一旦平面图确定下来，墙体和屋顶形式就围绕平面图形成。建筑效果趋向于体量的随意组合，而其中的部件则来自所选择的供应商目录图册，这些建筑在或多或少地营造传统社区意象的尝试中被随意进行布置。

在这一背景下，住宅被视为是产品单元，常常与工业产品相提并论。建造过程就是一条生产线，在计划预算之内，拥有一整套被设计用来生产住宅的特定操作过程——使用由开发商或建造商的采购部门（而不是由建筑师或设计师）所选定的预制部件。此外，建造商希望采用能够应用于每一栋住宅的重复性的细部装饰，以减少错误并加快建设速度。

那么，所面临的挑战就是要在标准化生产需求和建造技术之间找到一种工作方式，从而营建能够符合地方传统和环境条件的住宅，并形成高品质的建筑形式以建立受欢迎的社区形象。在这些能够批量、标准化生产的部件和那些需要特别进行设计从而与其场地相协调的独特场所的部件之间，必须求得平衡。

*官方批准和政治。*每一个工程项目都有赖于区划和规划的正式批准。也就是说，项目开展的方式必须在某种程度上得到公众的认可和支持。这种情况使开发商的处境变得更为复杂，因为在努力进行标准化生产、快速推进和应对瞬息变化的市场的同时，开发商还必须应对较大工程项目所在地的当地市民、当地行政管理者

社区模式

建成效果

社区模式

建成效果

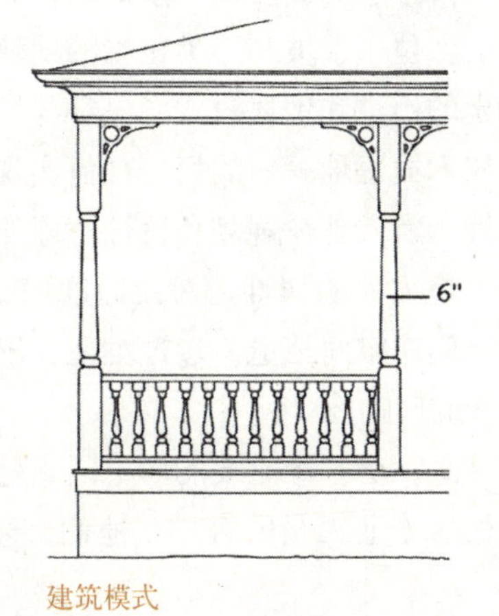

建筑模式

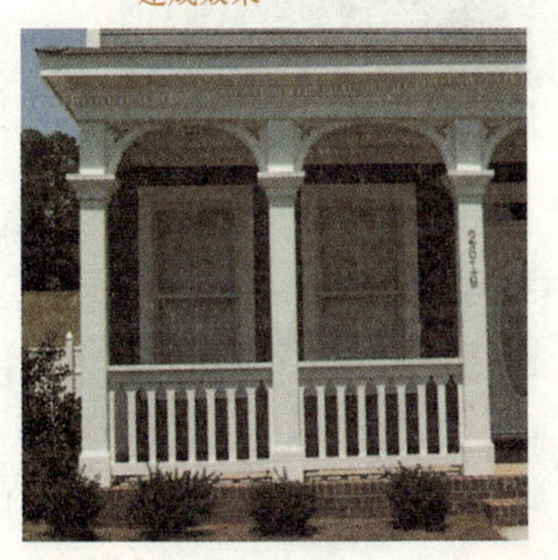

建成效果

范例

住宅设计

建成效果

社区模式

建成效果

和政治环境的干预。再一次地，必须要在给予公众信心的公共层次的要求和灵活性的要求之间取得平衡。地方官员通常要求承诺保证，但却很少关心时间改变所带来的机遇，不考虑市场会对开发行为产生什么样的影响。官方批准常常被视作是明确允诺所提出的全部计划内容。此外，许多现行的标准，例如区划条例、街道建设标准、停车要求规定等等，是与公众和规划当局表达的意望相悖。这就意味着需要修改许多标准，才能批准通过坚持塑造良好城市空间和当地模式——源自最好的传统邻里社区——的开发目标。关于已有标准的谈判和修改的新增工作层面，可能需要付出一定代价，并可能会导致预期良好的理想设计方案受挫。

现行的官方批准程序常常会导致开发商、政府官员和所在社区之间的对抗关系。提交给公共行政部门和规划官员的项目总体设计大多是处于完成状态的。当这种情况出现时，采纳或应对不同意见的余地就很小了，通常其后果就是紧张而艰难的协商谈判，而这是在开发团队已经花费了大量的努力和付出之后才发生的。无论其审批结果如何，这种标准的规划审批程序不可避免地造成某种对立。在这一情况下，对开发团队来说潜在风险非常大，而且要在成本高昂的政治环境中对审批的额外要求进行协商。

*物质环境。*项目场地周边的物质环境具有自身先决性的要求。对于邻里形式和建筑相互之间及其与场地之间关系来说，景观的形式和特征可能是一个重要的决定性因素。其中需要慎重考虑的问题包括：紧凑的街道与带有内部小巷的街区所形成的网格状布局，是如何在地形起伏的地区发挥作用的？位于上坡与下坡位置的地块是如何使用的？为满足地段总体设计所确定的街道和地块模式，保护与场地紧密相关的现存环境特征——例如成年树木或湿地——的目标如何实现？场地所具有的物质环境的约束条件或属性，常常是在规划和设计过程的初期中最不容易被认识的要素。这些物质环境方面的属性，对设计过程初期的场地三维认识而言是很重要的，因为它们在空间塑造、建筑类型、出入通道和平面布局的形式等方面都是关键性的影响因素。

*实施。*另一个重要的考虑因素是当地的实施能力。城市规划需要很多不同领域专业人员的合作，他们必须保持协调一致，并很好地完成任务。

因此，开发商就要面对一系列的问题：哪一家住宅建造商有能力保证建设质量和建设进度？基础设施（例如街道、市政设施和公共开放空间）如何建设？开发团队具有怎样的的潜力和经验？如何分阶段实施？本地设计人员的能力和资质怎么样？所有这些问题的答案在不同的地方也是各不相同的，但可以肯定的一点是，人才的质量并不是在所有的学科都一样。很多项目在设计中并没能清楚地理解建设实施程序是如何运作的。实施的方法和工具也因实施策略的不同而不同。考虑到对土地的处置能力和方法、施工协调、参与及控制等因素，很多制定出的规划不能成功地付诸实施。

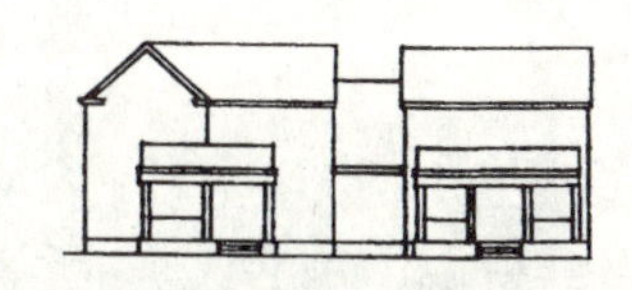

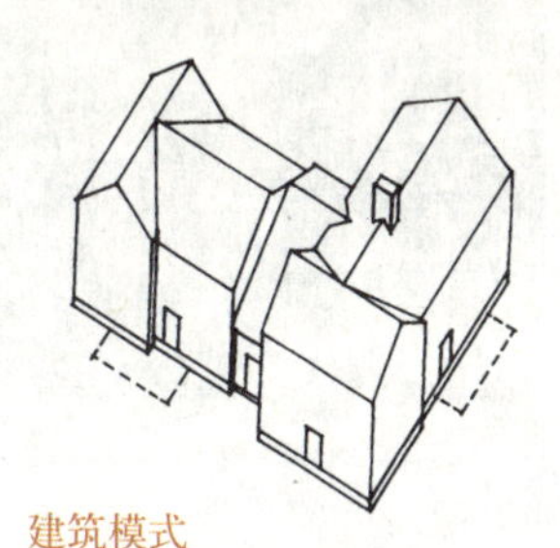

建筑模式

建成效果

社区模式

建成效果

社区模式

建成效果

通常情况下，项目的总体设计是充分考虑并经过官方批准的，地块被出售给建造者，并要求他们都能了解到如何使建设真正地与场地相适应。分类规划常常能保证在每一地块中建设住宅建筑，而没有考虑到每一条共用的街道或公园的特征和属性。建造者常常会因与建设区域不相协调的地块而使工程受阻。工程技术与景观设计相脱节，景观设计与城市设计相脱节，而城市设计又与住宅的设计相脱节，等等，等等。其后果就是，建筑物从未像最初预想的那样形成场所感。

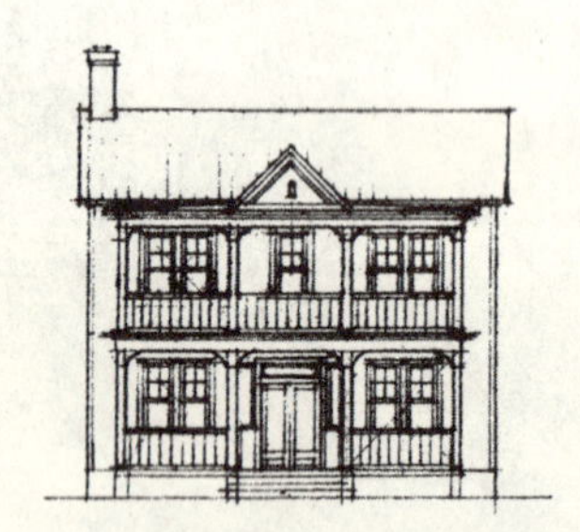
模式图则的可能效果

模式图则方法的益处

作为在所有建设参与者之间沟通开发目标的一种方法，模式图则能够提供一种解决上述问题的机制。传统的模式图则，例如建造者指南，能够对如何建造经过良好设计、符合正确建筑风格的建筑提供有益的建议。通过在场地、总体设计、实施策略和市场之间的发展图景或概念之间建立关键性的联系，现代模式图则对当前开发活动参与者所面临的更为艰巨的挑战进行了应答。这几乎以同样的方式应用于私人和公共的开发项目中，尽管后者常常会带来额外的复杂性。

模式图则对当前开发进程来说所具有的核心价值，是它能够在开发过程中将所有参与者联系起来，因为场地的发展图景被转化为已建成的环境景象。参与者包括开发商、职业销售人员、工程师、建筑师、管理和行政工作人员、公众、住宅建造者和媒体。

建成效果

在工作的早期阶段，这些专业人员的参与是非常重要的，便于资料的收集，同时也可以记录下当地的已有先例和现状条件。核心小组、访谈、野外研究活动和实地工作会议与问题研讨会，在开发过程中为从所有股东和参与者方面收集信息提供了多重机会。这建立起了一套关于设计意象和特征的共同价值标准。在开发过程的后期，则与所有这些参与的利益群体共同对总体设计进行严格的检验。在这一检验的过程中，总体设计通常能够得到改进，从而使场地与最初的设想和场所感相一致，并互为补充。

*市场。*模式图则最为成功的尝试，就是将对各种市场部分进行的认真分析和理解与以下要素联系起来：适当的建筑类型、尺度和造价，邻里特征，生活设施和单元类型的混合。模式图则的开发程序，汇集了规划方案的场地特征、预期的市场需求、建筑类型和尺度、景观设计和建筑语汇。规划方案的不断完善，是检验单元类型的混合、为营造方案中特定场所而对细部和要素进行推敲的过程的核心部分。一旦集合起这些不同的要素，就能通过场地总体设计获得满足要求的场所，也能够对建筑类型的混合和景观要素给予清晰说明。这种对场所的详细认知，在与公众和开发团队就建成环境特性方面进行沟通的过程中是非常有效的。预期的购买者、公共部门的官员和其他参与者，得以理解场地总体规划中各种要素所产生的特定属性。在模式图则中，概念和意象必须是被所有参与者所理解的，这也是开发计划获得市场成功所必不可少的。当以这种方式对其进行理解的时候，模式图则就将被视为有助于获得成功的有益工具，而不是被看作项目实施的障碍。

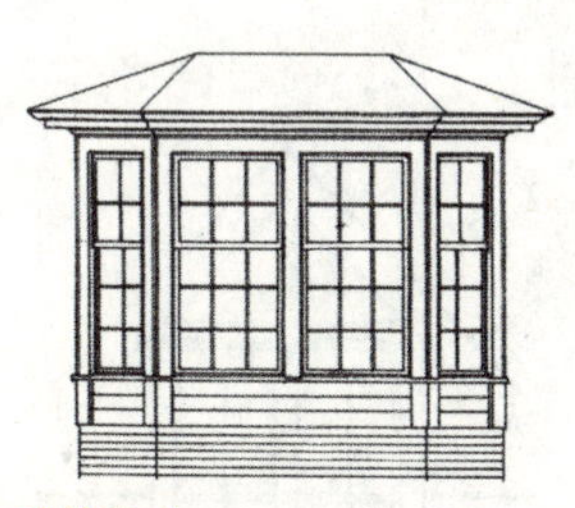
建筑模式

建成效果

可能的模式图则

建成效果

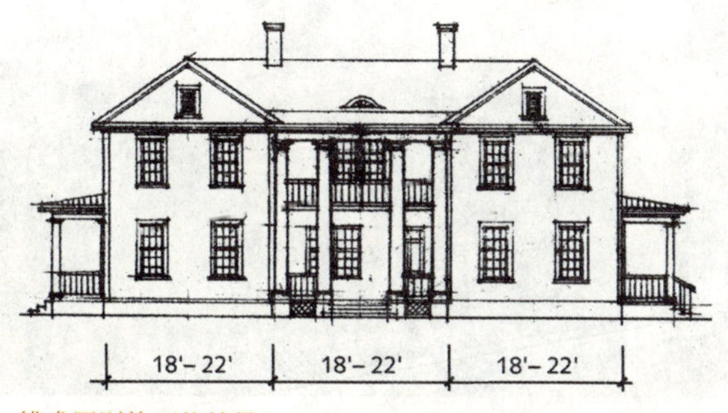

模式图则的可能效果

建成效果

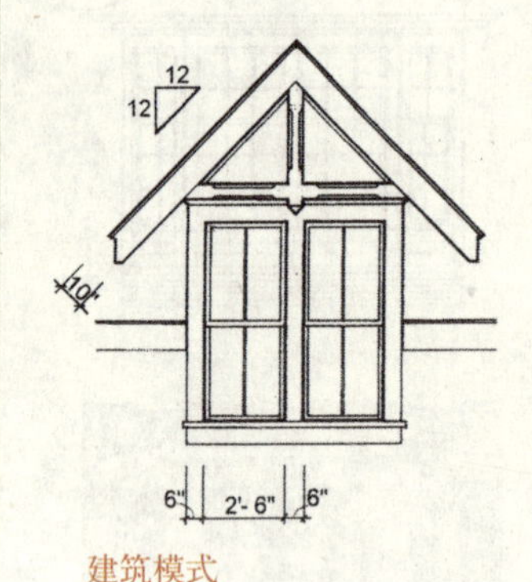

建筑模式

建成效果

*灵活性。*正如前面关于当前实践所进行的讨论那样，能够应对不同市场趋势和反应的灵活性，对于大多数开发努力来说都是非常重要的，特别是在有相当数量的单元要在一个较长时期内进行销售的时候。在模式图则的开发过程中，一般地，初始阶段是要详细开展工作的。社区模式通常是针对将进行为期 2 – 5 年的开发实施的特定区域而制定的。随后的阶段将在场地总体设计中提出同样的街道和开放空间模式，但可能会有建筑类型或单元大小的不同混合方式，这更多地考虑了在初始阶段过程中所显现出来的市场或开发偏好。通常，社区模式是在计划推进过程中形成的一系列章节。街道类型、公园、街区类型和建筑语汇等主要要素得以保持不变，而同时又保证了灵活性，能够应对不断发展变化的市场。

一般情况下，要对建筑类型重新进行界定，以顺应市场需求并形成宜人的场所属性。当共有产权住宅或公寓建筑被布置于邻里社区中时，它们可能具有大型住宅的特征，而当它们位于市区时则可能呈现出更多城市建筑的特征。场所特征并不一定会依赖于单元类型，相反地，认识正确特征的创造性过程，具有使市场力量与场所特性相适应的优势。UDA 所提出的城市组合部件方法，采取了在场地总体设计中确定街区和建筑类型的方式，从而具有灵活性和多样性。这种方法将场地总体设计中的特定部分——街道类型、街区、公园、建筑类型——视为各种部件的成套工具包，能够通过不同方式进行组合从而在整个规划中营造各不相同的场所。这种工具包的方法，使建筑类型能够进行交换或修改而无须改变街道和街区的模式，并使设计得以应对市场的变化。

社区模式

建成效果

模式图则的可能效果

建成效果

在营造令人难忘的邻里场所的过程中，公共空间的基本结构是最重要的因素。工具包方法包括了限定这些空间的建筑立面的标准和形象。例如，社区模式部分描述了空间特性和围绕其布置的建筑的特定模式。然而，它并没有指定某一特定单元或住宅类型。随着市场的变化，可以根据同样的模式布置不同的建筑类型，因而同时做到了满足变化的需求和保护场所的特征。

*标准化生产和地方建筑特征。*模式图则方法重新建立了使用特定地方建筑语汇的概念，从而营造了在场地总体设计中得以完善的意象和特征。涉及建造者和开发商的收集先例的过程和图则自身的版式，其目的都是建立地方传统的价值观念。模式图则对特定建筑语汇的关键要素进行了图解说明，从而使建造团队能够在设计和建设过程中具体落实这些策略。

下一步的工作，是在建造者的标准平面图中说明将建筑语汇最容易地应用于他们自己的“作品”中的途径。这种转化是通过与每一个建造者设计团队协作完成的，形成了一种积极的工作关系并在团队成员中建立信任。于是开发商也了解到只需对其标准化生产中的某一方面进行修改，并可以对成本和生产方法的相关问题进行评估。

这种方法常常使团队能够在传统的建造者设计或生产过程中作出重大改变。一改相当随意的工作方法，即室内关系的处理不考虑构成和集合元素，设计团队而是首先从模式图则中为特定风格提出的基本组合类型和相应的门窗搭配入手。平面关系的发展演变则是根据某一种基本的组合方式而做出的。详细设计的和实用性的元素——例如凸窗和门廊、贴面装饰和

社区模式

建成效果

建筑模式

建成效果

檐口——成为更加易于处理的一系列设计和建造问题。通常情况下，建造者将能够使用同样的平面形式建成至少两种不同的住宅。很多的传统住宅建筑都是由一些简单的形式构成的，又有其他的简单形式附加其上，即，主体建筑和两翼建筑。当屋檐、屋顶坡度、门窗风格以及贴面装饰和门廊的细部有所变化时，基本的组合方式常常能够发展出不同的建筑语汇。而且，由于组合工具包方法清晰地说明了哪些要素是固定不变的而哪些又是有所变化的，从而使建造过程更为容易。对于设计工作来说这是一种“土豆头先生”方法。它所产生的效果，是在严格的集成部件住宅建造过程中所完成的具有独特地方特色的、恰当的传统建筑。

*官方批准和政治。*对于在社区股东、行政领导、管理部门、建造者和开发团队之间达成一致意见来说，模式图则发挥着重要作用。对于很多开发项目非常关键的区划和规划批准，也是与开发过程融为一体的重要因素。对于抽象的区划标准和邻里的街道或公园特征之间的关系，是通过模型和相关透视图以三维方法进行说明的，从而使每个人都能够理解在基础模式和场所属性之间的联系。因此，当需要对法定规划进行变更或修改时，通过这种方式，能够在充分理解其效果的条件下完成这一过程。

建筑模式

城市组合部件方法再次发挥了作用。通过提取各种设计要素并将其分解为形象化的独立元素，使人们可能得以对每一元素进行评估和投资。例如，如果将街道和公园的设计与建筑设计一起进行整体构思，那么场所空间属性的效果就能够进行分离并分别提交审批。这种方法有助于解决因预期密度而引发的问题。一旦将所提出的建筑类型放在场地总体设计的场所文脉

建成效果

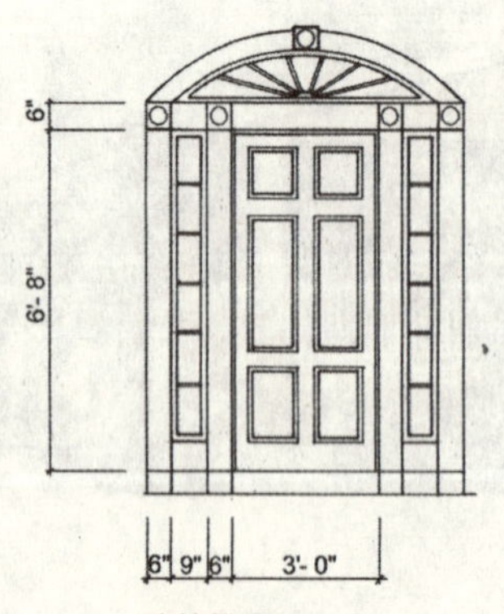

建筑模式

建成效果

环境中进行说明，那么关于每英亩土地上的单元数量或地块尺度的问题，其重要性就会变得次于最终模式和场所特征问题。对于向政府官员和公众来解释说明所提出的建筑类型和混合用途、街道尺度和类型的应用等问题来说，这是非常有帮助的。

*物质环境。*这里所描述的编制模式图则方法的一个核心部分，就是对场地总体规划进行检验和完善，从而满足地方前例、详细景观设计、地形条件、建筑语汇、建筑类型与混合用途的要求。与邻里特征、社区模式相关的模式图则章节，要满足以下要求：对单元面积和类型需求的市场预测，单元如何与场地相匹配，以及在平面图中这些要素的组合方式。这些元素通过带有能够准确反映建筑物、住宅类型和尺度的 1 英寸：20 英尺比例的模型进行检验。

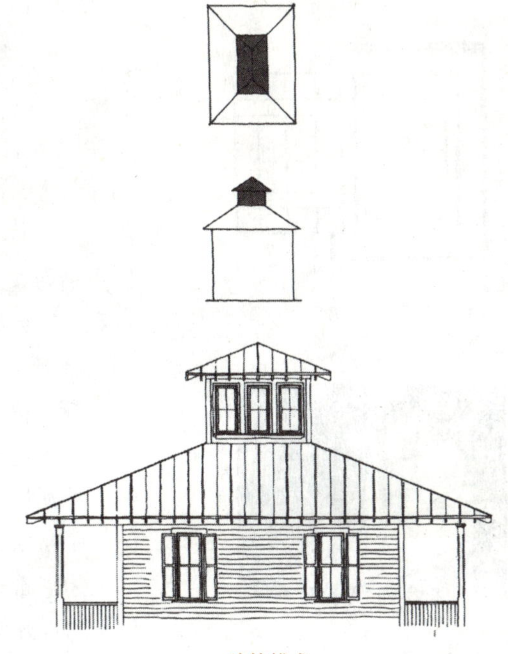

建筑模式

模型用于强化场地平面中特定场所的规划设计，并进行记录和说明。必要时要对场地平面图进行修改，以与检验结果相一致，并形成满意的空间品质以供模式图则作为范例和进行说明。

*实施。*模式图则使建造者、建筑师和开发、销售和营销团队能够理解所提出的场所特征。关于街道沿线的工程决策和景观设计决策，植物配置和路面铺装材料选择，建筑后退部分，路缘石类型和半径都可以在场所周边的文脉环境中看到，而不是作为孤立的部分。团队通过应用模型方法共同分析平面图中的要素，能够在最后的工程和实施过程中节约几个月的时间。这一过程为更详细地了解建造者和开发商在实现理想效果过程中所需要的东西打下基础。这样，每一地块和建筑所在场地也能够根据一定特征或适合的标准条件进行细化。每一处公园或街道可能会拥有一种特定的景观色卡，用以营造能够增加项目平面中不同场所魅力的独特环境感观。

建成效果

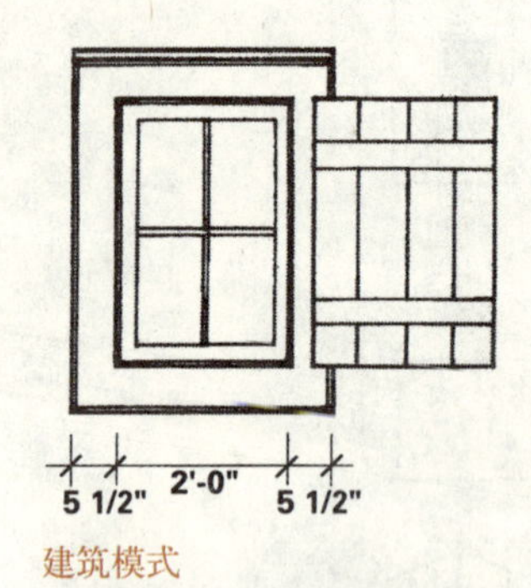

建筑模式

建成效果

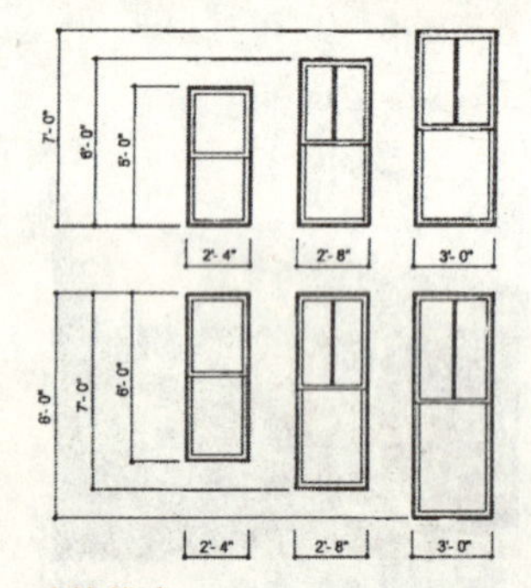

建筑模式

建成效果

建成效果的实例

本章以下的图片展示了在很多 UDA 模式图则中所包含的最初建筑模式和细部节点，以及一些实地拍摄的建成效果。这些作为图解的工程项目包括：

- 巴克斯特，一座混合用途的村庄，位于南卡罗来纳州的福特米尔，由克利尔·斯普林斯开发公司进行开发。巴克斯特混合应用了成品部件和定制部件。
- 莱奇斯，位于亚拉巴马州的亨茨维尔，是由亨茨维尔的莱奇斯公司开发、使用定制部件建造的居住性邻里社区。
- 帕克杜瓦拉，位于肯塔基州的路易斯维尔，是一项由路易斯维尔自治区、路易斯维尔住房当局、社区建造者与一家私人开发公司共同领导的名为“希望五号”的自发项目。帕克杜瓦拉是一项混合收益、混合资金的再开发项目。
- 沃特卡勒，位于佛罗里达州的狭长地带上，是由圣乔 / 阿尔维达开发的一处混合功能社区，根据库珀·罗伯逊所设计的项目总平面图采用定制部件建设。

对于任何既定项目，模式图则的目标都是确保建筑师和建造者能够投身于家园建设之中，因为每一个特定的场地都会具有他们所需要的房屋设计的工具，这些工具将营造出城市设计中预期的邻里社区和公共空间的属性和特点。

就像它们 19 世纪的前辈们一样，这些现代模式图则中的每一个，都被设计成为有效的市场工具。因为模式图则将建筑模式与给定开发项目的市场基调连为一体，模式图则使预期的住房购买者得以了解邻里社区的总体景象，并认识到他们的新居是与周边环境协调一致的。不像那些告诉人们不能够做什么的“规则”，模式图则帮助住房购买者展望他们能够做什么——他们新的梦想住房看起来会是什么样子。同样地，模式图则主动地提供建筑细部样本和建造房屋的方法，这能够激发人们的想像力，并支持邻里社区或城镇设计理念。

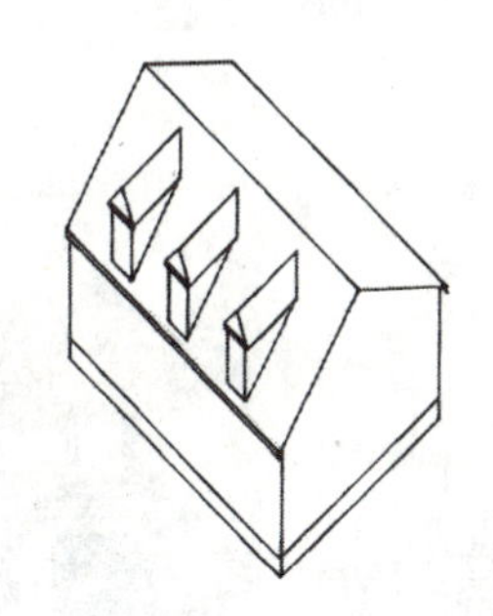

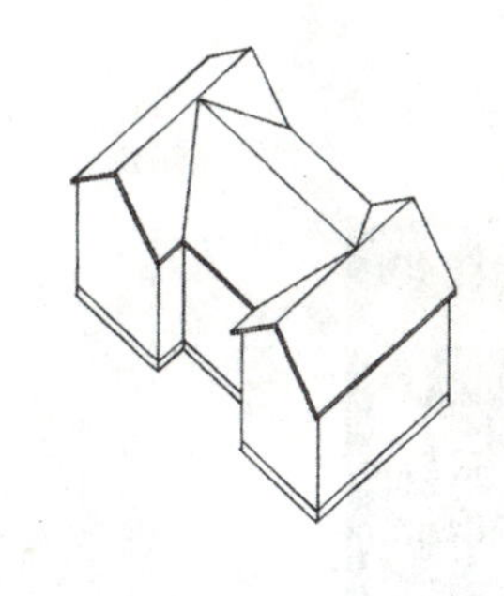

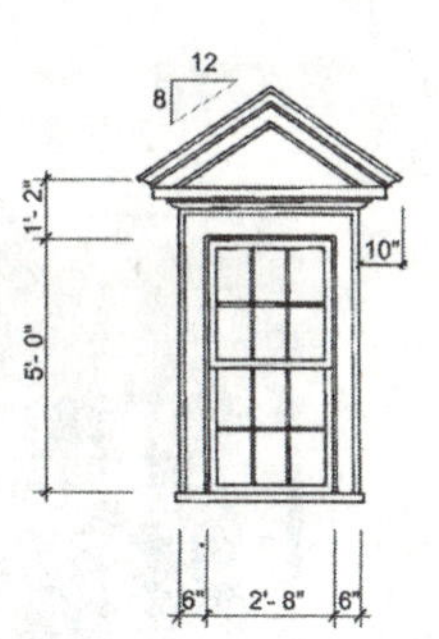

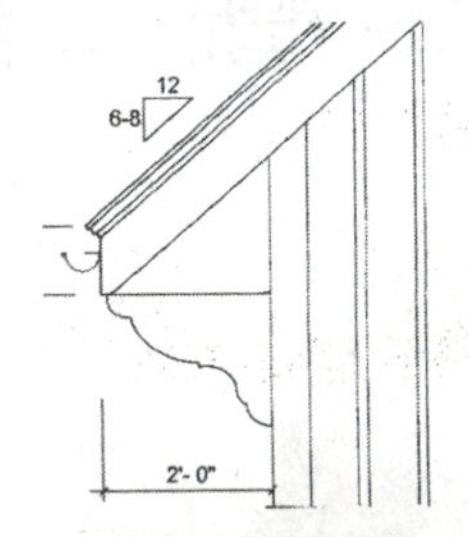

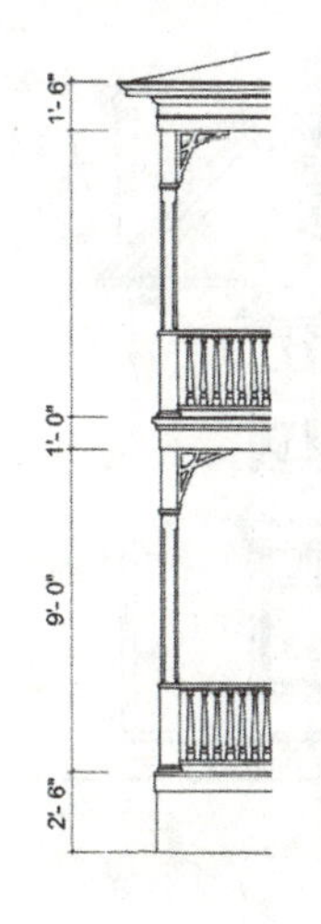

建筑模式和建成效果

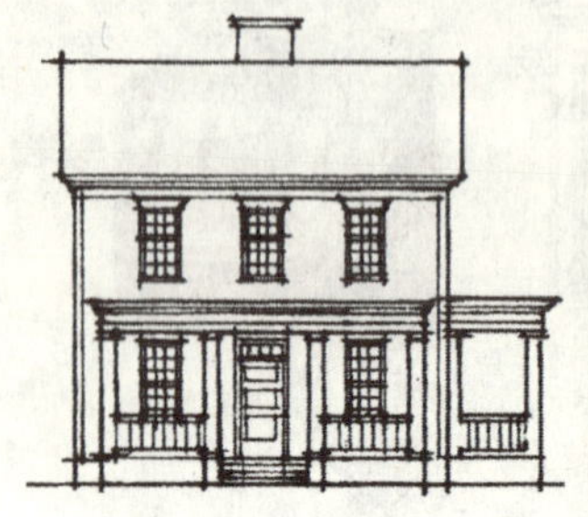

建筑模式和建成效果

第五章

编制模式图则的过程

编制模式图则的过程包括以下几个阶段：

- **第一阶段**：了解环境——过去与现在；
- **第二阶段**：形成图卡——记录相关特征；
- **第三阶段**：确定模式——社区、建筑和景观；
- **第四阶段**：制作成果——形成模式图则。

这有助于与美术设计师进行紧密协作，他们能够制定模式图则的版式，并在内容、设计、制作和印刷等问题方面辅助模式图则开发小组的工作。此外，一名职业撰稿人能够协助构思并编辑文本。通常这一过程需要花费大约6个月的时间，并包括以下步骤：

第一阶段　了解环境

过去与现在

最为有效的模式图则源自对场所地域特征的了解。因此，设计过程应从研究项目周边地域的传统城镇或邻里社区开始，包括限定了公共空间的住房和建筑物。很重要的一点是，要对客户和城市设计团队曾经用来营造推荐总体设计方案的意象和特征的范例给予密切关注。无论你是在合作进行一项总体设计还是在实施一项开发项目，都要确保了解记录了适当区域范围内（某些时候是全美国范围内的）与场地和项目的前景与特征相适宜的范例。

泰德沃特风格的建筑范例和细部设计

在与客户和开发或设计团队的初次现场会谈中，要努力去了解设计、范例、市场、阶段和建造过程。花上两到三天与你的客户共同了解区域内的范例。建立一支配备了照相机、测量卷尺和笔记本的工作小组。对区域进行调研、记录占优势的建筑类型、建筑风格和材料。仔细关注对于区域整体肌理来说最为基础的东西——它的主流风格——以及为不同邻里增添情趣的凸显外来的风格。要去了解当地社区的模式，对街道剖面、住宅退后部分、地块长度和宽度以及景观特征进行测量和拍照。对现状住宅和建筑类型要仔细地拍照和记录，确定它们的建筑风格、集成构造和关键的细部。以能表现住宅立面和街道空间的一系列剖面图的方法，来绘制记录每个邻里街道和公共空间的尺度与特质。

收集历史街区的卫星图片、已有邻里的城市地图、场地自身的平面图、与建造计划相关的市场信息、单元尺度等等。这些信息能够为项目提供出发点，并由此形成对项目所在地区文脉环境的理解与正确评价。

同时，借鉴很多开发商提供的标准平面图能够帮助你去了解他们进行工作的方式，并学会他们认为对赢得市场成功来说很重要的品质。通过这种方式，开发商或建造者的观点得以融入到这一过程之中。这有助于重新建立在建筑师和建造者之间曾经存在过的合作关系。当完成这些工作之后，就是回到你的办公室开始整理挑选制作图卡所可能需要的资料的时候了。

典型的街道剖面图

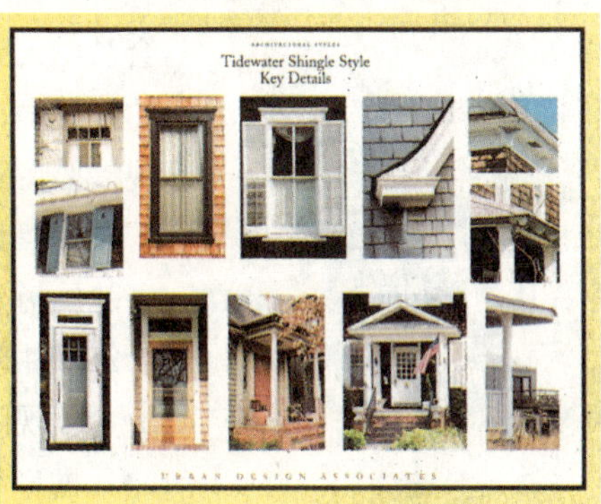

第二阶段　形成图卡

记录相关特征

第二阶段以对邻里社区的记录分类作为开始，既包括数字化资料，也包括复印资料。要去了解邻里的特征，根据场所对图片进行分类。准备邻里范例的场地图纸，并使其与街道和公园空间及其周边建筑后退的剖面图相一致。然后开始建立街区模式，在开发计划和适宜建筑的历史特征的基础上形成类型学图像。

这样，你就能够开始制作建筑图卡的草案初稿了。将照片的完整目录和与区域开发模式相关的历史文件，按风格或词汇表分为几大类。这一过程需要系列经过精选的图片资料，并对不同类型、主要元素和典型特征进行加工提炼。这包括使用照片、剖面图和选定为社区范例的邻里平面图来准备建筑物类型和建筑风格的范例图版。这些可以在与开发小组进行的第二次工作会议中使用，来检验图卡及其对目标社区的适用性。

范例研究

例如，伊斯特比奇是位于弗吉尼亚州诺福克市切萨皮克湾沿岸的一个新兴邻里社区，对于伊斯特比奇模式图则来说，UDA 设计小组研究了作为伊斯特比奇范例的东部滨海地区的城镇和邻里的特征与形式。该小组走访了 10 座城镇，对每座城镇中的住宅、建筑类型和建筑风格都进行了拍照和记录。从最初 8 种可能建筑风格的清单中，精简为适合于伊斯特比奇的 4 种风格（参见右侧图，从上至下）：泰德沃特木瓦风格、泰德沃特殖民复兴风格、泰德沃特维多利亚风格和泰德沃特工艺美术风格。

正如惯常做法那样，通过伊斯特比奇建筑风格的精选过程确定了这一地区看似独一无二的每一种风格或建筑语汇的关键标识要素。通常，在范例样本中找到的要素，被结合起来重点用于形成一系列连贯的指导方针和体现独特地方特征的基本属性要求。对于伊斯特比奇来说，很多提出的风格是从地区内的范例中推断而来的，并经过修订以强调早期殖民地的滨海和航运特征，在紧邻场地的区域内这一特征已被很好地恢复再现了。其重

建造一栋模式图则上的住宅

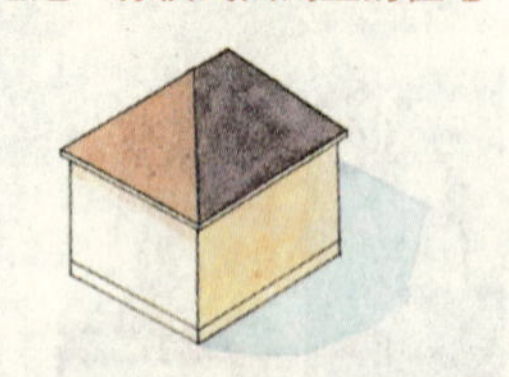

建筑主体

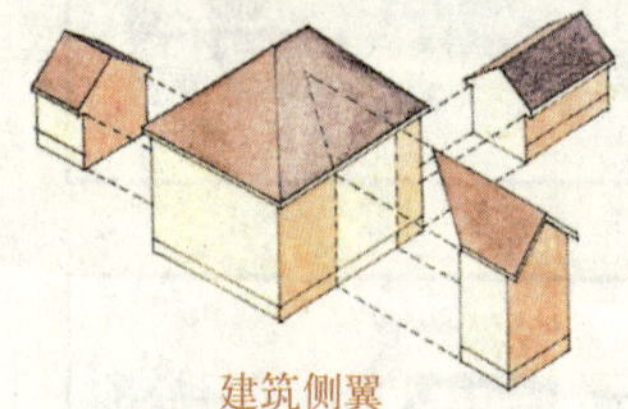

建筑侧翼

门洞和窗洞的构图布局

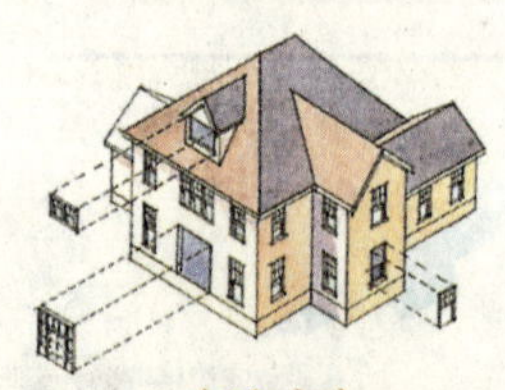

门和窗户

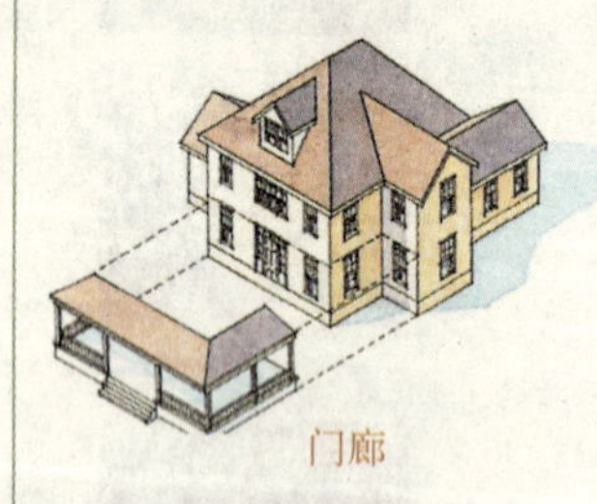

门廊

最后装配完成的住宅

点是将项目作为重新建立场地与滨海村庄起源原型之间联系的综合战略的一部分，而这一联系长期以来已被更为通行的开发和建设实践的模式所取代了。

在必要时，邀请专家参与特定建筑风格的工作，这对于获得详细的设计标准来说是非常有益的。一旦研究工作结束，你就可以开始在建议的市场类型基础上发展出一系列的住宅和建筑类型，并采用在每一种推荐风格中所发现的重要组合方式。要准备好轴测图，作为提供给模型制作者的信息资料。这些图纸应该包括重要的门廊要素和车库，这可能会、也可能不会被纳入到主要的住宅组合方式中去。在清晰说明每一种风格的组合方式的过程中，你将开始了解社区模式与建筑模式是如何结合的，并且你会发现需要在哪些地方开展更进一步的工作来完善你的设想。

一旦你使开发小组在图卡问题上达成一致意见，你就能够继续进行下一轮的对建筑风格特征所进行的更为集中的文件整理。让小组回到现场，对每一种选定风格的建筑细部和建筑要素拍照。其中包括：

- 体现门窗构图的正立面图；
- 檐口和屋檐；
- 窗和门的类型与布置；
- 门廊类型和细部；
- 组合方式和材料。

现在，你或许已经准备好来设计建筑图卡了，它将在随后进行的模式图则项目第三阶段中的现场模型研讨会中进行展示。让项目小组编辑所收集到的数据，并开始选择关键细部的典型范例，例如檐口、窗子类型、装饰元素、门廊栏杆、柱子类型、屋檐和柱上横楣、特殊门窗、材料、烟囱，以及其他建筑细部。在这一过程中要完成重要建筑细部的精确草图，并确定各种要素的尺度和比例。绘图标准的确定是以对模式图则方法的认同作为基础的，以确保在所有参与项目的人之间，包括景观建筑师、透视图制作者等其他协助者，达成一致性。

第三阶段　　确定模式

社区、建筑和景观

在第三阶段的开始，小组要准备一个活页的图则大纲，制作一张比例为 1 英寸：20 英尺的场地平面工作底图，以及各个类别中每种住宅原型的比例为 1 英寸：20 英尺的木质模型。在小组内部组装模型，并请全组一起进行讨论，指出每一处不足，并尽你所能在模型现场展示之前进行改正。

然后，将工作底图和成比例的模型带到社区中，并在现场模型研讨会中用来对设计方案进行检验。此外，还要绘制重要场所的透视图，这样将有助于你在研讨会中开展工作，从而在适宜的意象、地块模式和建筑功能混合方面获得一致意见。

研讨会对于客户来讲是非常具有启发性的，因为通常这是他们第一次能看到场地总体规划的三维图景。由于模型会暴露出在场地总体规划中所存在的不足，因而在模型研讨中常常就会进行修改完善。在研讨会议过程中要对社区模式、建筑模式和景观模式进行检查。要对以下要素进行评估和调整：建筑后退；场地划分；根据地块类型确定的特别设计标准，例如

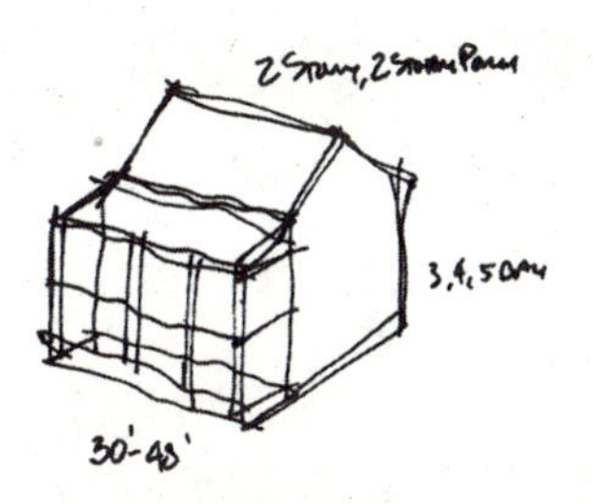

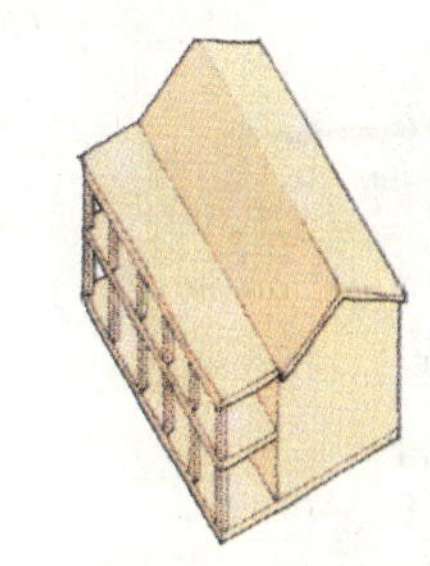

根据模型绘制的社区模式图

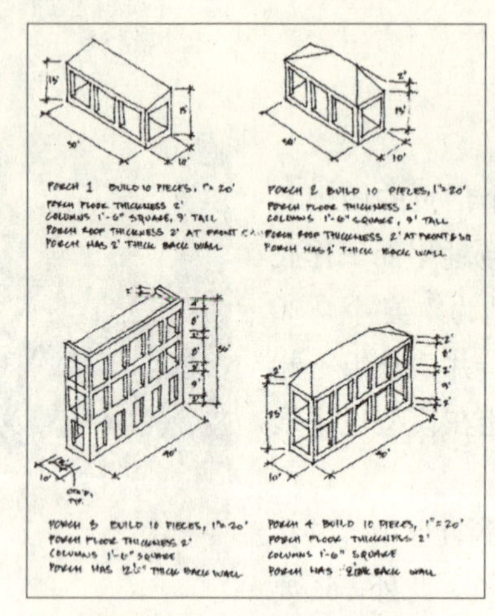

模型说明书

在转角住宅顶部的塔楼；街道定线；景观详细设计；公园、纪念性建筑和开放空间的设想；等等。这是一个工作量很大的过程，因为你需要一个地块一个地块地走遍整个场地，在工作底图上直接记录下每一处变更。这些记录文件成为对模式图则社区模式部分中的每一处特殊场所进行描述的原始资料。

请土木工程师、环境工程师、景观建筑师等各个方面的顾问参与开发小组的工作也很重要，这样在这一过程中所有小组成员都能够对各种场所的形式和特征获得充分的了解。通过在 1 英寸：20 英尺的比例尺上以三维方式完善规划设计，能够对工程部件进行评估和测试，并能明确对设计平面图所进行的新的调整或修改。同样，对设计平面图所进行的完善和说明也是综合过程的一部分，在其中每个人都能了解到各部分内容——建筑、景观设计、土木工程、市场营销和工程实施——是如何共同开展工作的。

在核心咨询小组之外，让可能的建造者加入到这一阶段的工作过程中是很有帮助的，这样他们能够参与并开始认识新的社区规划中诸多的不同方面，同时了解在塑造场所感的建筑、房屋类型以及使用功能的混合等问题之间存在的关系。这能帮助每一个人理解场所的更广阔图景，而且向他们提供提问的机会，去了解该过程是如何影响他们自己的操作方法和技术的，并开始明确认识到其公司所面临的机会。

在这一步的工作为模式图则的制作打下基础的同时，它也为完成场地总体设计的施工文件和最终的设计要素奠定了基础。需要重视的一点是，在总体设计中由开发小组设计的公共或市民共享的空间，是与从最初的场地总体设计设想中逐渐形成的建筑细部和设计要素相呼应的。要将注意力集中于在模式图则制定过程中所确定并进行图解的特定场所和建筑细部。同等重要的是，模式图则建筑导则与对街道、公园和开放空间的总体设计是互为补充的，这样对规划设计中场所特征进行的说明就能够融入到主要开发小组所完成规划设计的各种要素中。

第四阶段　制作成果

形成模式图则

在模型研讨会之后，你已在其中检验和完善了规划中的特定场所，现在是为完整的模式图则来制作更为详细的图板的时候了，要逐页地确定需要哪些图纸及应将其用于何处。通常情况下，由项目经理在与负责人进行协商后完成这些任务。

同时，最终的决定将明确哪些照片和图纸会被采用。每一本模式图则都可能会混合采用手绘图和工具线条图。在制作模式图则草稿的最初阶段，就要确定图纸的类型和不同建筑细部的比例，并在着手制图之前要与整个小组（包括小组之外对模式图则作出贡献的人）进行沟通。

由于小组与美术设计师和印刷专业人员进行紧密协作，在公司规定的模式图则版式、图纸与文字说明标准的要求下共同进行文件排版，从而使最终完成的模式图则的总体质量会得到很大的提高。

现在，是开始制作图则的时候了。以在研讨会所收集的反馈意见为基础，你的小组能够完成一份供客户进行评论的模式图则的完整草稿。

一旦草稿制定，并进行了小组内部的复查，包括从头到尾的校对，就该准备好提交给客户征求意见。在最终版本制作和分发之前，你也可以选择将其提交给建造者和建筑师进行检查。

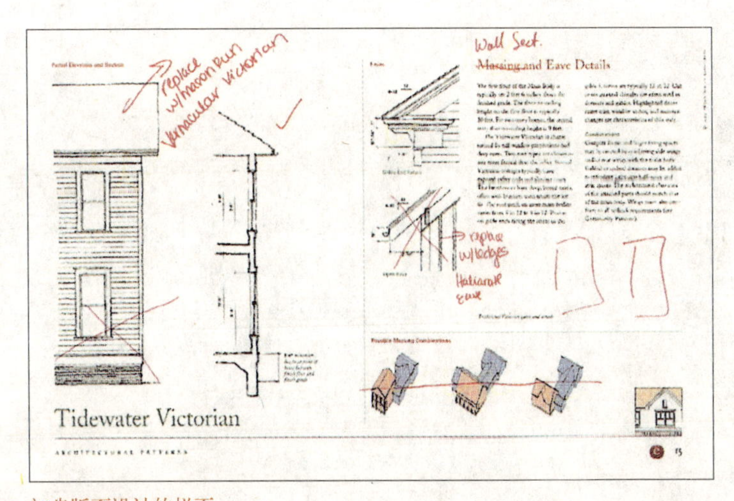

初步版面设计的样页

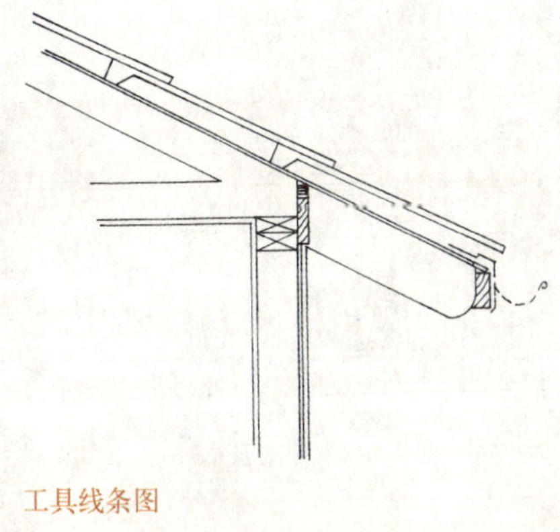

工具线条图

第二部分
UDA 模式图则范例

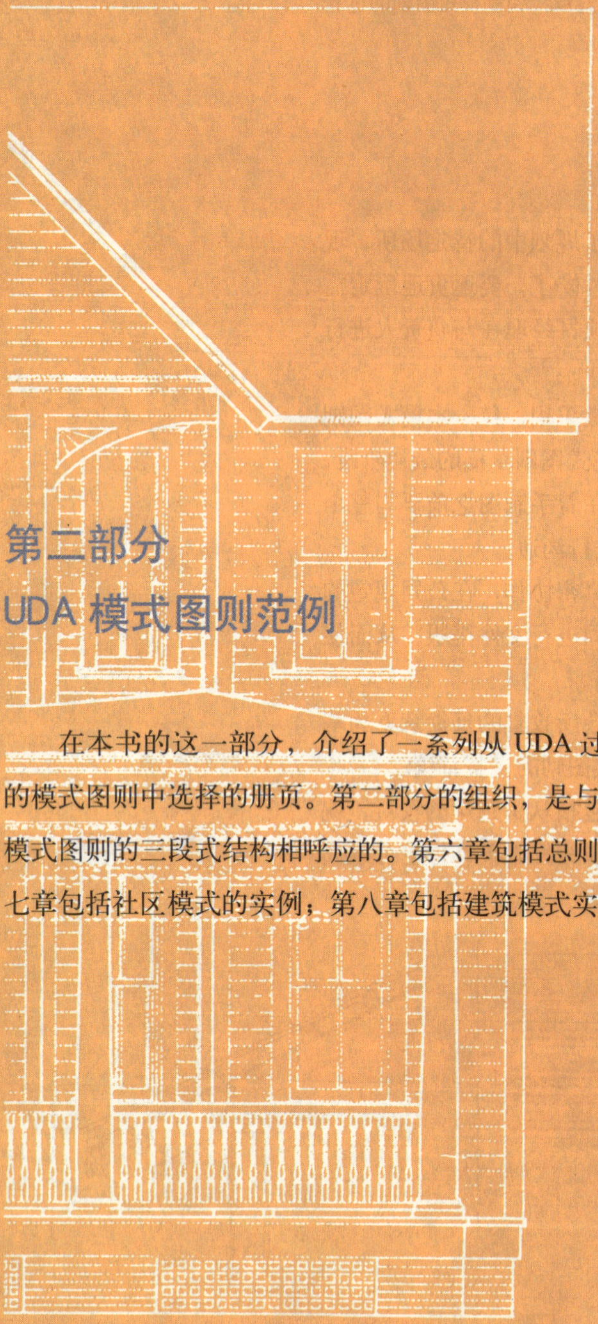

在本书的这一部分，介绍了一系列从 UDA 过去 10 余年中所完成的模式图则中选择的册页。第二部分的组织，是与在第三章中所推荐的模式图则的三段式结构相呼应的。第六章包括总则部分册页的实例，第七章包括社区模式的实例，第八章包括建筑模式实例册页的实例。

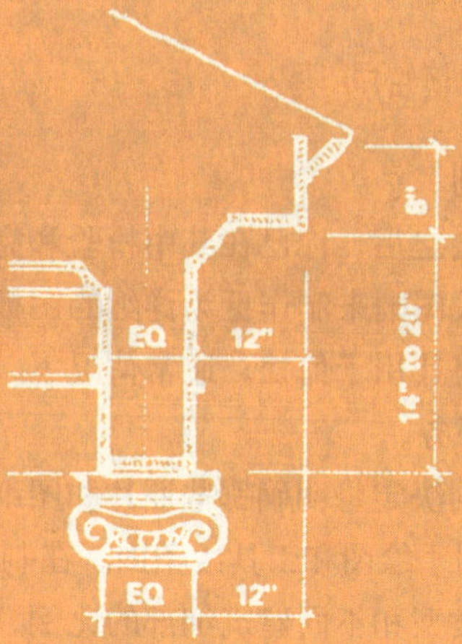

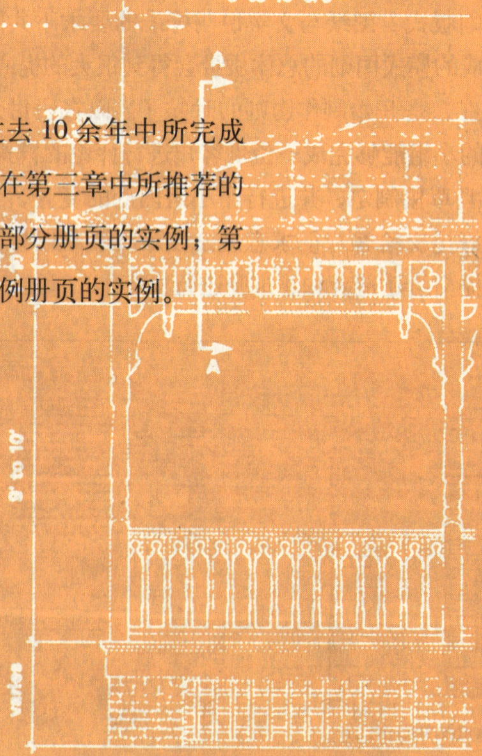

Section B-B

One-Story Porch Elevation

在第六章到第八章中所列举的模式图则册页范例，代表了各种项目类型，包括新城、古老城市中的新社区和在乡村环境中的开发实践：

- 伊斯特比奇，是位于弗吉尼亚州诺福克市欧申维尤邻里的一个开发项目。伊斯特比奇包括了从滨海岸地区到临海湾地区的场地范围，在中间还有一个传统邻里社区。邻里主题的确定，源自对沿东海岸地区泰德沃特市和大西洋城的海滨城镇的分析。
- 沃特卡勒，是位于佛罗里达州潘汉德尔地区的一座新城，建设了一系列与这座地处海洋和森林湖泊之间的城市的自然背景协调一致的公共空间。模式图则在从滨海地区，经过树林，直到湖泊的地域中营造了一系列的空间场所。
- 巴克斯特，是位于南卡罗来纳州福特米尔市的一座新城，规划从其公共空间属性和其住宅建筑风格两方面来延续南卡罗来纳高地的地方传统。
- 帕克杜瓦拉，是位于肯塔基州路易斯维尔市的一个邻里，已从低收入的公共住房地区转变为多样化、混合收入的邻里社区。在这个城市最受欢迎的场所的基础之上，模式图则创建了易于识别的社区和建筑模式，并通过借助其营造传统邻里，从而在成功的复兴路易斯维尔西部地区的过程中发挥了作用。
- 伊格尔帕克，位于北卡罗来纳州贝尔蒙特市的一项填充式开发项目，在这里模式图则包括一栋混合了居住与工作功能的建筑、由经修复的工厂改建的商业中心和住宅、新的联排住宅和每户独立的住宅。
- 达克尔山（比尔特摩），是位于北卡罗来纳州阿什维尔市附近比尔特摩的一个开发项目，在那里，开发形象是建立在对 19 世纪末建成的旅舍和度假胜地的建筑传统的详细了解基础之上的。
- 利伯蒂，是位于加利福尼亚州艾尔希诺湖市附近的一座新城，构想了一系列围绕一座广阔公园布置的邻里和在滨湖开垦土地之上的开放空间网络。建筑和邻里的特征，是通过对滨海和南加利福尼亚州城镇及邻里所进行的广泛研究确定的。

- 诺福克，位于弗吉尼亚州，在那里整座城市的模式图则，被设计用来帮助正在新建住宅或正在改造已有住宅的所有居民，以在他们的邻里中提升城市空间的品质。
- 梅森兰，是位于密歇根州门罗市的一个新建邻里，将对棕地的再开发和提供市场估价的可负担住房的目标结合起来，从而改善周边邻里环境。场地方案的设计的基础是对一系列场所的建设，其中很多是围绕着小型的邻里公园和一座设计用来以自然输送方式解决暴雨排放问题的长条形公园进行建设的。在继承遍布这座多元化的历史性城镇的建筑形式基础上，进行了新的填充式住宅的设计。
- 热尼特伊，位于巴黎东部的比西圣乔治，是一座拥有 2500 套住房单元、即将建成的规模宏大的新城。它所面临的挑战是在混合收入的新城中建立适于市场销售的形象，包括面向高端市场的独户住宅和相当比例的多户联排住宅。计划建造一条公园大道，将规划的热尼特伊公园延伸到城市边缘的开敞农田景观。在这条道路两侧布置着“别墅”，尽管它们实际上是联排住宅，但却具有在这一区域的城镇周边经常可以看到的大型别墅的形象。
- 伊斯特加里森，位于加利福尼亚州蒙特雷县的福特雷德，是在过去的军队营地上发展起来的新建社区。开发计划是在该区域内存在的早期教堂营地居民点和散布在谷地中的小型农业城镇的基础上形成的。1500 套非常密集的居住单元，包括独立式的和联排式的住宅，与艺术家工作室、保留下来的自然生境和区域内的娱乐设施融为一体。模式图则确立了一系列在场地周边城镇中存在的，并加以说明的住宅风格和建筑语汇。

第六章

总则册页

本书所倡导的模式图则的首要意图，是帮助设计师和开发商去建造能够塑造优美场所和城市空间的建筑物。为实现这一意图，重要的是形成对开发目的及其市场目标的充分理解。那样，所有参与者——无论是房屋建造者、建筑师、交通工程师、公园设计师还是政府官员——都具有对总体目标的共同理解。

因此，每一部模式图则的第一部分都应该提供对开发目标的总体评述，并包括能够表现开发建设完成时大部分重要空间属性的图像。总则部分提供了一个在其中能够表达各种特定模式的背景。

每一部模式图则都是各不相同的，因为它们应该是分别应对所提出的特定开发目标的。一些模式图则是专为新城设计的，在其中开发的市场前景是关键性要素。在大多数实例中，开发项目所在地的自然背景和区域传统，将成为开发活动和建筑主题的基础。对于在已建成城市中的开发项目来说，需要对地方建筑进行细致调查从而确定其特有的模式。

总则部分也向其读者说明如何去使用模式图则。

案例研究：伊斯特比奇

弗吉尼亚州诺福克市

伊斯特比奇，是作为弗吉尼亚州诺福克市欧申维尤地区总体规划的组成部分来进行开发的一个新建的填充式邻里。欧申维尤位于切萨皮克湾，地处沿海岸延伸约 4 英里长的狭长半岛之上。其周边地区主要在过去的 50

伊斯特比奇／泰德沃特的范例：（上图）马里兰州的安纳波利斯；（下图）北卡罗来纳州的伊登顿

年中进行了再开发，建造了毗邻欧申维尤北端海军基地的低品质的公寓住房和廉价的汽车旅馆，以服务于大量军队人员。

伊斯特比奇的发展设想，在使用功能的混合、建筑类型和尺度方面，将其描绘为作为真正的弗吉尼亚滨海居住区的新的邻里社区。紧邻地区的环境背景，在先例和社区特征的范例方面可供利用的东西很少。在确定植物配置、明确街道和公园布局方面，自然环境发挥了巨大的作用。在场地中要保护现有的成年海岸树木和植物类型。邻里特征的来源被确定为从北卡罗来纳到马里兰州东岸的沿东海岸地区，它们拥有共同的建筑类型、文化和物质属性以及环境背景。工作小组调查记录了很多村庄和城镇并将其作为提炼过程的一部分，从而在场地总体设计和从泰德沃特地区传统模式中发展而来的真实场所感之间建立联系。场地规划和开发计划确定了一系列具有特别意义的场所：滨临切萨皮克湾的地区、内陆地区的邻里、沿欧申维尤东岸普瑞提湖内湾的码头地区、位于作为进入伊斯特比奇通道的滨海大道处的公园入口。对场所范例所进行的研究，有助于建立场地中这些特殊空间的独特特征。

在滨海居住区的总体特征之外，通过对范例的研究形成了在整个地区所找到的最相符的建筑语汇的图卡。尽管拥有所有这些 19 世纪邻里建筑语汇的范例，诺福克在过去的年代里已丧失了很多更古老的滨海建筑类型。模式图则的总则部分提供了四种独特建筑语汇的图卡：泰德沃特殖民复兴风格、泰德沃特木瓦风格、泰德沃特维多利亚风格和泰德沃特工艺美术风格。另外，模式图则确定了混合使用的商业建筑的类型。伊斯特比奇总则部分，也描述了典型的伊斯特比奇住宅的基本部件以及它与其所处地块的关系。在这种海滨环境中，门廊和出挑的屋顶是很重要的，也因此被作为这些住宅的一个部件加以强调。其中还有一部分内容是关于如何使用模式图则去设计一栋与场地具有关系，并应用某种泰德沃特语汇的房屋。

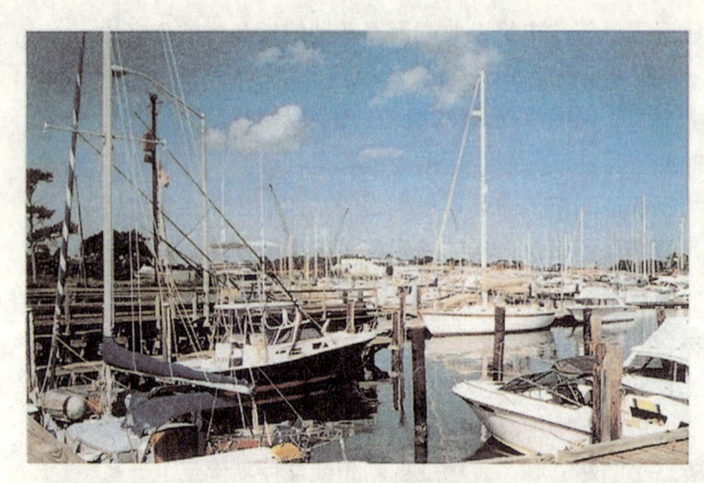

East Beach Character 伊斯特比奇风貌

INTRODUCTION

EAST BEACH IS A NEW NEIGHBORHOOD on Norfolk's Chesapeake Bay that draws upon southeastern building types and town planning practices to create a unique waterfront village rooted in the traditions of the region. It is intentionally and distinctly *Tidewater* in feeling, from its overall layout and landscape design to the details of its buildings, pathways and parks.

The plan of East Beach is a response to the historic pattern of neighborhood forms and specific natural features and contrasting qualities of the site. Pedestrian-scaled streets, hidden gardens, shuttered porches, narrow alleys and overhanging roofs have been brought together to provide a sense of familiarity, stimulation and ease.

This sense of wholeness is underscored by the interweaving of natural and built elements, each reinforcing an appreciation of the other. Mature shade trees and parks set the address for many intimate neighborhood streets while uses along Pretty Lake to the south provide a delightful contrast and a destination for residents. Residents and guests can walk from these neighborhoods out to the long stretch of preserved dunes and beaches along the Chesapeake Bay. Interspersed among and giving form to this distinctive local landscape are strongly vernacular Southern buildings of varying size, finish and color – all of which underscore the strong regional character of the place. The two pages that follow are samples of the regional precedents that help form the design of East Beach.

Norfolk, Virginia

Edenton, North Carolina

Edenton, North Carolina

Nantucket Postcard

Neighborhoods

THE CHARACTER AND QUALITY of the historic villages and towns along the East Coast of the United States have been studied carefully as a resource and guide to the planning and building of East Beach. The coastal character is expressed in the architecture which has been modified by local architects and builders over time to respond to the environment in subtle ways. It can be seen in the structure of the town, the street layout and public spaces, and in the landscape elements and materials. Towns and villages along the East Coast have both formal parks and courthouse squares as well as wonderfully landscaped local streets with a surprising variety of character. The green in Edenton, North Carolina is a great example of a public space surrounded by houses from different eras and a courthouse which looks east toward the sea. Many coastal towns have a main street that leads to the water as in Portsmouth, Virginia or Annapolis, Maryland.

Coastal neighborhoods have a variety of house types and architectural vocabularies. The Freemason and Ghent neighborhoods in Norfolk represent two distinct eras of building tradition, both of which are oriented to the water. Ghent is defined by a formal public edge along the inlet fronted by a mix of houses and civic uses such as the museum and opera. Freemason is characterized by narrow, cobblestone streets lined with a variety of attached townhouses and formal Colonial era houses. Many villages such as St. Michaels and Easton on Maryland's Eastern Shore have small cottages on narrow lots that give the village a delicate scale. Annapolis combines attached houses and mixed-use commercial buildings around the harbor to create an active and dynamic sense of place that is unique to the waterfront marina setting.

Historic settlements further north along Cape Cod and Nantucket are characteristic of the refined sense of place that results from the combination of the coastal landscape, neighborhood form and architectural materials.

The neighborhoods of East Beach will draw on these images and forms to establish a unique sense of living on two waterfronts in the Tidewater region.

Coastal Precedents　海滨先例

INTRODUCTION

A　2

Looking across the inlet in Annapolis

Window shopping in Annapolis

Marina and Main Street

MARINA PRECINCTS AND MAIN STREETS in towns and villages along the East Coast of the United States have a marvelous quality. They are diverse and exciting places to be. Annapolis sets the standard for a vibrant, mixed-use district tied to the image of sailing and the water. Buildings line the inlet facing the marina so that the activity around the boats and the water becomes a constant form of public theater that has attracted residents, merchants and visitors for over two centuries. The scale of the buildings is often three to four stories, with living units above shops, restaurants and offices. There are many forms of brick and clapboard buildings in Colonial, Federal and Victorian vocabularies. The ground floor is often a collection of diverse shopfronts that communicate the nature of the goods or services within. In total, the environment is dynamic and picturesque.

Typical Main Street architecture - Easton, Maryland

East Dover Street - Easton, Maryland

Nantucket Postcard

Coastal Precedents　海滨先例

INTRODUCTION

A　3

The Pretty Lake Marinas

Shore Drive Park Address

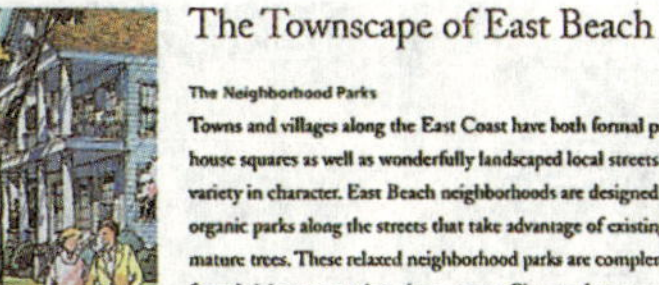

The Bay Front

East Beach Neighborhoods

The Townscape of East Beach

The Neighborhood Parks

Towns and villages along the East Coast have both formal parks and court-house squares as well as wonderfully landscaped local streets with a surprising variety in character. East Beach neighborhoods are designed around a series of organic parks along the streets that take advantage of existing landscape and mature trees. These relaxed neighborhood parks are complemented by more formal civic spaces such as the square on Pleasant Avenue and form a network of open spaces that link to both shorelines.

The Pretty Lake Marinas

Many coastal towns have a main street that leads to the water as in Edenton, North Carolina or Annapolis, Maryland. East Beach plans to have a marina precinct alive with a mix of residential and commercial uses in mixed-use buildings that add contrast and character to the inland residential neighborhoods. This precinct will have a continuous series of public spaces and thoroughfares that provide access to the waterfront for residents, slip owners and visitors.

The Bay Front

The neighborhood streets connect Pretty Lake to the Chesapeake Bay. Residents are never more than two blocks from the water living in East Beach. The Bay front will have a continuous public beach with access points at the ends of streets or through public greens and paths that open up to the spectacular views.

Shore Drive

The principal gateway into East Beach is along Shore Drive. The waterfront village meets Shore Drive across a continuous park. This park address connects the Pretty Lake mixed-use district overlooking the marinas with the stately housing designed in the tradition of an 'Admiral's Row' and a neighborhood shopping precinct at the entrance with Pleasant Avenue.

The Townscape of East Beach　伊斯特比奇的城市景观

INTRODUCTION　　A　4

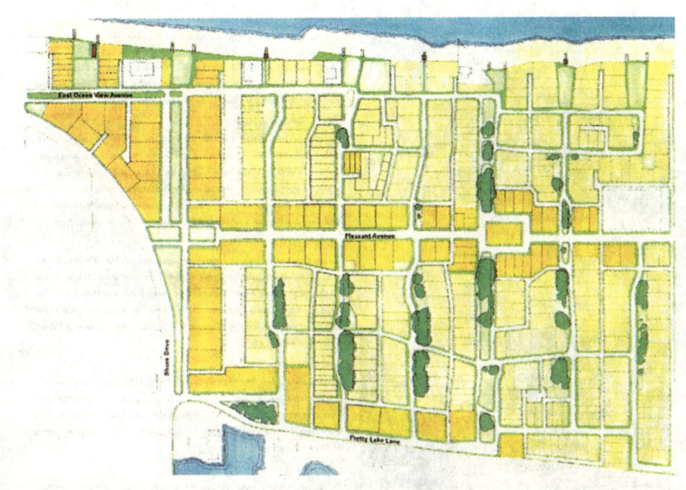

The Plan of East Beach

The Park Streets

The plan for East Beach features a series of unique addresses, each with a distinctive character and mix of houses. A series of informal park streets extends from the Chesapeake Bay to Pretty Lake. Mature trees that exist on the site become the focal points of the park streets and create a sense of a mature neighborhood. Each street has a different form, character and sequence of park spaces to create a rich inventory of neighborhood streets.

The Bay Front

The northernmost address is the Bay address with a mix of large and small houses flanking parks that look out to the Bay. The views from the ends of the park streets are preserved by the bay front parks that also provide public access to the beach. Houses along the Bay will feature deep porches and will build on the tradition of Tidewater Shingle Style waterfront architecture as the dominant image.

Pleasant Avenue

Pleasant Avenue is the heart of the new neighborhood and provides a strong identity for East Beach. A village square, adjacent to Shore Drive, anchors Pleasant Avenue creating a memorable address. The square will be lined with a mix of residential over shops at Shore Drive, manor houses, and a mix of house sizes in the westernmost blocks.

Pretty Lake Avenue

Pretty Lake Avenue is the primary address running east-west along the marina precinct. Near the Shore Drive intersection, the intended character of Pretty Lake is that of 'Little Annapolis' - a reference to the scale, character and mix of uses and buildings found in historic marina districts like Annapolis. This location and mix of uses will create a vibrant place to live, work and shop in East Beach.

The Plan of East Beach　伊斯特比奇平面图

INTRODUCTION　　　5

Tidewater Vernacular Architecture

THE TIDEWATER INFLUENCE in East Beach is clear in its architecture. Houses are simple, low-key and defer to one another and to the indigenous qualities of the landscape. The regional building traditions, which over the years have evolved to take advantage of shade and capture breezes, will be seen in the porches, overhanging eaves, shuttered windows, and screened doors, the traditional use of shingle and clapboard siding, and the picket fenced yards and gardens. The architectural goal is a simple elegance derived from well proportioned massing and fenestration, a rich color palette and details that are derived from the building traditions throughout the region.

Tidewater Vernacular Architecture 泰德沃特乡土建筑

INTRODUCTION

6

Tidewater Colonial Style

Tidewater Shingle Style

Tidewater Victorian Style

Tidewater Arts & Crafts Style

The Houses of East Beach

EAST BEACH HOUSES WILL DRAW on four primary architectural languages that have a unique regional and coastal character appropriate to this site along the Chesapeake Bay. These four languages include:

Tidewater Colonial Houses

These houses have roots in the Colonial and Classical traditions of the region. Later Colonial Revival houses derived their forms from more expressive Classical motifs with Ionic and Doric order columns and entablatures on the porches, deeper eaves and cornices, and a wider variety of house massing and window and door elements. The coastal adaptation of Colonial Revival features deep porches and a more relaxed composition of windows and doors.

Tidewater Shingle

Houses designed in this style have roots in the country's New England coastal villages. Houses are generally simple, elegant forms clad in cut shingles. In the South, many of these houses were built with deep porches and windows under shade to protect from the summer sun. Windows, doors, porches and trim can have either simple colonial trim details or Victorian era proportions and details typically painted in white.

Tidewater Victorian

In many towns, these Victorian houses are the principal 'spice' elements in a neighborhood. Steeply pitched gable roofs facing the street, deep porches and decorative trim combine with vertical proportions to create an endearing style. The coastal variations include many full façade, one- and two-story porches as well as deep eaves and ornate porch trim.

Tidewater Arts & Crafts

Arts & Crafts houses were based on the English tradition of summer cottages and became popular in this country in the late nineteenth century. Deep eaves, robust porch elements and shaped rafter tails are signature elements of this language. Windows tend to be wide in proportion and combined to take advantage of the light in living areas. An asymmetric composition and massing is part of this vocabulary.

East Beach Architecture 伊斯特比奇建筑

INTRODUCTION

7

An East Beach House
Simple, dignified massing with large porches and overhanging roofs

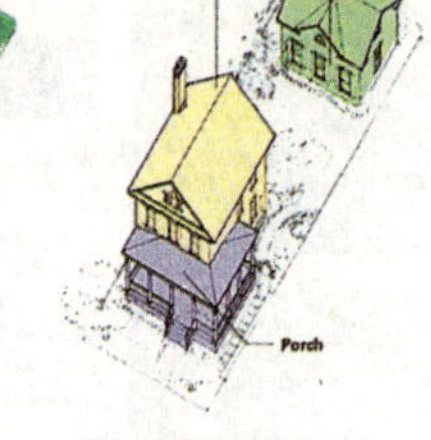

Elements of an East Beach House
The Main Body is the largest and most visible element with the most specific design requirements. Side or Rear Wings, Porches, and Out buildings provide a wide range of options for homebuilders.

East Beach Roof Types

Side Gable House

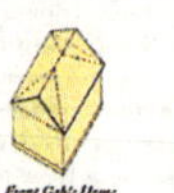

Front Gable House

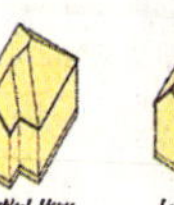

Gable-L House

L-Shaped House

Gambrel House

Mansard House

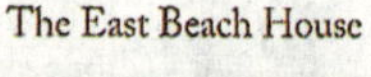

The East Beach House

EAST BEACH HOUSES WILL CREATE the backdrop for the many distinct addresses within the neighborhood. As in traditional Southern towns, the houses define the character of the public space and reflect the individual composition of the private realm behind the porch or front door.

In these traditional neighborhoods, the front portion of the house is the most public and must be responsive to the character of the neighborhood and the adjacent houses. The landscaping of the front yard, the setbacks from the street, the size and placement of the house on the lot and the front porch are all shared elements that form the public realm.

The houses in East Beach are based on the traditional vernacular architecture of the East Coast, using regional house types with style elements applied. The four house styles for East Beach are defined by the character and shape of the Main Body which can draw from any of the six types shown at left.

Principal Elements

The East Beach House includes these principal elements:

The **Main Body** of the house, which is the principal mass and includes the front door.

Side or Rear Wings, which are one or two stories connected to the Main Body. These optional additions have smaller massing than the Main Body and are set back.

Porches are encouraged on the Main Body of the house. These include full-façade front porches, wraparound porches and side porches. Some architectural styles have inset porches into the Main Body of the house.

Out Buildings are optional structures that include carports, garages, storage, carriage buildings, and work studios. Typically, Out Buildings must be placed behind the Main Body.

Towers, Cupolas and Widow's Walks are optional elements that allow distant views from certain lots.

The East Beach House 伊斯特比奇住宅

How To Use This Pattern Book

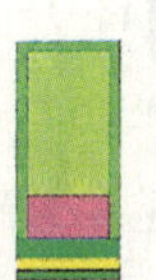

A typical Cottage lot

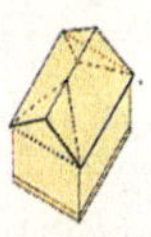

Typical Main Body massing for a Single Cottage

The East Beach Pattern Book will guide the development of neighborhoods and houses to fulfill the original vision described in the master plan. The Pattern Book has three principal sections: *Community Patterns*, which illustrates neighborhood character; *Architectural Patterns*, which establishes the architectural vocabulary and elements that may be used; and *Landscape Patterns*, which sets palettes and standards for the various lot types within the different ecological zones of East Beach.

Step 1　Selecting an East Beach Lot

The *Pattern Book* along with the *Lot-Specific Community Patterns* (separate document) should be used in the very beginning of the process of selecting the lot for your East Beach home. Different lots have different setback requirements. Each lot also has particular requirements for the location of porches and publicly oriented façades. Refer to the *Lot-Specific Community Patterns* to find the lot that best suits the size and layout of the house you plan to build. The *Community Patterns* section in this book will provide a sense of what the different locations within East Beach will be like as places.

Step 2　Shape and Size

The basic mass of the house will determine the general location of the programmatic elements. *The East Beach House* on page A-8 describes the massing pieces: A Main Body, the Porch, Side or Rear Wings, and Out Buildings. The *Lot-Specific Community Patterns* determines specific requirements for setbacks, porch locations and other special conditions related to specific lots. Each Architectural Style section describes the basic massing types found in the precedents for each vocabulary. The layout of rooms should be designed to fit into the massing types found within the particular style you are designing. The roof types are part of this overall massing description.

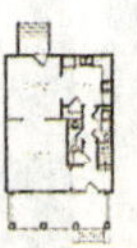

A sample Single Cottage plan

Window and door placement diagram

Standard window type and detail

Step 3　Room Layout and Location of Windows and Doors

The window and door spacing is related to both the shape and the style of the house. It is important that all sides of the house have correctly composed façades. Each section on architectural vocabularies describes the basic elements for each of the four design vocabularies that are found in traditional Tidewater architecture — Colonial Revival, Arts & Crafts, Victorian, and Shingle. Typical window and door compositions are illustrated as part of the massing illustrations for each style. Typical window and door proportions, trim details and special window or door elements are illustrated on a separate page within each section.

Step 4　Porch Design

Porches are important to the character of the neighborhoods. The massing of the front porch is specific to each house type and distinct within a particular vocabulary. The location and design elements of porches on the site are described on a designated page for each vocabulary in the *Architectural Patterns* section. Additional porch requirements for particular lots may be described in the *Lot-Specific Community Patterns*.

An example of a traditionally detailed porch

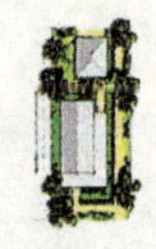

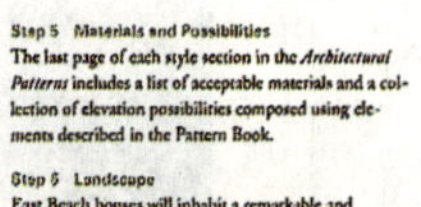

Step 5　Materials and Possibilities

The last page of each style section in the *Architectural Patterns* includes a list of acceptable materials and a collection of elevation possibilities composed using elements described in the Pattern Book.

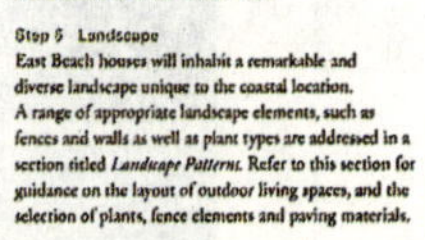

Step 6　Landscape

East Beach houses will inhabit a remarkable and diverse landscape unique to the coastal location. A range of appropriate landscape elements, such as fences and walls as well as plant types are addressed in a section titled *Landscape Patterns*. Refer to this section for guidance on the layout of outdoor living spaces, and the selection of plants, fence elements and paving materials.

Examples of Possibilities from each of the style vocabularies

How To Use This Pattern Book 怎样应用这本模式图则

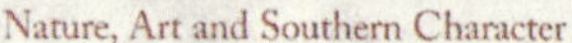

Nature, Art and Southern Character

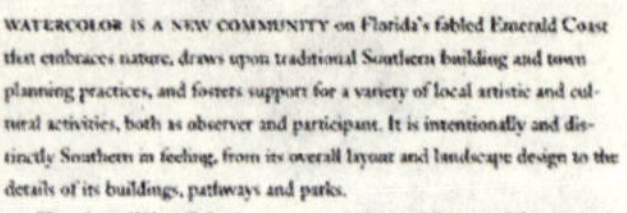

WATERCOLOR IS A NEW COMMUNITY on Florida's fabled Emerald Coast that embraces nature, draws upon traditional Southern building and town planning practices, and fosters support for a variety of local artistic and cultural activities, both as observer and participant. It is intentionally and distinctly Southern in feeling, from its overall layout and landscape design to the details of its buildings, pathways and parks.

The plan of WaterColor is a response to the specific natural features and contrasting qualities of the site and to the best aspects of traditional vernacular place making found in the American South. Pedestrian-scaled streets, scented gardens, shuttered porches, narrow alleys and overhanging roofs, vivid as well as pale colors, deep shades and bright surfaces have been brought together to provide a sense of familiarity, stimulation and ease.

This sense of wholeness is underscored by the interweaving of natural and built elements, each reinforcing an appreciation of the other. Marshes, creeks and wooded frontages around the quietly reflective waters of Western Lake provide a variety of complementary but contrasting settings to the long stretch of dunes and dazzling white beaches. Interspersed among and giving form to this distinctive local landscape are strongly vernacular Southern buildings of varying size, finish and color—all of which underscore the strong regional character of the place.

Nature, Art and Southern Character 自然、艺术和南方风貌

INTRODUCTION

1

The Landscape of WaterColor

THE LANDSCAPE OF WATERCOLOR is extraordinary, a rare and magical configuration of plant communities found only along this stretch of the Florida beachfront. Here the dark, leaf-stained freshwater lakes and waterlily sloughs lie next to the aquamarine surf of the Gulf of Mexico. Two beautiful ecosystems, adjacent yet intact, create between them a remarkable, diverse environment, where one encounters many distinct and identifiable plant communities, including dry upland pine stands, freshwater marshes, cypress depressions, beach dunes with coastal scrub, and sawgrass needle-rush wetlands. Numerous endangered or threatened plant and animal species inhabit these overlapping ecosystems. Because it is such an exceptional environment, and is wooded throughout, very special care has been taken to preserve both the existing vegetation and the animal habitat as integral parts of the new community.

Wind, temperature, sea-salt, soil and, most importantly, water, determine what can grow in WaterColor. And because water is central to everything about this environment, it is embraced, protected and celebrated at WaterColor.

The Landscape of WaterColor 沃特卡勒的景观

INTRODUCTION

2

WaterColor's Town Center

A view of The Lawn

Western Lake Drive

An intimate neighborhood street

The Townscape of WaterColor

The Parks

In the best Southern communities, parks tend to be associated with the most prominent public buildings. Thus, at WaterColor, a framework of public spaces—large and small public parks and squares—defines the plan, with the largest of these open spaces, The Lawn, serving as the front door to the community. This broad public space is not only the heart of WaterColor, but provides a long 'water axis' that connects the beach front and main entrance to the highest point of land on the other side of Western Lake. The lawn also creates a visual corridor through the site, from gulf to lake and unites the wooded upland with the white beaches of the Gulf, integrating two quite different places in a direct and powerful way.

The Main Residential Drive

Many Southern towns have a main street, a communal spine, on which the largest and most prominent homes are situated. Western Lake Drive can be seen as a 'land axis'—it links the three residential areas that surround Western Lake, connecting the entrance lawn at the western end of WaterColor to County Road 395 to the east.

The Neighborhoods

WaterColor is comprised of neighborhoods that use a regional palette of landscape and architecture. Houses are oriented toward the street with deep front porches that convey a sense of neighborhood and civic responsibility. Regardless of their size, houses are unpretentious and defer to the landscape and the street. Low fences or hedges provide a subtle delineation between the public zone of the street and the semi-private zone of the front yard and porch. The predominant public image is of shaded porches nestled within a richly textured native landscape.

The Townscape of WaterColor 沃特卡勒的城市景观

The Architecture of WaterColor

THE VERNACULAR 'SOUTHERNNESS' of WaterColor is clear in its architecture. Houses are simple, low-key and defer to one another and to the connective tissue of the landscape. The regional building traditions, which over the years have evolved to take advantage of shade and capture breezes, will be seen in the porches, 'dogtrot' passages, overhanging eaves, shuttered windows, and screened doors, as well as the traditional use of wood siding, metal roofs and exposed rafter tails. The architectural goal is a simple elegance derived from well proportioned massing and fenestration, a rich color palette and details that catch the sun and create shadows—a sinuous profile on a rafter tail, a delicately turned porch post, the subtle sheen of a cast bronze door knob.

The Architecture of WaterColor 沃特卡勒的建筑

Town Center – The heart of WaterColor, with shops, civic buildings, and residences arranged around The Lawn.

Park Row – Cottages overlooking Grayton Beach State Park.

Sunset Ridge – Tall houses with two-story porches look across the street to the lake, past lakeside houses set back in the trees.

Rose Garden Mews – An intimate street that connects the lake to the beach. A garden surrounded by cottages forms its center.

The Lake District – A gently curving road, lined with prominent houses, that serves as the main residential street.

The Cottage District – An intimate street, defined by the front porches of cottages, connects the lake to the beach.

Rainbow Row – A block-long street of colorful cottages.

Sunrise Ridge – Larger houses along a common green have views to the lake, while smaller cottages face the tennis courts.

Beach Lane – Cottages connected by a pedestrian path have direct access to the beach.

Addresses within the First Phase

THE CHARACTER OF SOUTHERN TOWNS and neighborhoods that one most admires did not happen by accident. Towns and neighborhoods were laid out according to standard surveying practices. Civic spaces were defined, lots were created and building setbacks established. Houses and buildings filled in the plan over the course of many years, the massing and setbacks following what had come before.

Similar tools are used to describe distinct addresses within the community. While houses employ a consistent architectural language throughout the community, each address will be defined by its particular arrangement of house types, use of landscape elements and building coloration.

Phase I is surrounded by water, spanning from the beach to Western Lake.

One of WaterColor's addresses is Rose Garden Mews, with cottages surrounding a public garden.

Sunrise Ridge is an address characterized by larger houses facing a common green.

Addresses within the First Phase 第一阶段里的地址

INTRODUCTION　A　5

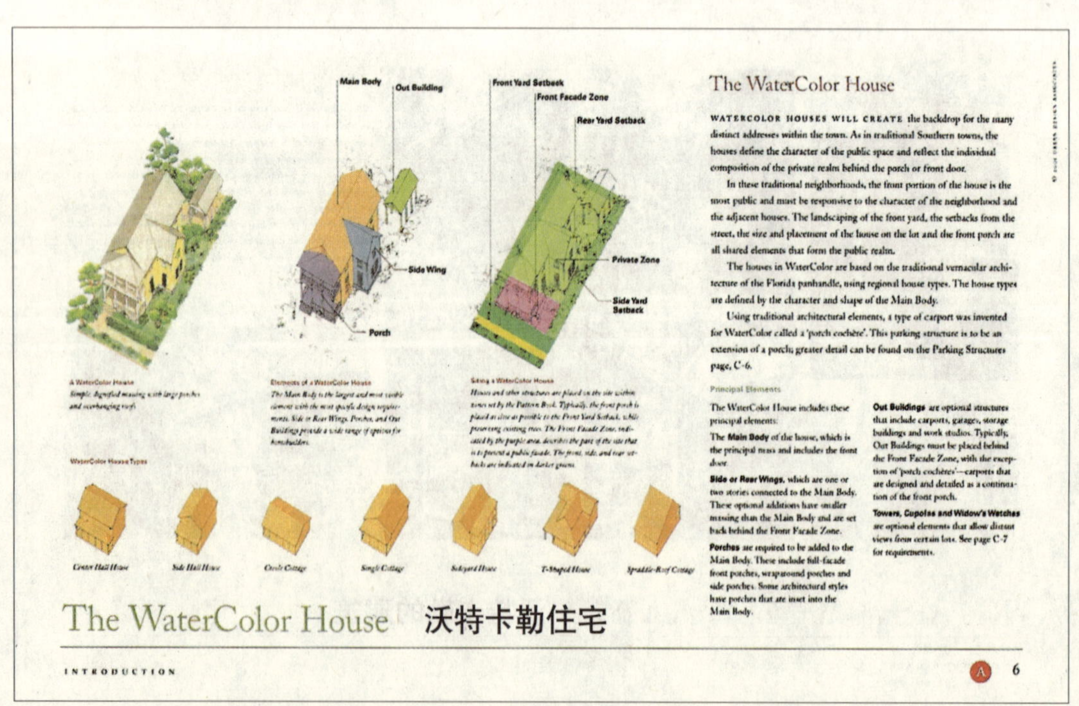

A WaterColor House

Simple, dignified massing with large porches and overhanging roofs.

Elements of a WaterColor House

The Main Body is the largest and most visible element with the most specific design requirements. Side or Rear Wings, Porches, and Out Buildings provide a wide range of options for homebuilders.

Siting a WaterColor House

Houses and other structures are placed on the site within zones set by the Pattern Book. Typically, the front porch is placed as close as possible to the Front Yard Setback, while preserving existing trees. The Front Facade Zone, indicated by the purple area, describes the part of the site that is to present a public facade. The front, side, and rear setbacks are indicated in darker green.

WaterColor House Types

Center Hall House　Side Hall House　Creole Cottage　Single Cottage　Sideyard House　T-Shaped House　Spraddle-Roof Cottage

The WaterColor House

WATERCOLOR HOUSES WILL CREATE the backdrop for the many distinct addresses within the town. As in traditional Southern towns, the houses define the character of the public space and reflect the individual composition of the private realm behind the porch or front door.

In these traditional neighborhoods, the front portion of the house is the most public and must be responsive to the character of the neighborhood and the adjacent houses. The landscaping of the front yard, the setbacks from the street, the size and placement of the house on the lot and the front porch are all shared elements that form the public realm.

The houses in WaterColor are based on the traditional vernacular architecture of the Florida panhandle, using regional house types. The house types are defined by the character and shape of the Main Body.

Using traditional architectural elements, a type of carport was invented for WaterColor called a 'porch cochère'. This parking structure is to be an extension of a porch; greater detail can be found on the Parking Structures page, C-6.

Principal Elements

The WaterColor House includes these principal elements:

The **Main Body** of the house, which is the principal mass and includes the front door.

Side or Rear Wings, which are one or two stories connected to the Main Body. These optional additions have smaller massing than the Main Body and are set back behind the Front Facade Zone.

Porches are required to be added to the Main Body. These include full-facade front porches, wraparound porches and side porches. Some architectural styles have porches that are inset into the Main Body.

Out Buildings are optional structures that include carports, garages, storage buildings and work studios. Typically, Out Buildings must be placed behind the Front Facade Zone, with the exception of 'porch cochères'—carports that are designed and detailed as a continuation of the front porch.

Towers, Cupolas and Widow's Watches are optional elements that allow distant views from certain lots. See page C-7 for requirements.

The WaterColor House 沃特卡勒住宅

INTRODUCTION　A　6

How to Use this Pattern Book

The WaterColor Pattern Book will guide the development of neighborhoods and houses to fulfill the original vision described in the master plan. The Pattern Book has three sections. *Community Patterns*, which establishes guidelines for placing the house on its lot and defining neighborhood character; *Architectural Patterns*, which establishes the architectural patterns and elements that may be used; and *Landscape Patterns*, which sets palettes and standards for the various lot types and ecological zones.

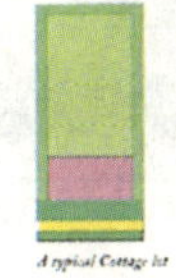

A typical Cottage lot

Step 1　Selecting a WaterColor Lot

The Pattern Book should be used in the very beginning of the process of selecting the lot for your WaterColor home. Some lots require two-story houses; other lots require one- or one-and-a-half-story houses. Different lots have different setback requirements. Each lot also has particular requirements for the location of porches and publicly oriented facades. Refer to the Community Patterns section to find the lot that best suits the size and layout of the house you plan to build.

Typical Main Body massing for a Single Cottage

Step 2　Shape and Size

The basic mass of the house will determine the general location of the programmatic elements. The Pattern Book addresses the appearance of the house and the yard from the street or public space. The portions of the house that are not visible from the public areas can accommodate a broader range of elements and may not follow the compositional patterns found on the public faces of the house.

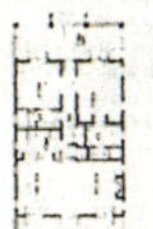

A sample Single Cottage plan

Window and door placement diagram

The WaterColor House on page A-6 describes the massing pieces: Main Body, Porch, Side or Rear Wing, and Out Building. House Types are determined by the shape of the Main Body and the location of porches. The Address pages specify which house types are permitted on each lot, and whether the house is to be one story or two.

Depending on the house type, the Main Body will be a simple rectangular volume oriented perpendicular to the street or parallel to the street, or a simple T-shape, of one or two stories. The Main Body massing for each house type is described on page C-2.

Step 3　Room Layout and Location of Windows and Doors

The window and door spacing is related to both the shape and the width of the house. It is important that all sides of the house that are exposed to public view have correctly composed facades. On facades with porches, window and door spacing should relate to porch bays. Some possible facade compositions are shown on page C-2. Window and door designs are described on page C-5.

Step 4　Porch Design

Porches are important to the character of the neighborhoods. The massing of the front porch is specific to each house type. The location of porches on the site is described in the Building Placement pages (B-3 through B-8). Additional porch requirements for particular lots are described in the Address pages (B-9 through B-14). Porches must be at least 8 feet deep. Page C-4 specifies column types, handrail and eave profiles to use in designing the porches.

An example of a traditionally detailed porch

Step 5　Materials

As outlined in Section C-7, Materials and Possibilities, WaterColor houses will use materials traditional to this region: wood board-and-batten or drop siding, metal roofs and brick or stucco pier foundations.

Step 6　Details

Before lumber yards and building products manufacturers took over the design of windows, doors and columns, builders distinguished themselves by inventing signature eaves, handrails and columns. The Architectural Patterns section (Section C) documents some of the typical details found in the best traditional examples of Southern vernacular architecture.

Typical window head/sill profile

Step 7　Color

Each address within WaterColor will be described by a particular color palette. Refer to Section E when selecting the colors for your house.

Step 8　Landscape

WaterColor houses will inhabit the landscape with the least possible disruption to the site. Refer to Section D for guidance on the layout of outdoor living spaces, and the selection of plants and paving materials.

How to Use this Pattern Book　怎样应用这本模式图则

INTRODUCTION

7

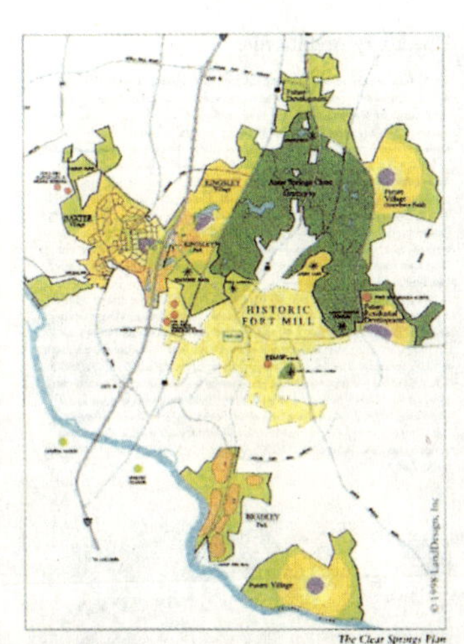

The Clear Springs Plan

Clear Springs and Upcountry Traditions

Clear Springs, a 6,200-acre development being undertaken by the Close family in Fort Mill, South Carolina, embodies a remarkable concept for how to make a great place to live, work, raise children, enjoy the natural environment and participate in building community. Clear Springs is a constellation of villages, each with its own distinctive character surrounding historic Fort Mill. Neighborhoods are designed in the tradition of South Carolina's Upcountry towns complete with tree-lined streets, neighborhood parks, and gracious houses with architectural details such as deep front porches that face the street and convey a sense of neighborhood and civic place.

The Close family has dedicated over 2,300 acres surrounding Fort Mill as an environmental preserve and greenway. The Anne Springs Close Greenway features over 26 miles of hiking and horseback riding trails, three lakes, the Leroy Springs recreation complex and restored historic buildings. Each of the villages is linked to this greenway system and residents will enjoy access to this resource as well as to a wide variety of civic parks and recreational amenities.

The design of each village is based on the treasured legacy of Upcountry towns and villages that developed during the nineteenth century and into the early part of the twentieth century. These places are admired today for their character and quality of architecture. The pattern of development is an expression of the democratic ideals of civic responsibility and participation. Each neighborhood, street, park, or public space is designed using the regional palette of landscape and architecture ensuring a continuation of the best traditions and sense of identity that is unmistakably Upcountry.

The Anne Springs Close Greenway provides over 2300 acres of preserved natural environment for the residents of Fort Mill and the Clear Springs villages.

Clear Springs neighborhoods and houses will be designed in the tradition of the region. The Springs' Founders Home in Fort Mill is an example of Victorian-era architecture found throughout the region.

Clear Springs Development Company, LLC reserves the right to change the plans, landscaping, and buildings at their sole discretion and without notice. All renderings, plans and maps are artists' conceptions and are not intended to be an exact depiction of the landscaping or buildings.

斯普林斯公司和内陆传统

Urban Design Associates

Introduction · Baxter　A-1　Clear Springs & Upcountry Traditions

Neighborhood streets have a variety of lot widths, setbacks, and landscape

Upcountry Towns and Villages

Neighborhoods

One of the most consistent features of the Upcountry neighborhoods is the pattern of houses on a street. Big houses are mixed with small houses; large lots will often interrupt a pattern of narrow lots; setbacks may vary from one side of the street to the other. There are seldom fences between properties and yards often flow into one another. The predominant image is of white, frame houses sitting within a deep green band of trees and lawn. Streets are generally narrow with parking on both sides. Sidewalks with trees and lawns of varying widths give neighborhoods an informal and diverse character. This loose fit, predicated on pockets of generous lawns within relatively modest lot sizes, creates the charm of the small town.

Main Streets

Main Streets have a unique character in the Upcountry region. The pattern is a series of two- and sometimes three-story buildings, often attached but many times separated slightly to form the fabric of the downtown. The vocabulary is a mix of Victorian-era detailing and proportion as well as a more classical building vocabulary. The most successful towns have centers. The square in Chester is an example of a Main Street that ends at the top of the hill in an enclosed space.

Civic Buildings

Davidson, Chester and York all have excellent examples of civic architecture. In towns of this scale, there are a series of small-scale buildings such as the town hall, the courthouse, the library and institutional buildings such as the masonic halls and churches. These are typically set within the normal pattern of lots and blocks with the exception of a city hall or courthouse sometimes fronting on a public space as in Chester. These buildings are scaled down to reflect the character of the single-family house with the exception of the churches, which historically use spires and bell towers to mark the skyline and provide orientation. The dominant style for the public buildings is Classical.

Parks and Open Space

In the smaller towns, the parks tend to be associated with either the most prominent public buildings as formal squares or front lawns or as community parks with recreation facilities within them. There is also the rural edge – the distinct line between farmland, woodlands and the compact pattern of houses within a village. Charlotte's best neighborhoods, developed in the 1920's, have a sophisticated sequence of neighborhood parks. This pattern can be integrated into the small town to create another layer of richness in the Upcountry character.

Neighborhood parks bring nature into the heart of the community

Gracious lawns are a hallmark for corner lots

A classic Upcountry retail street

Urban Design Associates

Introduction · Baxter　A-2 内陆市镇和村庄 Upcountry Towns & Villages

Upcountry Classical-Adams/Federal Style

Upcountry Colonial Revival with full front porch

Upcountry Architecture

The historic settlements in the Upcountry region of North and South Carolina were largely built by Irish, Scotch and German settlers migrating from Pennsylvania through the Shenandoah Valley. These settlements were frontier settlements in the late 1700s and early 1800s. They were agricultural in nature and grew into villages and towns serving regional populations. Today the predominant architectural styles found in Upcountry towns include Classical, with houses designed in the Adams/Federal and Greek Revival styles. These include a mix of red brick and clapboard exteriors with slate and metal roofs. During the early 1900s, Colonial Revival houses were built that provided a vocabulary complementary to the Classical-era styles. Colonial Revival is also a pervasive style. The Victorian-era houses typically add 'spice' to the neighborhood fabric. They were largely built as either large Italianate mansions in urban locations, or Carpenter Gothic cottages in village neighborhoods or farmhouses.

This palette of architectural styles has limitless variations and creates a varied and interesting neighborhood character. This mix is the distinguishing character of neighborhoods in towns like York and Chester and sets them apart from later developments where the vocabulary is more eclectic and discordant. Houses built in Clear Springs will be designed in one of the three architectural vocabularies. While there are pre-approved house plans available to builders and homeowners, this *Pattern Book* ™ contains a series of design guidelines for architects to use when designing individual houses in one of the three styles. Since the *Pattern Book*™ is not an exhaustive or definitive manual for each style, architects and designers may also draw on historic Upcountry precedents as a basis for new house designs.

Upcountry Classical Revival Style

Upcountry Victorian

Upcountry Colonial Revival

Upcountry Victorian Carpenter Gothic Cottage

Houses in Baxter will be designed in one of the three Upcountry styles.

Urban Design Associates

Introduction · Baxter　A-3　建筑先例 Architectural Precedents

Phase II Plan of Baxter

The Pattern Book and Baxter Neighborhoods

The character of the towns and neighborhoods that we most admire did not happen by chance. The builders used pattern books from England and the colonies to design and build sophisticated houses and civic buildings without the aid of architects in areas remote from the urban centers. Towns and neighborhoods were laid out according to set surveying practices. Civic spaces were defined, lots were created and building setbacks established. Houses and buildings filled in the original plan over many years. While the architectural styles changed over the years, the common elements of streets and public open space as well as setbacks and massing followed what had come before.

The building of Baxter, the first village to be built in the Clear Springs Plan, will also be guided by similar tools. Each village in Baxter will have its own distinct character formed by the rolling topography, woodlands, the design of the streets and neighborhood parks and the placement and design of houses on the lots. This *Pattern Book* describes guidelines for placing houses on lots and for designing and building houses in the Upcountry traditions. The *Community Patterns* section of the book describes the range of setbacks for houses and ancillary structures, the width of streets, parks and location of sidewalks and tree lawns for each neighborhood. The common areas of streets and parks are carefully designed as shared gardens and civic spaces – some formal and some informal. Indigenous plants, traditional landscape treatments found in historic Upcountry towns and neighborhoods serve as models for the design of Baxter. The *Architectural Patterns* section describes the range of architectural styles found in Upcountry neighborhoods with key elements defined for architects and builders.

The following pages provide a sense of what the finished neighborhood character might look like using both the architectural styles described in the Architectural Patterns section and the setbacks and landscape elements described in the Community Patterns section.

View along Colonel Springs Way

Baxter Square

Typical Baxter Neighborhood Street

Houses look out across neighborhood parks

Urban Design Associates

Introduction · Baxter　　A-4　　场地 The Site

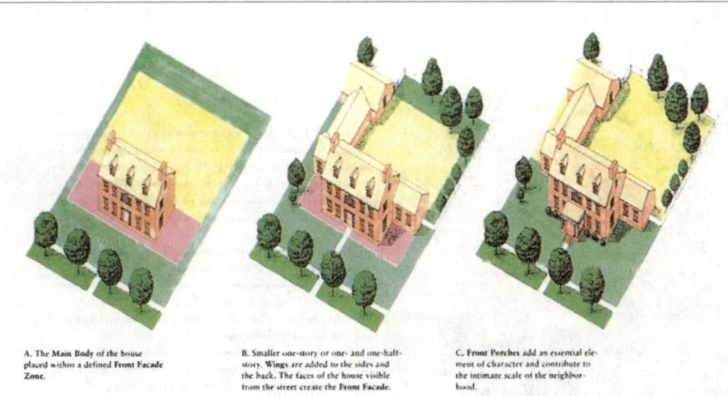

A. The **Main Body** of the house placed within a defined **Front Facade Zone**.

B. Smaller one-story or one- and one-half-story. **Wings** are added to the sides and the back. The faces of the house visible from the street create the **Front Facade**.

C. **Front Porches** add an essential element of character and contribute to the intimate scale of the neighborhood.

Baxter Houses

Building a house in Baxter is part of building a neighborhood. Each house contributes to the overall character and quality of a particular street or park address. Houses will relate to one another within a given address – setbacks will be consistent, landscaping will be coordinated, and massing will be similar.

The public face of the house, the **Front Facade**, is an important element that defines the neighborhood character. Porches, windows and doors, correct proportions and traditional 'detailing' are important facets that contribute to the character of Baxter. The private areas of the back yard will be screened from public view by the siting of the house and the relationship between houses.

The **Front Facade** of each house includes all the sides of a house that are visible from the street or public spaces.

Traditional houses have three major elements that make up the house form including:

- The **Main Body**, the largest mass of the house, contains the front door and has a composed window and door pattern.
- **Side wings** are one- or one-and-one-half stories set back from the front facade of the Main Body. These are typically smaller in scale than the Main Body.
- **Porches** can include full-facade, one- and two-story structures, side porches on Colonial Revival houses, or smaller porticos surrounding the front door.

Corner houses have public facades on two streets. These houses use side wings and fences to create a formal composition on the side street, and to screen the private world of the backyard from public view. Each lot has zones within which the Front Facade of the house can be placed. These are described on page B-2.

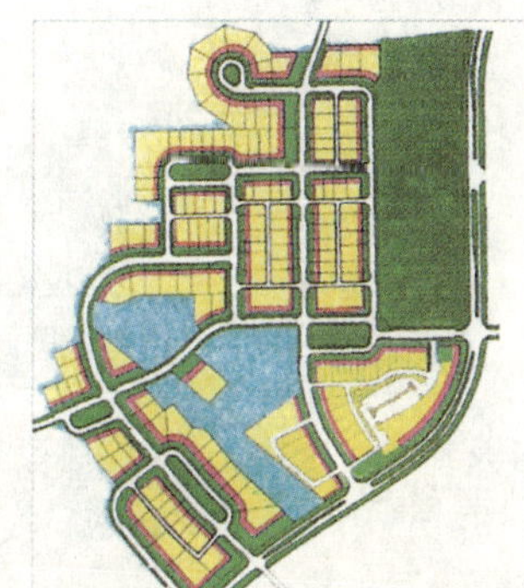

Partial plan of Baxter

Urban Design Associates

Introduction · Baxter　　A-5　　巴克斯特住宅 Baxter Houses

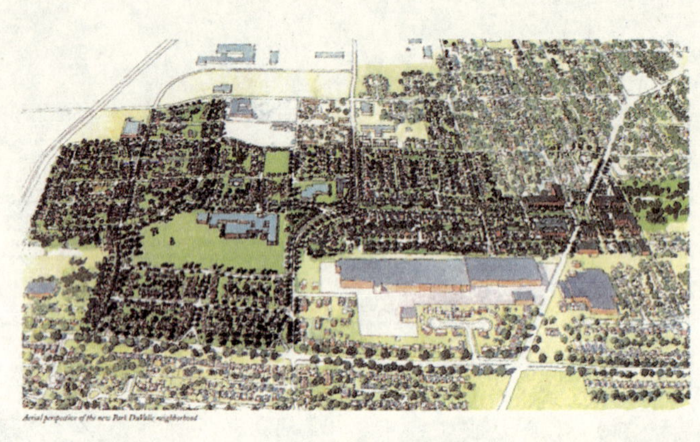

Aerial perspective of the new Park DuValle neighborhood

The New Park DuValle Neighborhood

THE NEW NEIGHBORHOOD IS DESIGNED to become a mixed-income, mixed-use, compact and pedestrian-friendly community with many activities of daily life within walking distance.

The Park DuValle neighborhood is comprised of several distinct Addresses, each with its own character. It includes a wide range of housing types and a sequence of public open spaces and parks linking the parts of the neighborhood together and to nearby amenities such as Algonquin Park. A new front door for the entire neighborhood will be established with single-family houses along Algonquin Parkway and at several points along Wilson Avenue.

Within the neighborhood are a series of institutions and community facilities, including a Village Green. Lined with buildings that combine public uses, retail shops, and residential uses, the Village Green will become a new heart for the neighborhood. The system of parks and parkways provide appropriate settings for civic and religious buildings that will be reinforced as anchors for the community.

The Park DuValle Plan　帕克杜瓦拉平面图

INTRODUCTION

 1

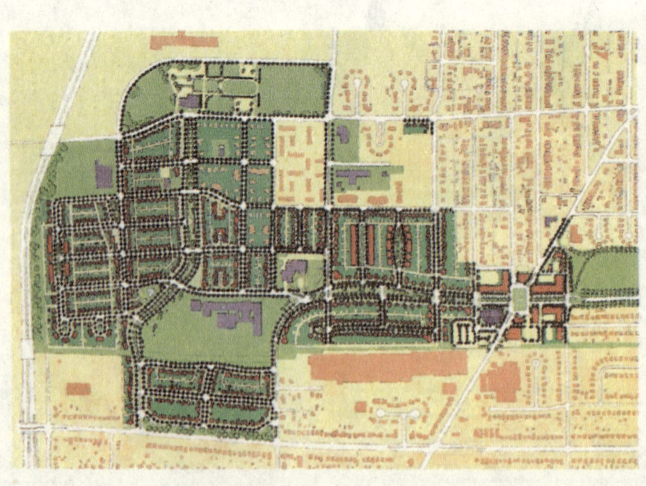

The Plan as Framework

THE PLAN PROVIDES AN INTERCONNECTED network of streets and public open space. They define blocks within which a wide variety of individual lot types are accommodated. Placement of alleys and front yard set-backs are fixed but can accommodate lots of varying widths. The plan will be developed over time, in market conditions which may change. Therefore, this flexible system of blocks and lots provides the capacity to respond to changing conditions.

In order to create marketable, mixed-income development, however, it is essential to create a series of attractive Addresses, each with its own character. The plan calls for developing these Addresses by following Louisville's great traditions of neighborhood development. The architecture of individual houses and the image of Park DuValle's public spaces should recall the best and most stable of Louisville's neighborhoods.

Research in preparing this Pattern Book identified a number of well known residential 'Addresses' in Louisville which have served as models for the guidelines. Three of them are illustrated on the following pages: Old Louisville, Olmsted's Parkway designs, and the Cherokee Triangle area.

© 2001 URBAN DESIGN ASSOCIATES

The Plan as Framework　作为框架的平面图

INTRODUCTION

A 2

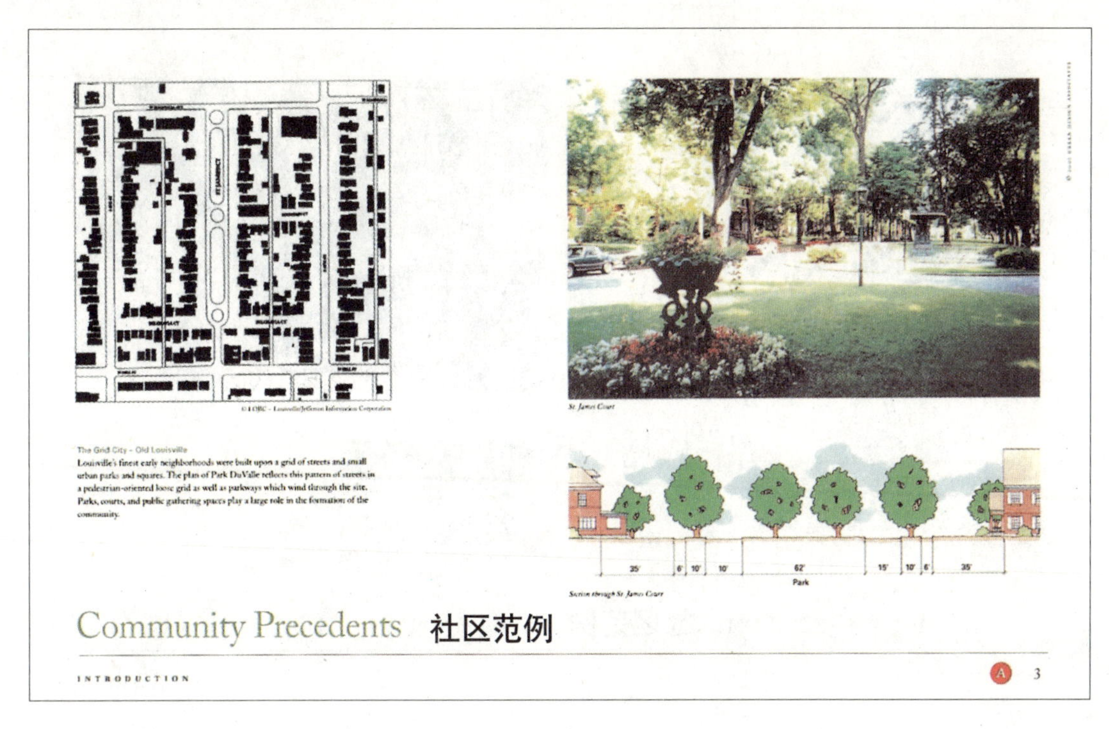
© LOJIC – Louisville/Jefferson Information Corporation
St. James Court
35'
6'
10'
10'
62'
Park
15'
10'
6'
35'
Section through St. James Court
The Grid City – Old Louisville
Louisville's finest early neighborhoods were built upon a grid of streets and small urban parks and squares. The plan of Park DuValle reflects this pattern of streets in a pedestrian-oriented loose grid as well as parkways which wind through the site. Parks, courts, and public gathering spaces play a large role in the formation of the community.
Community Precedents
INTRODUCTION
A
3

社区范例

Cherokee Parkway
Algonquin Parkway
30'
9'
6'
20'
40'
40'
6'
9'
Section through Cherokee Parkway
Olmsted's Parkways
The Parkways form an armature of green space linking Louisville's three major Olmsted-designed parks. New park drives in Park DuValle, which reflect this traditional linkage of green spaces, wind through the site and link to Algonquin Parkway as well as Algonqion Park.
Community Precedents
INTRODUCTION
A
4

社区范例

Cherokee Triangle, Crescent Hill and The Highlands

The establishment of Cherokee, Iroquois, and Shawnee Parks at the turn of the century provided an attractive setting for the development of new 'suburban' neighborhoods. Built largely between 1890 and 1930, Cherokee Triangle, Crescent Hill, and The Highlands neighborhoods most clearly express the influence of Olmsted's design principles in their picturesque street layout and expanses of community green space.

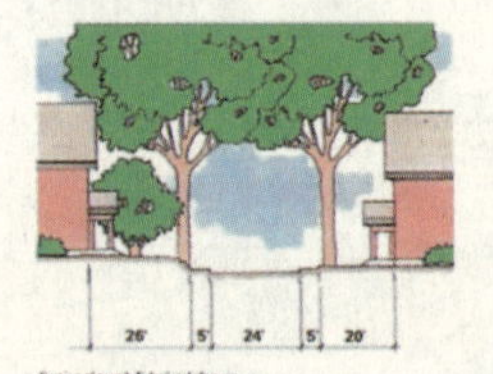

Section through Edgeland Avenue

Community Precedents 社区范例

The Park DuValle Pattern Book

THE PARK DUVALLE PATTERN BOOK contains design guidelines for both community character and architectural character. The section that follows, *Community Patterns*, describes the general principles for placing houses on their lots, as well as specific requirements for lots within the different neighborhoods. These principles include setbacks, overall massing of the house, locations for fences and ancillary structures and access from driveways or alleys. The third section, *Architectural Patterns*, describes the palette of architectural styles for Park DuValle and includes guidelines for designing the parts of the house that are visible from a street or public space.

Once a lot has been selected, the guidelines for placing the house on the lot can be determined by turning to the page that describes the general conditions for the lot type and then the page that describes specifics for the particular neighborhood in which the lot is located. For instance, if you decided to build a Victorian house on a Cottage Lot along Von Spiegel Avenue, you would first turn to B-2 to determine the general setbacks and width of the house. You would then turn to B-12 to determine what specific variations or additional site guidelines might apply.

The specific house design would then be developed or selected in accordance with the Victorian style described in the Architectural Patterns section.

The Park DuValle Pattern Book 帕克杜瓦拉模式图则

View of renovated mill building and new live/work buildings from the proposed town square park

Colonial Revival house in Belmont

European Romantic house in Charlotte

New residential street lined with Colonial Revival, European Romantic and Craftsman houses

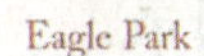

Eagle Park

EAGLE PARK IS A NEW VILLAGE designed in the tradition of the best small towns and villages found throughout this region in North Carolina. There is an informal quality to the neighborhood character with a mix of different houses that range in size and character. Streets will have an intimate feel with narrow lanes and porches facing front gardens and raised yards. Tree-lined streets will create a shaded, quiet atmosphere that transforms the neighborhood street into a shared 'outdoor room.' Service lanes provide access to garages behind the houses. Eagle Park features a series of neighborhood parks and squares each with its own character and sense of place. Residents are never more than a block away from a park.

The western neighborhoods in the village will feature a restored mill with loft housing and offices as part of the village center which will have live-work houses and attached houses combining to create an active precinct to complement the neighborhoods. The Pattern Book serves as a guide to the creation of special neighborhoods and houses designed and built to make an authentic place to live, work and play.

Overview of Eagle Park 伊格尔帕克总则

INTRODUCTION  1

Arts & Crafts House

Informal Garden in Pinehurst

Belmont Front Yards

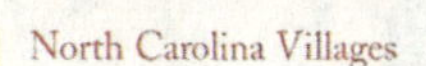

North Carolina Villages

THE CHARACTER OF villages and small towns in the region provides precedent for the neighborhood design in Eagle Park. The design of the new district surrounding the historic mill in Eagle Park draws on the commercial centers of villages like Belmont, Davidson, Southern Pines and Pinehurst. These precedents combine the small-scale character and softness of a residential street address with commercial uses. The treatment of sidewalks, landscape and building character will be reflected in the design for Eagle Park. The neighborhoods of Eagle Park are also drawn from these small towns and villages as well as historic neighborhoods in Charlotte. The character found in Belmont neighborhoods features front lawns raised above the street, often with a sloped lawn or short wall. Small-scale wrought iron fences and hedges reinforce the pattern of a seperation between the public street and the front yard. Many neighborhoods have cottages combined with larger houses on corners and along the Main Street. Many houses were influenced by the romantic periods of domestic architecture available from Pattern Books and catalogs early in the twentieth century.

Pinehurst

Historic Sanborn map of Belmont, South Carolina

Regional Precedents　地区范例

INTRODUCTION A 2

A Colonial Revival Precedent

A Craftsman Precedent

A European Romantic Precedent

A Mixed-Use Building Precedent in Belmont

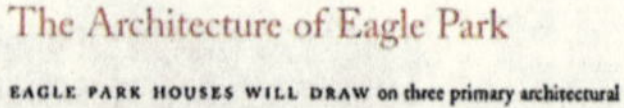

The Architecture of Eagle Park

EAGLE PARK HOUSES WILL DRAW on three primary architectural languages that have a unique regional character appropriate to the context of Belmont. These three languages include:

Belmont Colonial Revival Houses

These houses have roots in the Colonial and Classical traditions of the region. Later Colonial Revival houses derived their forms from more expressive Classical motifs with Ionic and Doric order columns and entablatures on the porches, deeper eaves and cornices and a wider variety of house massing and window and door elements. The regional adaptation of Colonial Revival features deep porches and a more relaxed composition of windows and doors.

Belmont Craftsman Houses

Arts & Crafts houses were based on the English tradition of summer cottages and became popular in this country in the late nineteenth century. Deep eaves, robust porch elements and shaped rafter tails are signature elements of this language. Windows tend to be wide in proportion and combined to take advantage of the light in living areas. An asymmetric composition and massing is part of this vocabulary.

Belmont European Romantic

Houses designed in this style have roots in the country's interpretation of English and European cottages around the first quarter of the twentieth century. Houses designed in this Romantic style became hallmark images for aspiring homeowners. In the South, many of these houses were built as interpretations of the original stone or stucco precedents found in England using cut shingles and clapboard siding. There are many brick examples with half-timbered accents as well. Houses are generally simple, elegant forms with asymmetric compositions and a variety or casement or double-hung windows.

A Belmont Colonial Revival House

A Belmont Craftsman House

A Belmont European Romantic House

Eagle Park Architecture 伊格尔帕克建筑

INTRODUCTION

3

The overall Master Plan for North Ducker showing neighborhoods surrounding the golf course

North Ducker will have a system of streets and open space similar to those found in Biltmore Forest

NORTH DUCKER LIES just south of the historic Biltmore House and is part of the original 125,000 acre estate established by George Washington Vanderbilt. Ducker Mountain is a geographic landmark in the region encircled by the French Broad River and rising over 600 feet above the river valley. From the mountain, you experience spectacular views over the rugged and beautiful western North Carolina landscape including the estate grounds, Pisgah National Forest and the town of Biltmore Forest. North Ducker continues the heritage of carefully designed landscapes and villages commissioned by the Vanderbilt family.

North Ducker encompasses 1,000 acres of woodland forest crossed with streams. It has been designed to accommodate houses tucked into the forest overlooking an 18-hole championship golf course. Mountain lanes wind up the slopes along the original trails that followed courses of minimal disturbance to the mountain terrain. Houses on North Ducker will be designed to take advantage of the views while preserving the woodland character of the mountain site. The mountain lanes, private drives and garden landscaping are all elements that build on the inherited patterns of the Olmsted-inspired plan for Biltmore Forest.

Biltmore Arts & Crafts

Biltmore Colonial Revival

Biltmore European Romantic

North Ducker 北达克尔

INTRODUCTION

A　I

伊格尔帕克和达克尔山（比尔特摩）

Parks in Biltmore Forest

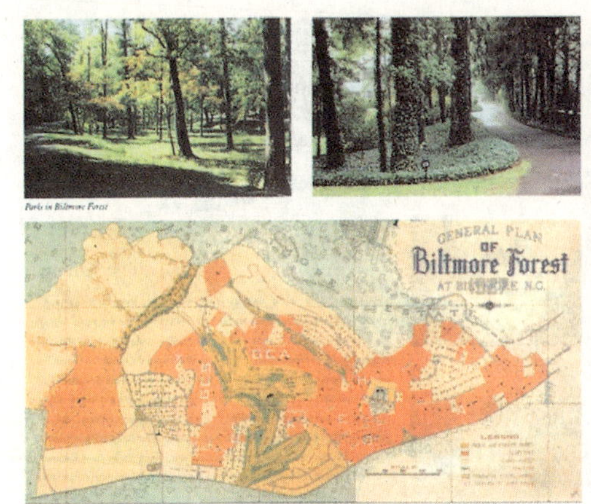

Plan of Biltmore Forest

Houses in Biltmore Forest

View of the terrace overlooking the golf course at the Biltmore Country Club

IN THE EARLY 1900S, the Vanderbilt family commissioned the design of a village adjacent to the estate, Biltmore Forest, which eventually became incorporated as a town. The Frederick Law Olmsted-inspired plan for the village was based on concepts of creating organic, garden neighborhoods nestled into the woodland foothills flanking the estate. A compact and elegant golf course, designed by Donald Ross, creates a social and recreational center to the plan. A village center was constructed adjacent to the Forest neighborhoods to provide neighborhood shopping and services. Biltmore Forest provides a strong precedent for the design of North Ducker. Narrow, winding streets are positioned to take advantage of the natural topography and define a network of private parks and woodland vistas that create a strong sense of place reflective of the western North Carolina landscape.

The use of low stone walls as edging for many of the streets complements the Arts & Crafts light fixtures and the dense forest landscape flanking the streets. Houses are set deeply into their lots with significant stands of trees in a natural woodland setting acting as a green veil between the private realm of the house and the more public realm of the street or park. The best house sites have a series of landscaped terraces to create a proper setting for the house on a steeply sloped lot. Driveways are narrow and unobtrusive. Often they are flanked by stone walls to provide a gracious arrival and minimal disturbance to the forest landscape. The golf course creates open vistas for residents across the rolling hills. These elements form the basis for the North Ducker addresses.

The Biltmore Forest Legacy 比尔特摩森林遗产

Portrait of Olmsted painted by John Singer Sargent on display in the Biltmore House

The Legacy of the Biltmore Estate

THE RUGGED BEAUTY OF ASHEVILLE'S rural terrain instantly bewitched George Washington Vanderbilt, and it is here that he decided to fulfill his long time vision of an estate that was to be modeled after the immense land baronies he had seen in Europe. They not only served as cherished retreats, but also as self-supporting businesses. It had been suggested to Vanderbilt that commercial forestry represented a good investment, so he purchased 125, 000 acres, and began taking the first steps toward his goal.

Vanderbilt called the estate 'Biltmore' – from Bildt, the Dutch town where his ancestors originated, and 'more', an old English word for open, rolling land. He then commissioned architect Richard Morris Hunt and landscape architect Frederick Law Olmsted to collaborate on the estate's design. For Olmsted, this was his last private commission, and it is said to have been a summation of all his ideas.

Vanderbilt later hired Gifford Pinchot to help restore the forests of Biltmore at Olmsted's request. Pinchot was one of the first people in the United States to start thinking about forest health and management. His work helped to establish the Biltmore School of Forestry and what is now part of Pisgah National Forest as the Cradle of Forestry in America.

Much of Biltmore's lush woodland, the first forest in the country to be managed scientifically, was planted in the 1890s; today it covers some 4,500 acres. Although some improvements have been made to improve the quality of soil, water and wildlife habitats, the harvesting of mature trees, as well as selective thinning and pruning, remain a vital part of current forest management practices.

Landscape Precedents of Biltmore Estate 比尔特摩地产景观范例

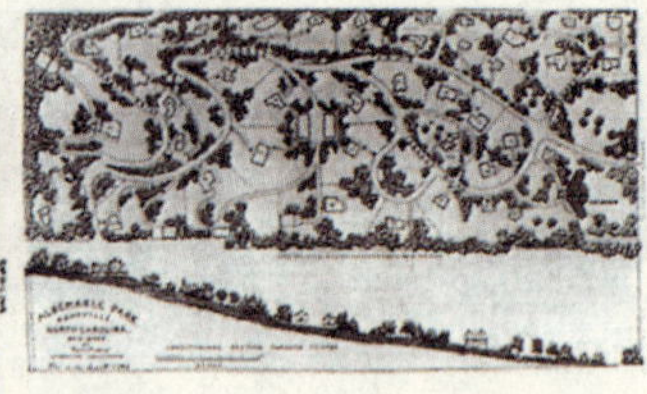

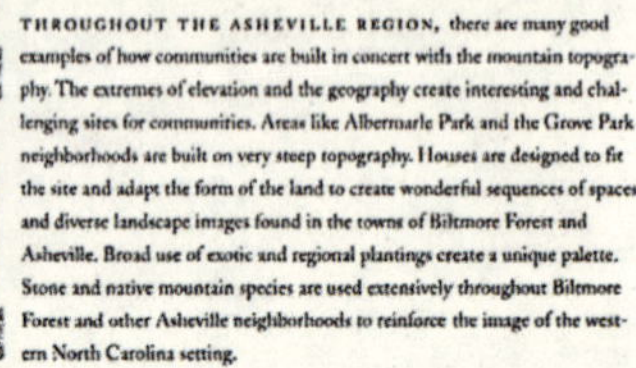

The Neighborhoods of Asheville

THROUGHOUT THE ASHEVILLE REGION, there are many good examples of how communities are built in concert with the mountain topography. The extremes of elevation and the geography create interesting and challenging sites for communities. Areas like Albermarle Park and the Grove Park neighborhoods are built on very steep topography. Houses are designed to fit the site and adapt the form of the land to create wonderful sequences of spaces and diverse landscape images found in the towns of Biltmore Forest and Asheville. Broad use of exotic and regional plantings create a unique palette. Stone and native mountain species are used extensively throughout Biltmore Forest and other Asheville neighborhoods to reinforce the image of the western North Carolina setting.

Landscape Precedents 景观先例

INTRODUCTION

4

Mountain Shingle Precedent

Classical Precedent

Mountain Shingle Precedent

European Romantic Styles with English and French Elements

English Romantic Precedent

Colonial Revival Precedent

Mountain Architecture

ONCE THE RAILROAD CONNECTED Asheville to the east coast network of cities and towns in 1880, the region became a magnet for people drawn to the mountain environment for recreation, health benefits and respite from the urban centers in the northeast. With this growth and influx of people, the character of the region's architecture was transformed rapidly in the late 1800's with such projects as the Biltmore Estate. The prominent architect Richard Morris Hunt drew on strong European influences for his work and brought with him an interest in the current architectural influences developing in the northeastern region of the country. Architects such as Richard Sharp Smith, who came to Asheville to supervise the Biltmore construction, contributed many civic and residential buildings to the Asheville region. A unique architectural language that used combinations of different styles and architectural elements developed within the region. This trend is evident in many of the area's historic neighborhoods such as Montford and Albemarle Park. Significant variations of the Arts & Crafts style and the European Romantic styles are found throughout Asheville and the surrounding region. Classical and Colonial Revival houses provide a balance to the more exotic architectural varieties.

North Ducker houses will be designed using the incredible variety found in four dominant style families: the Biltmore Mountain Shingle style, the Biltmore Colonial Revival, the Biltmore English Romantic style, and the Biltmore Classical style. The Pattern Book establishes design patterns for houses in each of the style categories. Each style builds on the unique elements found in the inherited stock of fine quality houses in the region.

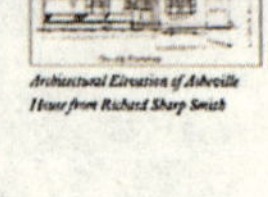

Architectural Elevation of Asheville House from Richard Sharp Smith

Architectural Precedents 建筑范例

INTRODUCTION

5

The Southern Appalachians

THE SOUTHERN APPALACHIANS contain extraordinarily diverse landscapes and rare forest types. As a result of this inherent natural beauty, the Southern Appalachians have been attracting visitors for centuries. In the late 19th century, Asheville was a popular health resort due to the availability of mineral springs, a pleasant climate and fresh air. Today, as the Blue Ridge Parkway and the Great Smoky Mountain National Park traverse the Appalachians, millions of visitors are exposed to the unique natural wonders of the region.

The Appalachian Mountains rise to a maximum of 6,684 feet, and diminish to less than 1,000 feet. As a result of this elevation change, highly varied climates and environments appear. The lower regions are filled with mixed deciduous woods throughout the dry zones and pines in the shaded, moist environments. As the topography rises, oak/hickory forests flank the slopes of the mountains leading to boreal and transitional forests on the ridge tops, representing almost every eastern forest type in America.

At the heart of the Southern Appalachians, these habitats converge seamlessly allowing visitors to hike from a deciduous forest in the lowlands to a spruce-fir forest on the ridge-tops to habitats at or near sea level, all within a matter of hours. However, these forest types are not easily distinguishable due to the moderating effects of the humidity level.

Resulting from the variation and multitude of forest types, the southern end of the Blue Ridge Mountains in North Carolina possesses the greatest number of tree species in North America. Many of the species proliferate in record proportions and are among some of the oldest in North America.

Landscape Character of North Ducker 北达克尔景观风貌

INTRODUCTION  6

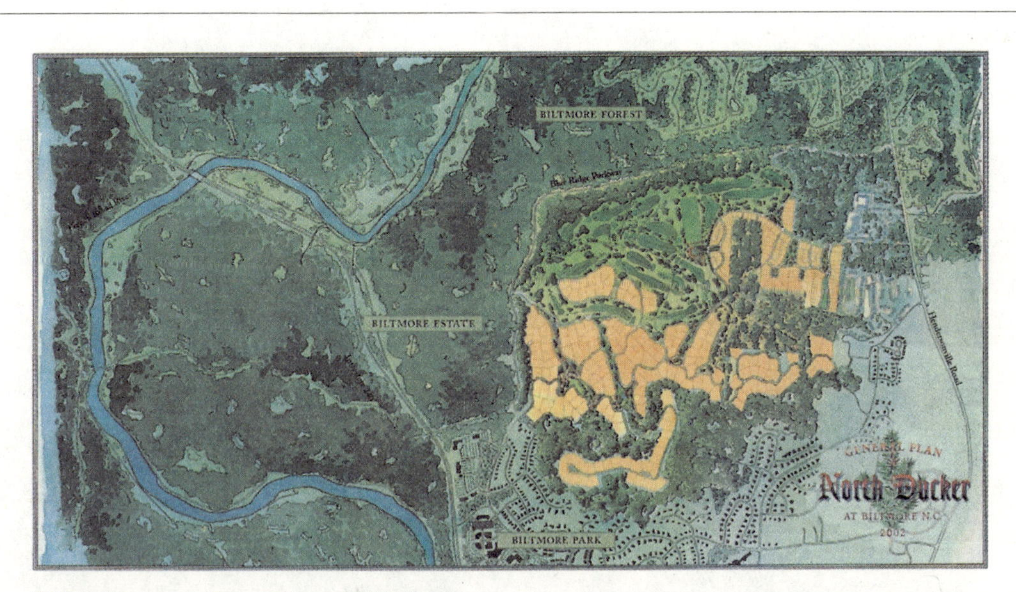

The Site in Context 场地背景

INTRODUCTION  7

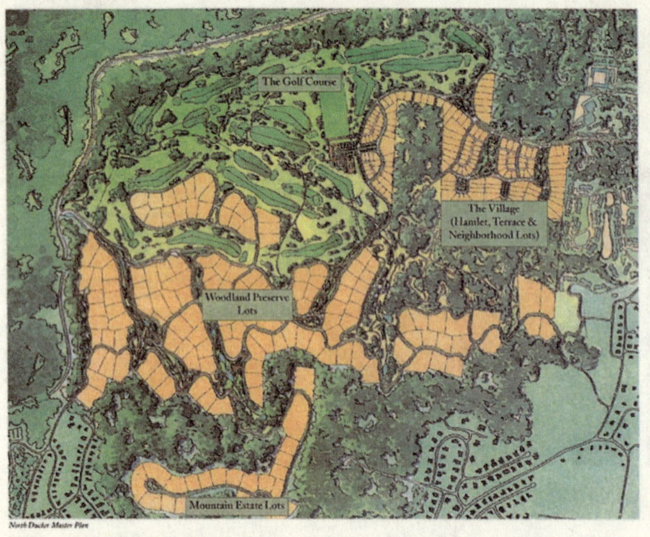

North Ducker Master Plan

A Site in Nature

THE DESIGN OF THE NORTH DUCKER community is both a response to the inherited quality and character of the neighborhoods in Biltmore Forest and the diverse and unique landscape of the mountain. The design principles include the celebration of the natural stream corridors, as preserved parks linking the mountainside neighborhoods to the golf course which traverses the lowland areas of the site adjacent to the Blue Ridge Parkway. Access to the mountain slopes is patterned after the pre-existing trails and roadways to minimize disturbance to the woodland environment. Houses sited on the slopes are carefully planned to preserve the woodlands and minimize grading and clearing. These neighborhoods are organic in form and follow the natural contours of the slope. Specific design criteria for how houses are designed to meet these slopes are detailed in the Landscape Patterns section of this Pattern Book. A village centered on the Clubhouse and recreational amenities offered by the Club builds on the traditional pattern found in the original Biltmore Village. Houses front tree-lined, neighborhood streets and parks on land that is suited for a village environment. A range of lot and house sizes are provided in the North Ducker Plan as well as a diverse selection of neighborhoods, each with its own character and charm.

Architectural styles are also carefully chosen for North Ducker. These styles are found throughout the Asheville region and are particularly appropriate in Biltmore Forest and the North Ducker environment.

North Ducker Master Plan 北达克尔总平面图

INTRODUCTION

A 8

Looking across Lake Elsinore toward the Santa Ana Mountains

The overall Master Plan for Liberty showing the various villages around the lake

LIBERTY IS A 2900-ACRE DEVELOPMENT nestled in the Santa Ana mountains along the eastern shoreline of Lake Elsinore. This 3300-acre natural lake creates a spectacular setting for a series of villages that will look out over the lake toward the mountains. The villages are designed on principles of historic town planning and environmental conservation. Each village is linked by a 1300-acre preserved open space that will have wildlife preservation areas, protected marshlands, and recreation amenities such as golf courses and walking trails.

The design of each village is based on the tradition of Southern California resort communities and towns. Lake Elsinore was a prominent resort area in the early 1900s and attracted visitors to the remarkable views and lake setting. The Liberty Master Plan, developed by Cooper, Robertson & Partners, builds on the character and unique quality of Inland Empire towns, coastal villages, and historic Southern California resorts. The neighborhoods, streets, boulevards, parks, and town center are connected to the shoreline, the parklands, and the view of the mountains so that every resident enjoys the natural amenities. Each village is designed with a village center containing shops, restaurants, offices, and apartments within walking distance of the surrounding residential neighborhoods.

Liberty neighborhoods will have a character similar to historic settlements such as this neighborhood in Orange.

Classic

Craftsman

Victorian

European Country

Monterey Ranch

Spanish Revival

Liberty and California Traditions 利伯蒂和加利福尼亚传统

INTRODUCTION

A 1

Streets in the Belmont Shores neighborhood take you to the water. The blend of streetscapes and private gardens creates a distinctly nautical character.

Colonial Revival houses sit comfortably next to Craftsman, Spanish Revival, and European houses with front yards that are gardens for the street.

The downtown historic district in Riverside contains a variety of mixed-use buildings designed in the Spanish Revival and Mission styles.

Plan of historic neighborhood in Riverside

Inland Empire and Coastal Towns

SOUTHERN CALIFORNIA HAS A VARIETY of town forms that serve as precedents for Liberty. One distinct form can be found in coastal towns, such as Coronado Island, Corona Del Mar, Venice, Long Beach, and Balboa Island, where narrow streets and nautical imagery lead you to the water's edge. These are quirky and delightful environments with a variety of different house types, streetscapes, and water edge conditions. Some neighborhoods are built on canal frontages, such as Venice and Naples. Others, such as Balboa, create another kind of public address along the beach. A common trait is the informal nature of the street and yard configuration and the sense of being near the water, even on inland streets.

Another form is represented by valley settlements such as Riverside, Orange, and Pasadena — established neighborhoods with a distinct character quite different from that of the Coastal precedents. Many of these places developed at the turn of the century and have marvelous neighborhoods from the early 1900s. Places like Madison Heights in Pasadena and Floral Park in Orange are models of beautiful public spaces and street designs that create settings for a varied collection of houses of all different sizes, styles, and prices. The common element is the unique mix of landscape, garden, and regional adaptation of architecture. Some neighborhoods were built at one time as compositions in a single style, like the bungalow courts in Craftsman or European Romantic styles. Others are a diverse collection of styles built over many years.

Southern California Towns and Neighborhoods 南加利福尼亚市镇和社区

INTRODUCTION

2

American Classic houses include California adaptations of Colonial Revival.

California Craftsman houses from the early 1900s

Victorian-era farmhouses and cottages

European Country styles use English and French elements.

Spanish Revival houses from the early twentieth century

Monterey Ranch houses are a unique California house type.

Southern California Architecture

THE TRADITIONAL ARCHITECTURE of Southern California neighborhoods includes a wide variety of styles built in different time periods and with unique regional adaptations to styles as well as housing types. While Spanish Revival architecture has been a signature style for so many recent developments throughout the region, the traditional towns and villages have a much more varied palette of styles and materials. Craftsman houses, Victorian farmhouses, Monterey Ranch houses, and varieties of American Colonial houses dominated many early town and neighborhood patterns in the Inland Empire as well as the coastal towns. This variety and diversity has been lost in recent years and will be re-established in Liberty. Six styles have been adapted and re-introduced. These include: Liberty Classic, which draws on Colonial Revival precedents; Liberty Craftsman; Liberty Victorian; Liberty European Country; Liberty Monterey Ranch; and Liberty Spanish Revival.

Architectural Precedents 建筑范例

INTRODUCTION

3

Perspective views of various street and park addresses in the North Neighborhood

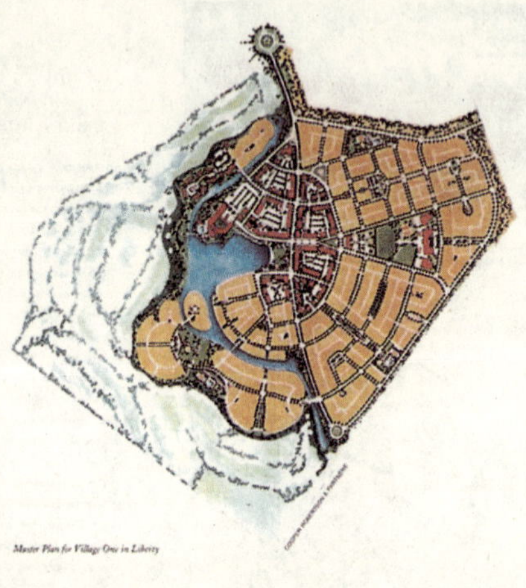

Master Plan for Village One in Liberty

The Pattern Book and Liberty Neighborhoods

THE CHARACTER OF SOUTHERN CALIFORNIA towns and neighborhoods that we most admire did not happen by accident. Early in this century, builders used Pattern Books from England and later plan books and magazine publications such as *Pencil Points* and *The Bungalow* as resources to design and build sophisticated houses in emerging Southern California settlements. Towns and neighborhoods were laid out according to standard surveying practices. Civic spaces were defined, lots were created, and building set-backs established. Houses and buildings filled in the original plan over the course of many years. While the architectural styles changed according to current fashion, the massing and set-backs followed what had come before.

The building of Liberty will be guided by similar tools. *The Liberty Pattern Book* is a device developed for guiding the development of neighborhoods and houses consistent with the original vision described in the Master Plan. The Pattern Book has three sections: *Community Patterns*, which establishes guidelines for placing the house on its lot and defining neighborhood character; *Architectural Patterns*, which establishes architectural precedents for six particular styles in Liberty; and *Landscape Patterns*, which sets palettes and standards for the various lot types and house styles.

The Site 场地

INTRODUCTION

4

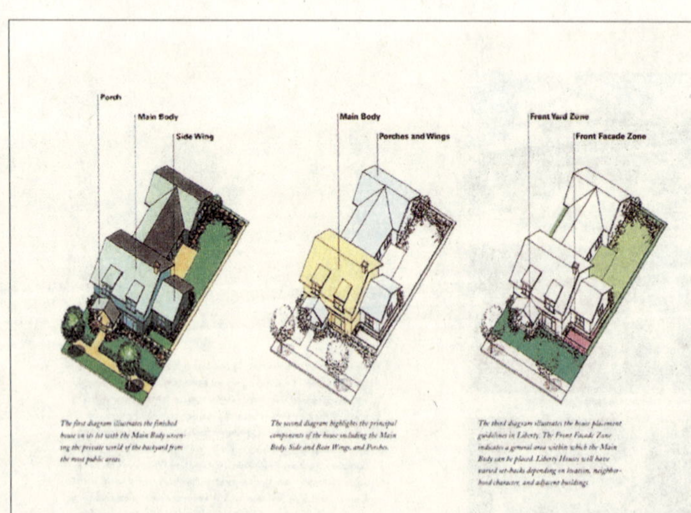

The first diagram illustrates the finished house on its lot with the Main Body screening the private world of the backyard from the most public areas.

The second diagram highlights the principal components of the house including the Main Body, Side and Rear Wings, and Porches.

The third diagram illustrates the house placement guidelines in Liberty. The Front Facade Zone indicates a general area within which the Main Body can be placed. Liberty Houses will have varied set-backs depending on location, neighborhood character, and adjacent buildings.

Liberty Houses

LIBERTY HOUSES WILL CREATE the backdrop for the many distinct neighborhoods within the village. As in traditional California towns and villages, the houses define the character of the public space and signal the individual character of the private realm behind the porch or front door.

In these traditional neighborhoods, the front portion of the house is the most public and must be responsive to the character of the neighborhood and the adjacent houses. The landscaping of the front yard, the set-backs from the street, the size and placement of the house on the lot, and the front porch, courtyard, or verandah are all shared elements that form the public realm.

Three Elements

The Liberty House includes three principal elements.

The **Main Body** of the house, which is the principal mass and includes the front door.

Side Wings, which are one or one-and-one-half stories connected to the Main Body. These are set back from the front facade of the Main Body the same depth as the length of the wing.

Porches are typically additive to the Main Body. These include porticos, side porches, full-facade front porches, or wraparound porches. Some architectural styles have porches that are inset into the Main Body.

The Liberty House 利伯蒂住宅

INTRODUCTION

5

Purpose of the Norfolk Pattern Book 诺福克模式图则的目的

Norfolk has a rich architectural heritage that has created a collection of neighborhoods, remarkable for their diversity and unique character. The architectural style of the houses varies from neighborhood to neighborhood, especially in the traditional neighborhoods built between 1850 and 1950. In recent years, the distinctly different quality of the traditional architectural styles has been affected by the mass production of houses that seem the same wherever they are located. Also, homeowners often have a difficult time finding builders, architects, or materials and components that are in keeping with the period and detailing of their original house.

The Department of Planning and Community Development has commissioned *A Pattern Book for Norfolk Neighborhoods* to provide a resource for homeowners, builders, and communities as they repair, rebuild and expand their houses and preserve their neighborhoods. From remodeling a front door, adding a wing to your house, building a new house, to building a whole new housing development, you will be able to find the appropriate patterns to help guide the process of designing and building in ways that are consistent with the traditional Norfolk architecture and are compatible with the neighborhood character.

Overview of the Norfolk Pattern Book 诺福克模式图则的总则

This Pattern Book is organized in four sections: The Overview, Neighborhood Patterns, Architectural Patterns, and Landscape Patterns. Each section is designed to provide key information to help you make design and site planning decisions about a planned renovation or new house construction. The Neighborhood Patterns section provides a description of the various types of Norfolk neighborhoods by era. Building setbacks, the character of the streets, landscaping, and architectural diversity are described for each type. This gives owners a sense of what key elements to look for when planning to build or renovate a house in one of these neighborhoods.

The Architectural Patterns section presents guidelines for building or renovating a traditional Norfolk house within a specific architectural style. Six different traditional styles found throughout the Norfolk neighborhoods are illustrated with key details, materials and shapes to help owners determine the appropriate design elements for their house.

The Landscape Patterns section illustrates specific examples of fencing, walls, paving, and driveway types for Norfolk houses.

An Appendix listing materials resources and reference materials is also included.

Neighborhood Patterns

Architectural Patterns

Landscape Patterns

SECTION A

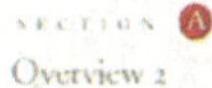

Overview 2

Purpose and Overview of The Norfolk Pattern Book 3
How To Use The Norfolk Pattern Book 4

SECTION B

Neighborhood Patterns 6

Norfolk Neighborhoods 7
Neighborhood Patterns 8
Nineteenth Century Neighborhoods 10
Early Twentieth Century Neighborhoods 12
Twentieth Century Post-War Neighborhoods 14

SECTION C

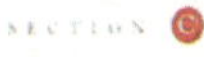

Architectural Patterns 16

Norfolk Architectural Styles 17
Building a Norfolk House 18
Renovations 20
Additions 22
Transformations 25
Garages & Other Ancillary Structures 28
Norfolk Classical Revival 30
Norfolk Colonial Revival 36
Norfolk European Romantic 42
Norfolk Arts & Crafts 48
Norfolk Victorian 54
Norfolk Coastal Cottage 60

SECTION D

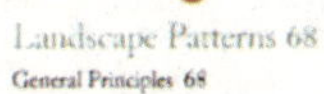

Landscape Patterns 68

General Principles 68

SECTION E

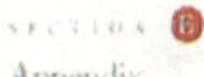

Appendix

Material & Component Manufacturers 70
Resources & Glossary 71
The History of Pattern Books 72

A Norfolk neighborhood pattern

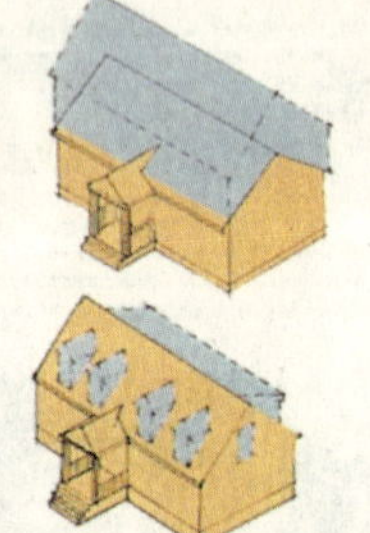

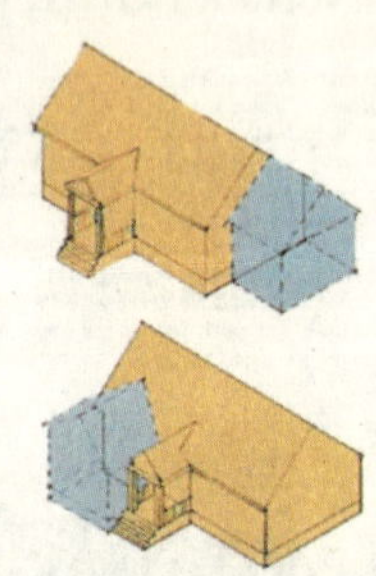

Designing and Renovating a Norfolk House

How To Use The Norfolk Pattern Book 如何应用诺福克模式图则

The following is a step-by-step outline which describes how to use the Pattern Book for both homeowners who are interested in renovating or adding on to their house, and individuals who are interested in constructing a new house.

STEP 1 Identify Your Neighborhood

Whether you own an existing house or are building a new house, refer to the Neighborhood Patterns section of the Pattern Book (pages B6 through B15) and review the three eras of neighborhood building.

If you already own a house, select the era which your neighborhood most closely resembles. Read about the individual components – such as the typical front yard depth, streetscape character, house spacing, landscape treatments (both public and private) – that define your neighborhood.

If you are searching for a lot on which to build your new house, the Pattern Book can also be helpful. The Neighborhood Patterns section provides an overview of the unique characteristics of each era of neighborhood building and a listing of many Norfolk neighborhoods that fall within each era. This introduction can direct you to the neighborhoods that have characteristics that interest you.

Refer to directly to Step 4 if you are constructing a new house. Otherwise, continue to Step 2.

STEP 2 Identify the Architectural Style of your House

Once you've familiarized yourself with the era of your neighborhood, identify the architectural style that most closely resembles your house.

The Overview of Architectural Styles in the beginning of the Architectural Patterns section (pages C16 and C17) describes in visual form the predominant architectural styles found in Norfolk. The Table of Roof Pitches on page C22 in the Additions and Renovations section might also be helpful.

If your house does not have an identifiable style or is a mix of two styles, select one for it that would work best with its massing and height.

STEP 3 For Additions and Renovations

For information on appropriate means of modifying your house (whether historic or post-war) refer to the Additions and Renovations section (pages C22 and C23). This section describes strategies for adding on extra rooms or garages as well as changing or replacing exterior components such as windows, doors and materials.

STEP 4 For New Construction

If you are planning on constructing a new house, please refer to the "Building a Norfolk House" and "Siting a Norfolk House" sections (pages C18 through C21).

The "Building a Norfolk House" section outlines the step-by-step process of composing a Norfolk House and relates the individual elements, such as windows, doors, and porches, to the architectural style sections (as described in Step 5).

"Siting a Norfolk House" explains how to locate your house, garage, ancillary structures, and landscaping on your lot in a manner appropriate to the neighborhood context.

Also refer to the "House Lot Diagram" which is shown for every era of neighborhood in the Neighborhood Patterns section. The diagram describes the typical "zones" of a house lot, such as front yard, front facade, side yard, and private zone, all of which vary depending on the era.

STEP 5 Review the Architectural Style Sections

Six architectural styles found in Norfolk are documented in the Pattern Book: Classical Revival, Colonial Revival, European Romantic, Arts & Crafts, Victorian, and Coastal Cottage.

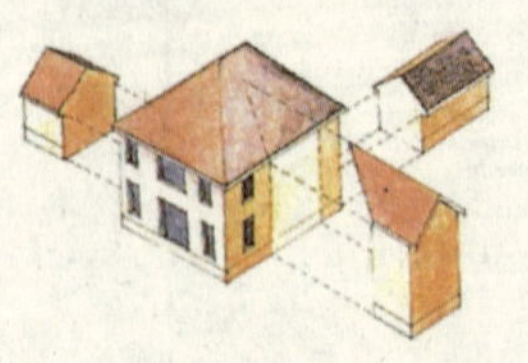

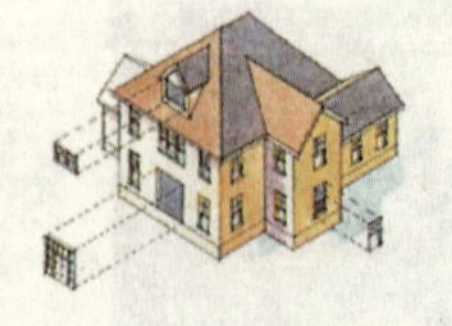

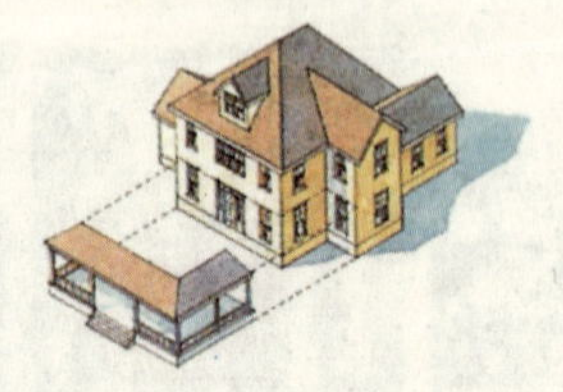

Assembling the elements of a Norfolk house

Character sketch of a Norfolk Arts & Crafts house

Massing and composition diagrams

HISTORY & CHARACTER PAGE

The first page of every architectural style section begins with a brief description of the style and its history. Photos of relevant examples of the style in Norfolk have been documented and are shown along with the essential qualities of each style. A partial elevation drawing and measured cross-section relay the critical dimensions and elements of the facade.

MASSING & COMPOSITION PAGE

This page describes the basic massing types or shapes of houses found in the Norfolk precedents for each architectural style. Each massing type is shown as a 3D image with a corresponding elevation diagram showing potential additions. The layout of rooms should be designed to fit into the massing types found within the particular style you are designing. The roof types are part of this overall massing description.

WINDOWS & DOORS PAGE

The window and door spacing is related to both the shape and the style of the house. Typical window and door compositions are illustrated as part of the massing illustrations for each style. Typical window and door proportions, trim details and special window or door elements are illustrated on a separate page within each section.

PORCHES & CHIMNEYS PAGE

Porches are essential elements of the character of many Norfolk neighborhoods. The location and design elements of porches are covered on this page. The massing of the front porch is specific to each house type and distinct within a particular style.

Chimneys are a key element in the composition of the elevation for several styles. Massing and details such as chimney caps are outlined on this page.

MATERIALS &D APPLICATIONS PAGE

This page of each style section in the Architectural Patterns includes a list of acceptable materials and their application. Also included on this page are hand-drawn elevation "possibilities" composed using elements described in the Pattern Book to illustrate the end result achieved if one follows the guidelines of the Pattern Book.

GALLERY OF EXAMPLES PAGE

This last page of each style section contains both a collection of photos of Norfolk houses in that style as well as detail photos of porches, doors and windows.

STEP 6 Review the Materials & Components List in the Appendix

Please review the list of materials and components such as doors, windows, columns, and moldings. Keyed to the appropriate architectural style, the list can serve as a reference or resource when searching for the appropriate building supplies from local sources.

STEP 7 Review the Reference List in the Appendix

For those who are interested in learning more about Norfolk's residential architecture, architectural styles in general, Norfolk's history, or available resources from the City of Norfolk, this list provides a handy reference.

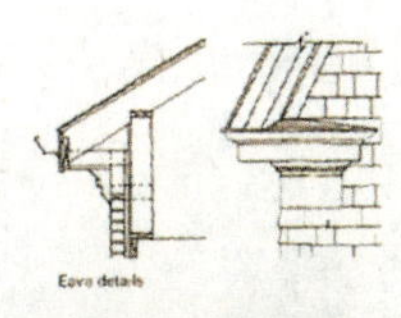
Eave details

Style examples

Identifying or selecting a porch

Identifying or selecting a window

Material options example

沃特卡勒的建成效果

第七章

社区模式册页

社区模式通常包括总体规划设计的两个重要要素——与其所在地块相关的不同建筑类型的基本设计参量，以及确立场地中不同场所特征的导则。一旦开发小组确定了不同建筑类型和地块大小的具体计划，并将其纳入总体设计的最终定稿中，模式图则就可以详细说明重要的建筑后退、覆盖范围，以及与地块的长宽相关的住宅和建筑的相对尺度。通常情况下，这些导则为在地块中安排建筑物提供了一个灵活的框架，从而能够在不同的位置和环境中进行相应的变化。在本章所介绍的范例中，导则以“地块类型——总体条件”的标题出现在册页之中。这些条件可以说是对社区模式章节第二部分所介绍的特定设计或后退标准的补充，在该章节也对邻里场所进行了描述。

邻里场所的设计是作为通过三维模型来检验场地总体设计的结晶。在这一部分的设计过程中，设计和开发小组参加了历时 2 至 3 天的一系列工作会议，来完善规划设计中的重要场所。这一项工作包括确定有助于营造规划确定的预期效果或空间特征的特定条件。特别的建筑要素、建筑后退、建筑类型或景观因素都可以被确定下来以营造这些独特的场所。这些条件被纳入到为单个地块和建筑物制定的模式图则导则中。每一处不同的场所都是借助透视图、典型建筑后退平面图和街道剖面图来进行描述的，以表达这些要素布置之间的基本关系。

社区模式建成效果，巴克斯特，南卡罗来纳州福特米尔市

案例研究：巴克斯特

南卡罗来纳州福特米尔市

巴克斯特模式图则的社区模式部分确定了一系列住宅地块的类型，其中包括2到3单元的叠拼住宅、联排住宅、45英尺 ×100英尺的小型别墅用地、55英尺 ×120英尺的大型别墅用地和70英尺 ×120英尺的住宅用地。一些地块类型——小型别墅、大型别墅和住宅——拥有前车道和服务小径的条件。这些地块类型构成了主要的独户住宅类型；同时在规划方案中也包括了模式图则地块类型以外的公寓、生活/工作单元和混合用途的建筑。

巴克斯特是作为南卡罗来纳州“内陆”地区的传统村镇来进行设计的。在这一地区的村镇具有不受拘束、富于变化的住宅类型模式，在同一条街道中混合布置着大型和小型的住宅建筑。而且，在这里也看不到惯常联排住宅建筑类型的传统，而是具有更多的城市环境。巴克斯特的联排住宅应作为一种与内陆邻里的不拘小节的特征相一致的建筑类型出现。另外，巴克斯特的规划设计还必须进行修改，以反映这种在同一条街道上各种尺寸的用地和住宅混合布置的特征。在传统模式中对建筑后退和人行道位置的安排具有一定的随意性。这种建筑后退环境的混合安排，成为对巴克斯特社区模式部分进行完善和引导发展的组成部分。模式图则中对联排住宅进行了详细说明，从而使它们既可以作为大型住宅也可以用来塑造中心公共空间，并呈现出在该地区很多小城镇中所具有的小型学院建筑的特征。社区模式也为巴克斯特的不同邻里确定了景观特征和建筑意象。每一处场所和每一处地块都明确了特定的建筑后退和/或所需要的建筑要求。

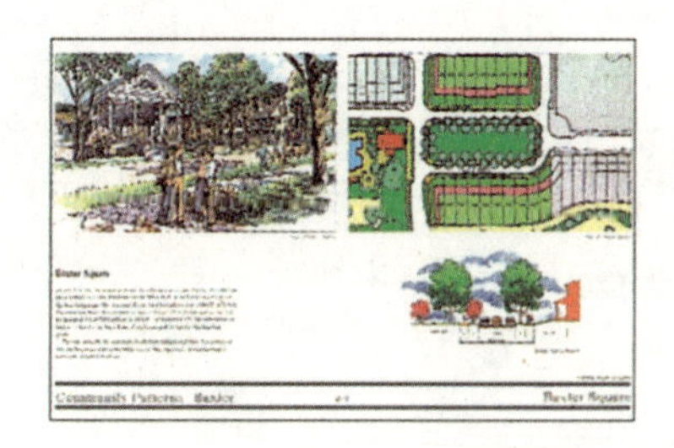

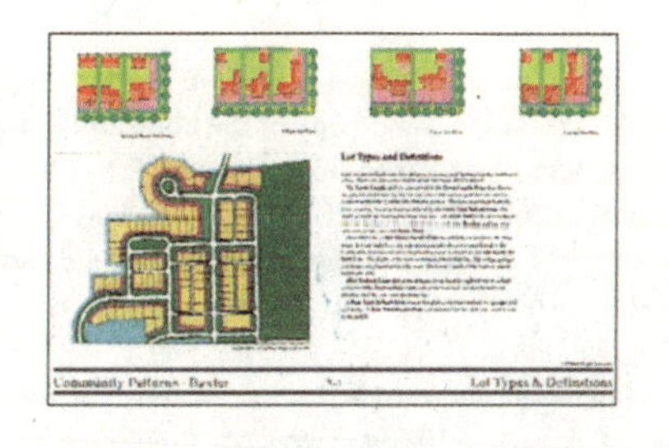

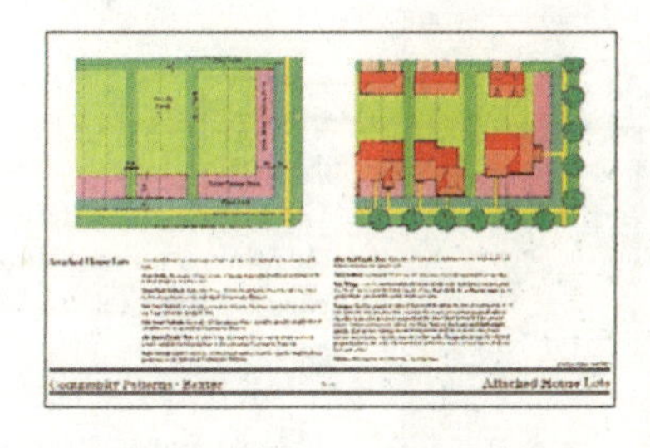

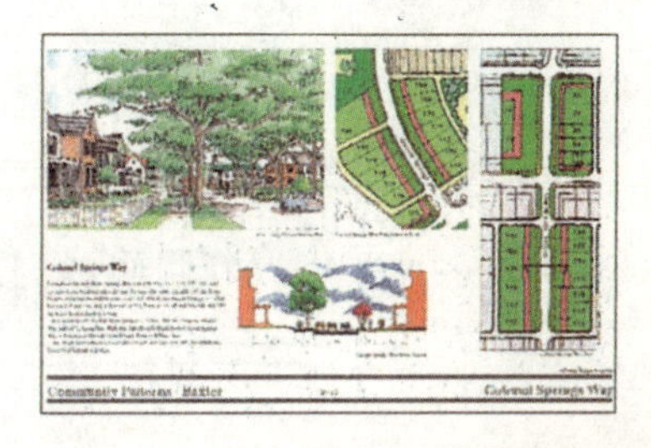

View of Baxter Square

Baxter Square

For lots 115–117 the minimum Front Yard Setback is 25 feet. For lot 115, the Side Street Setback is 15 feet minimum for the Main body of the house and 8 feet for the detached garage. The minimum Front Yard Setback for lots 118–120 is 20 feet. The minimum Front Yard setback for lots 122 and 123 is 35 feet and for lot 121 the minimum Front Yard setback is 28 feet. For corner lot 123, the Side Street Setback is 17 feet for the Main Body of the house and 10 feet for the detached garage.

For lots 149–158, the minimum Front-Yard Setback is 20 feet. For corner lot 149, the Minimum Side Street Setback is 10 feet. Two-story (minimum height) houses are required on all lots.

Community Patterns · Baxter 巴克斯特社区模式 B-8

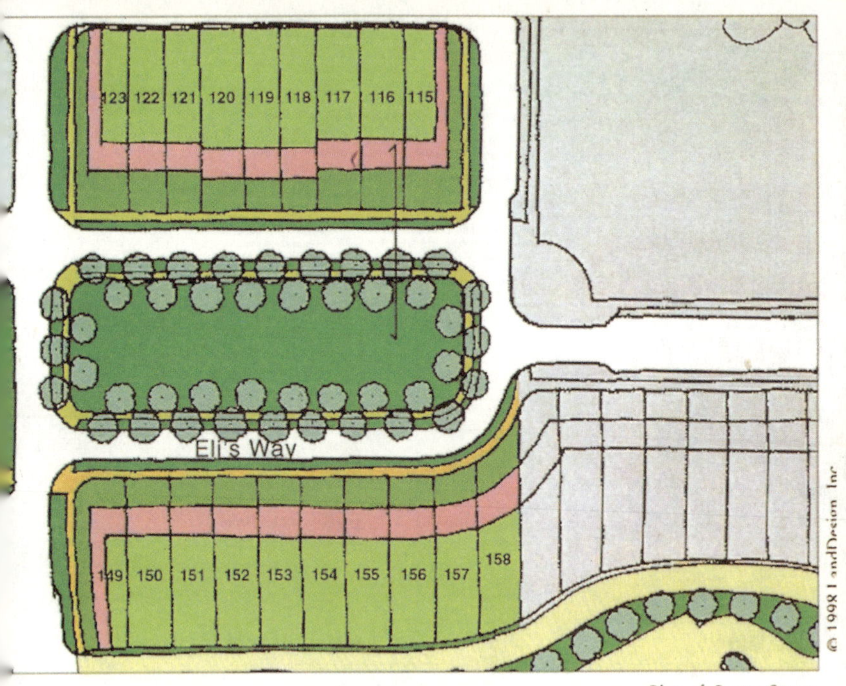

Plan of Baxter Square

Baxter Square Section

巴克斯特广场 Baxter Square

巴克斯特

南卡罗来纳州福特米尔市

巴克斯特的特征是参考了南卡罗来纳“内陆”地区各处的小城镇而确定的。这一特殊地区的独特属性要归因于其农业特征、土壤和植物种类、文化迁移和居住模式、以及随之形成的居住特征。巴克斯特广场是作为多户住宅和社区中心建筑的用地来进行建设的。在这一地区的小城镇中没有建设联排住宅或较大型的多户住宅的传统。广场的设计反映了南卡罗来纳州的奇罗或北卡罗来纳州的戴维森等地的市镇特征。社区中心是参照小型市政厅设计的，而住宅被设计为可以随时间推移增建的大型住宅。

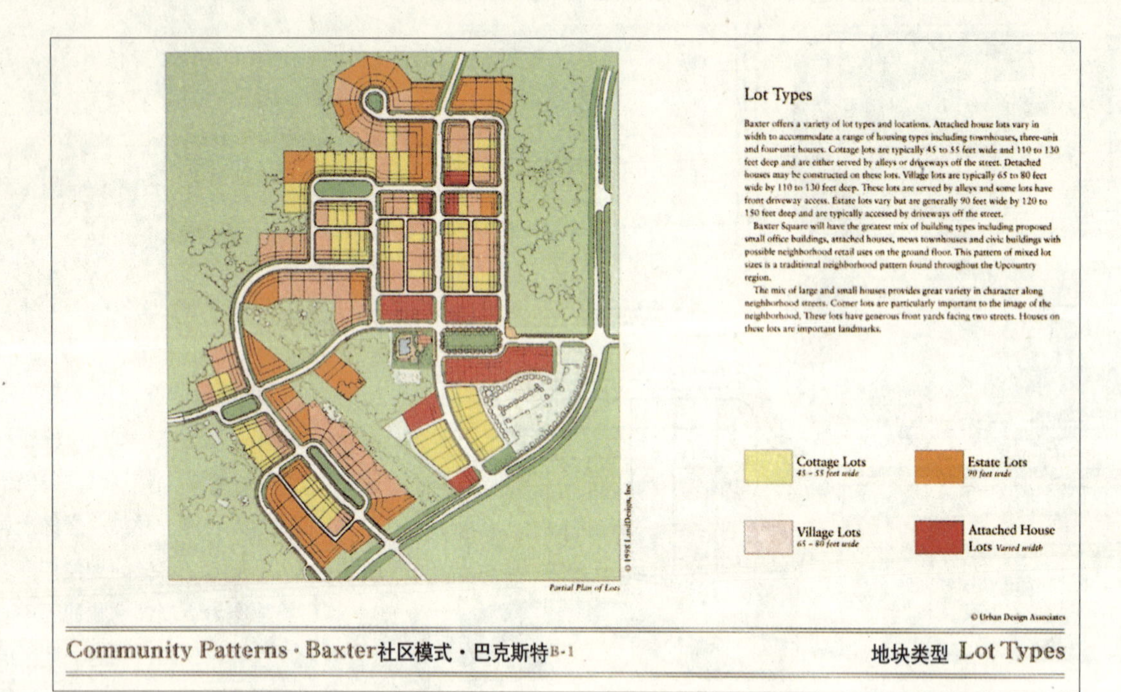

Lot Types

Baxter offers a variety of lot types and locations. Attached house lots vary in width to accommodate a range of housing types including townhouses, three-unit and four-unit houses. Cottage lots are typically 45 to 55 feet wide and 110 to 130 feet deep and are either served by alleys or driveways off the street. Detached houses may be constructed on these lots. Village lots are typically 65 to 80 feet wide by 110 to 130 feet deep. These lots are served by alleys and some lots have front driveway access. Estate lots vary but are generally 90 feet wide by 120 to 150 feet deep and are typically accessed by driveways off the street.

Baxter Square will have the greatest mix of building types including proposed small office buildings, attached houses, mews townhouses and civic buildings with possible neighborhood retail uses on the ground floor. This pattern of mixed lot sizes is a traditional neighborhood pattern found throughout the Upcountry region.

The mix of large and small houses provides great variety in character along neighborhood streets. Corner lots are particularly important to the image of the neighborhood. These lots have generous front yards facing two streets. Houses on these lots are important landmarks.

Cottage Lots *45 - 55 feet wide*

Estate Lots *90 feet wide*

Village Lots *65 - 80 feet wide*

Attached House Lots *Varied width*

Partial Plan of Lots

© 1998 LandDesign, Inc

Community Patterns · Baxter社区模式·巴克斯特B-1　　地块类型 Lot Types

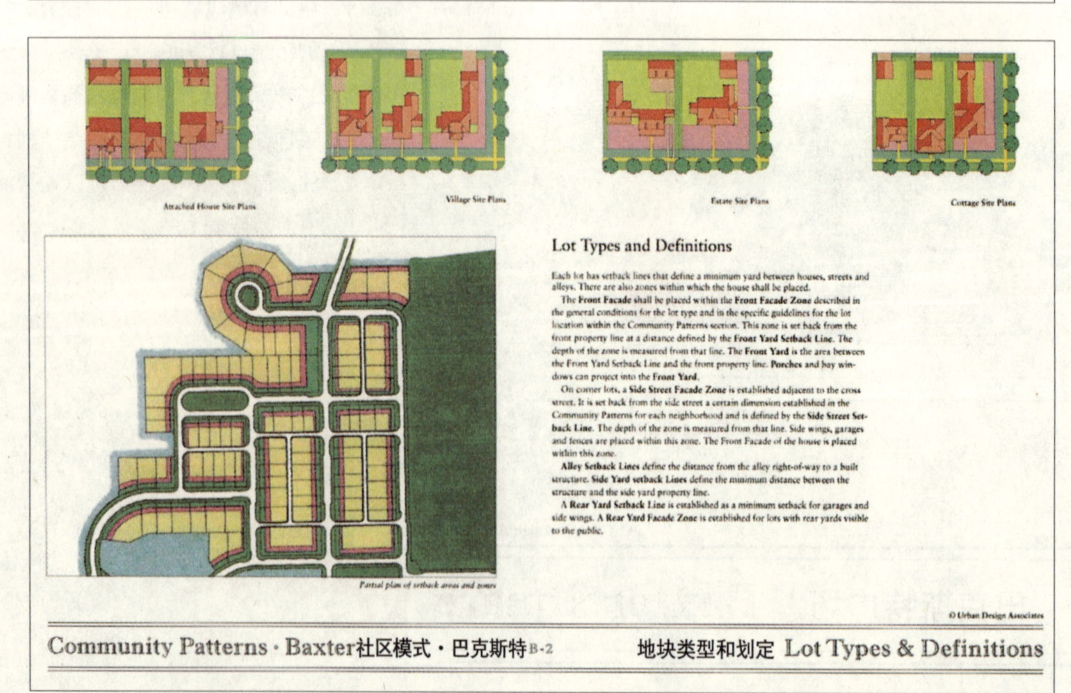

Attached House Site Plans　Village Site Plans　Estate Site Plans　Cottage Site Plans

Lot Types and Definitions

Each lot has setback lines that define a minimum yard between houses, streets and alleys. There are also zones within which the house shall be placed.

The **Front Facade** shall be placed within the **Front Facade Zone** described in the general conditions for the lot type and in the specific guidelines for the lot location within the Community Patterns section. This zone is set back from the front property line at a distance defined by the **Front Yard Setback Line**. The depth of the zone is measured from that line. The **Front Yard** is the area between the Front Yard Setback Line and the front property line. **Porches** and bay windows can project into the **Front Yard**.

On corner lots, a **Side Street Facade Zone** is established adjacent to the cross street. It is set back from the side street a certain dimension established in the Community Patterns for each neighborhood and is defined by the **Side Street Setback Line**. The depth of the zone is measured from that line. Side wings, garages and fences are placed within this zone. The Front Facade of the house is placed within this zone.

Alley Setback Lines define the distance from the alley right-of-way to a built structure. **Side Yard setback Lines** define the minimum distance between the structure and the side yard property line.

A **Rear Yard Setback Line** is established as a minimum setback for garages and side wings. A **Rear Yard Facade Zone** is established for lots with rear yards visible to the public.

Partial plan of setback areas and zones

Community Patterns · Baxter社区模式·巴克斯特B-2　　地块类型和划定 Lot Types & Definitions

巴克斯特

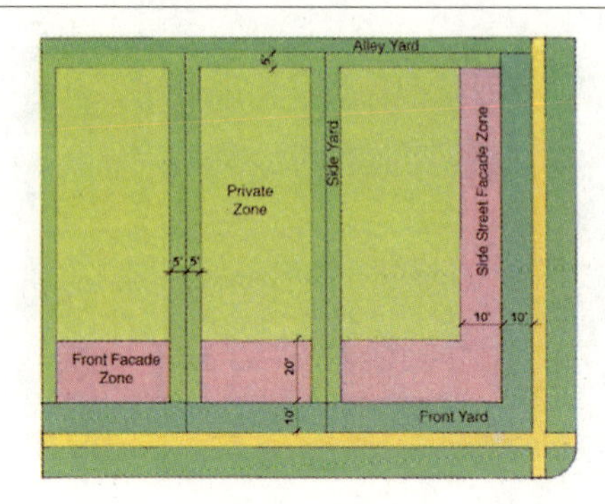

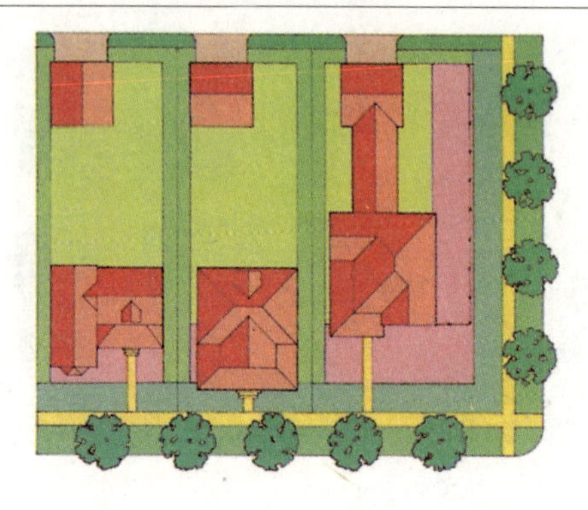

Cottage Lots

Cottage Lots are typically 45 to 55 feet wide by 110 to 130 feet deep. They may vary in size from lot to lot depending on location.

Main Body Width: Generally 32 feet or less within the Front Facade Zone.

Front Yard Setback: 10 feet except where noted by specific neighborhood guidelines in the individual Community Patterns.

Side Yard Setback: Minimum setbacks for all structures are 5 feet.

Side Street Setback: Generally 10 feet except where noted by specific neighborhood guidelines in the individual Community Patterns.

Side Street Facade Zone (Corner Lots): Generally 10 feet except where noted by specific neighborhood guidelines in the individual Community Patterns.

Front Facade Zone: Generally 20 feet except where noted by specific neighborhood guidelines in the individual Community Patterns.

Alley Yard Facade Zone: Generally 20 feet where indicated in the individual Community Patterns for specific lots.

Alley Setback: 15 feet for all structures from center line of alley.

Side Wings: 1 or 1½ stories within the Front Facade Zone. Side wings should generally be set back from the Front Facade of the Main Body by a distance equal to, or greater than, one-half the width of the side wing.

Garages: Shall be placed at either 5 feet from the property line or a minimum of 15 feet from the rear property line. Garage doors may be oriented perpendicular to the alley. Lots with driveway access from the street shall have garages placed behind the front facade of the Main Body of the house and shall require specific plan review. Garage doors facing streets shall be no wider than 9 feet. Garage doors facing the alley may be 18 feet wide. Garage doors may be oriented perpendicular to the alley. On corner lots with alley access, garage doors shall not face side streets.

Porches may extend into the Front Yard Setback.

Community Patterns · Baxter 社区模式 · 巴克斯特 B-3 小别墅地块 Cottage Lots

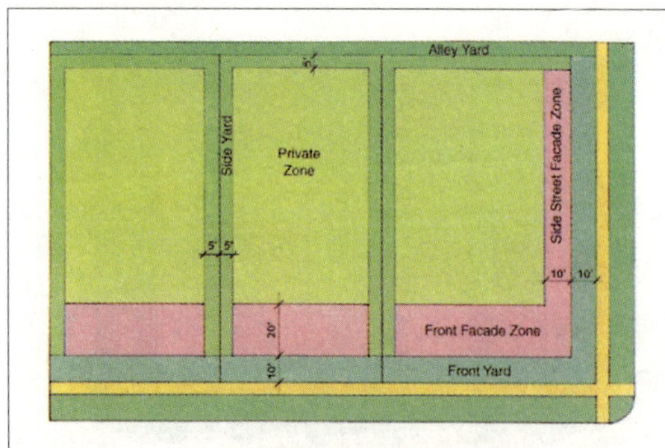

Village Lots

Village Lots are typically 65 or 80 feet wide by 110 to 130 feet deep. They may vary in size from lot to lot depending on location.

Main Body Width: Generally 40 feet or less.

Front Yard Setback: Generally 10 to 30 feet except where noted by specific neighborhood guidelines in the individual Community Patterns.

Side Yard Setback: Minimum setbacks for all structures are 5 feet.

Side Street Setback: Generally 10 feet except where noted by specific neighborhood guidelines in the individual Community Patterns.

Side Street Facade Zone (Corner Lots): Generally 10 feet except where noted by specific neighborhood guidelines in the individual Community Patterns.

Front Facade Zone: Generally 20 feet except where noted by specific neighborhood guidelines in the Community Patterns.

Alley Setback: Generally 15 feet for all structures from the center line of the alley.

Rear Yard Setback. Generally 5 feet for all structures.

Side Wings: 1 or 1½ stories within the Front Facade Zone. Side wings should generally be set back from the Front Facade of the Main Body by a distance equal to, or greater than, one-half the width of the side wing.

Garages: Shall be placed at either 5 feet from the property line or a minimum of 15 feet from the rear property line. Lots with driveway access from the street shall have garages placed behind the front facade of the Main Body of the house. Garage doors facing streets shall be no wider than 9 feet. Garage doors facing an alley may be 18 feet wide. Garage doors may be oriented perpendicular to the alley. On corner lots with alley access, garage doors shall not face side streets.

Porches may extend into the Front Yard Setback.

Community Patterns · Baxter 社区模式 · 巴克斯特 B-4 大别墅地块 Village Lots

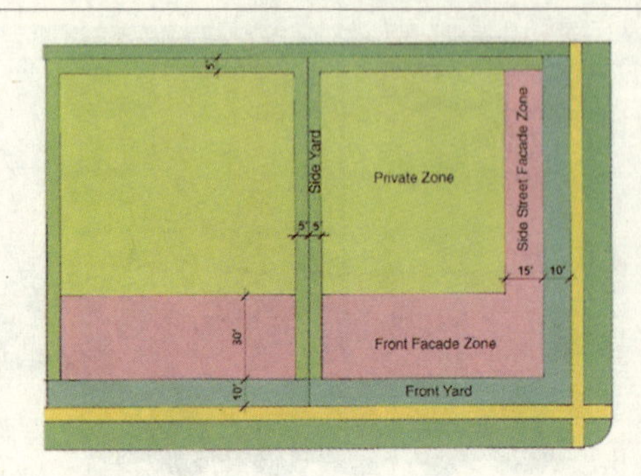

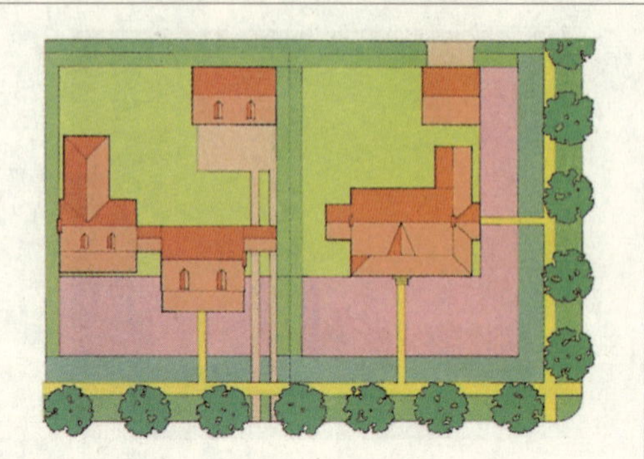

Estate Lots

Estate Lots are typically 90 feet wide by 120 to 150 feet deep. They may vary in size from lot to lot depending on location.

Main Body Width: Generally 48 feet or less.

Front Yard Setback: Generally 30 feet except where noted by specific neighborhood guidelines in the individual Community Patterns.

Side Yard Setback: Minimum setbacks for all structures are 5 feet.

Front Facade Zone: Generally 30 feet except where noted by specific neighborhood guidelines in the individual Community Patterns.

Alley Setback: Generally 15 feet for all structures from the centerline of the alley.

Rear Yard Setback: Generally 5 feet minimum for all structures.

Side Wings: 1 or 1½ stories within the front facade zone. Side wings should generally be set back from the Front Facade of the Main Body by a distance equal to, or greater than, one-half the width of the side wing.

Garages: Shall be placed at either 5 feet from the property line or a minimum of 15 feet from the rear property line. Lots with driveway access from the street shall generally have garages placed behind the front facade of the Main Body of the house and shall require specific plan review. Garage doors facing streets shall be no wider than 9 feet. Garage doors facing an alley may be 18 feet wide. Garage doors may be oriented perpendicular to the alley. On corner lots with alley access, garage doors shall not face side streets.

The maximum width of a garage with doors facing the street is 24 feet.

Porches may extend into the Front Yard Setback.

Community Patterns · Baxter 社区模式　B-5　房产地块 Estate Lots

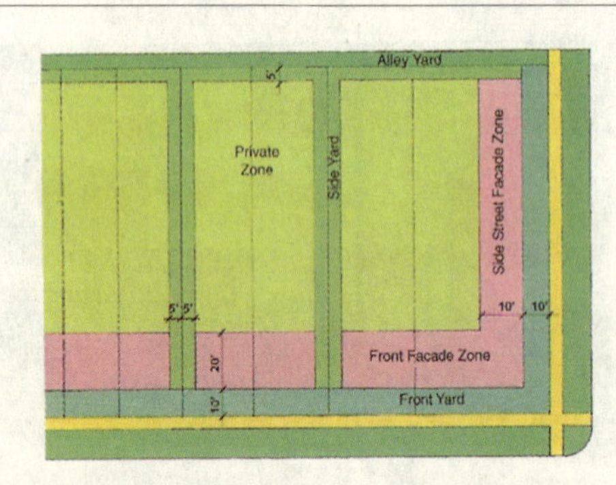

Attached House Lots

Attached House Lots may vary in size from lot to lot depending on location and type.

Main Body: Generally 50 feet or less of facade is permitted without a change in the vertical plane of at least 6 feet.

Front Yard Setback: Generally 10 to 25 feet except where noted by specific neighborhood guidelines in the individual Community Patterns.

Side Yard Setback: Generally, minimum setbacks between attached unit structures are 5 feet from the property line.

Side Street Setback: Generally 10 feet except where noted by specific neighborhood guidelines in the individual Community Patterns.

Side Street Facade Zone (Corner Lots): Generally 10 feet except where noted by specific neighborhood guidelines in the individual Community Patterns.

Front Facade Zone: Generally 20 feet except where noted by specific neighborhood guidelines in the individual Community Patterns.

Alley Yard Facade Zone: Generally 20 feet where indicated in the individual Community Patterns for specific lots.

Alley Setback: Generally 15 feet for all structures from the centerline of the alley.

Side Wings: 1 or 1½ stories within the front facade zone. Side wings should generally be set back from the Front Facade of the Main Body by a distance equal to, or greater than, one-half the width of the side wing.

Garages: Shall be placed at either 5 feet from the property line or a minimum of 15 feet from the rear property line. Garage doors may be oriented perpendicular to the alley. Lots with driveway access from the street shall generally have garages placed behind the front facade of the Main Body of the house and shall require specific plan review. Garage doors facing streets shall be no wider than 9 feet. Garage doors facing the alley may be 18 feet wide. Garage doors may be oriented perpendicular to the alley. On corner lots with alley access, garage doors shall not face side streets.

Porches may extend into the Front Yard Setback.

Community Patterns · Baxter 社区模式　B-6　毗连住宅地块 Attached House Lots

巴克斯特

View of Founders Street from North Sutton Road

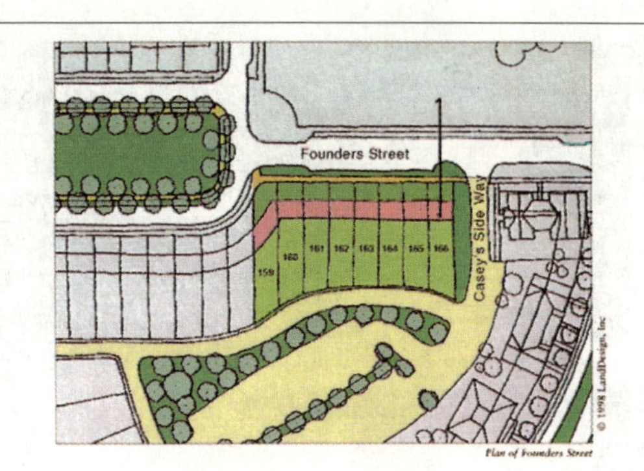

Plan of Founders Street

Founders Street

Houses along Founders Street look out across the park and frame the arrival into the heart of the neighborhood. For lots 159–166, the minimum Front Yard Setback is 15 feet. The minimum height for houses along Founders Street is two stories. Double-height front porches are strongly encouraged.

Founders Street Section

© Urban Design Associates

Community Patterns · Baxter 社区模式·巴克斯特B-7　　创始者大街 Founders Street

The close provides an intimate setting for residences along Sonny's Way

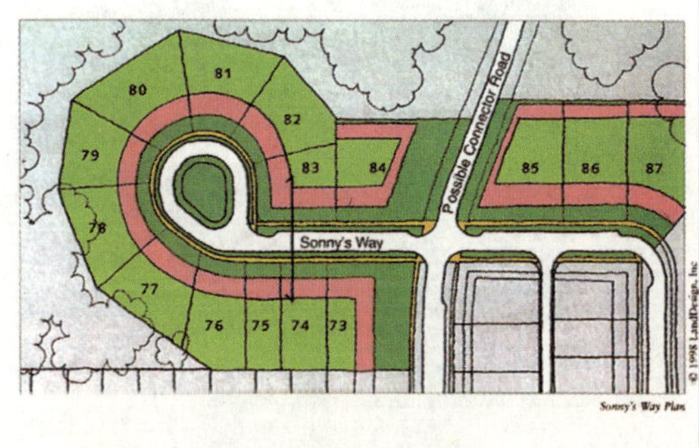

Sonny's Way Plan

Sonny's Way

For typical lots on Sonny's Way, lots 74–83, the Front Yard Setback is 20 feet. Houses on these lots shall have a one-and-one-half story or two-story Main Body and are strongly encouraged to have two-story porches. For lot 73, the Front Yard Setback is 40 feet. For lots 85–87, the Front Yard Setback is 15 feet and a 3 foot high front yard fence shall be built at the front property line. No portion of the house, including the porch, can be built closer than 5 feet from the property line.

For lots 74–83 and 85–87, two-car garages are permitted to be built in the front yard provided the garage doors are perpendicular to the street and face a paved motor court in front of the Main Body of the house. A front yard fence, wall or hedge shall be built at the property line.

For corner lot 84, the Front Yard Setback is 20 feet and the Side Street Setback is 60 feet. For lot 85, the Side Street Setback is 20 feet. For both lots the Side Street Facade Zone is 20 feet deep. If the connector road is eliminated, the setbacks for 84 and 85 shall be revised.

Sonny's Way Street Section

© Urban Design Associates

Community Patterns · Baxter 社区模式·巴克斯特B-9　　宝路 Sonny's Way

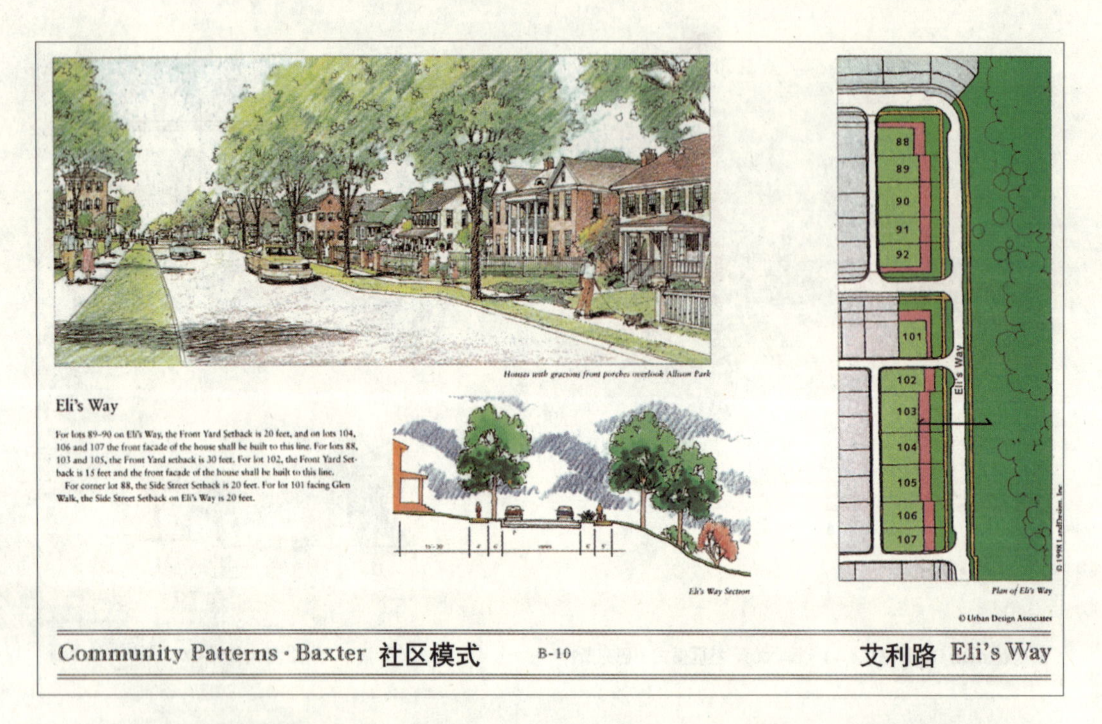
Houses with gracious front porches overlook Allison Park
Eli's Way
For lots 89–90 on Eli's Way, the Front Yard Setback is 20 feet, and on lots 104, 106 and 107 the front facade of the house shall be built to this line. For lots 88, 103 and 105, the Front Yard setback is 30 feet. For lot 102, the Front Yard Setback is 15 feet and the front facade of the house shall be built to this line.
For corner lot 88, the Side Street Setback is 20 feet. For lot 101 facing Glen Walk, the Side Street Setback on Eli's Way is 20 feet.
Eli's Way Section
88
89
90
91
92
101
102
103
104
105
106
107
Eli's Way
Plan of Eli's Way
© Urban Design Associates
Community Patterns · Baxter 社区模式
B-10
艾利路 Eli's Way

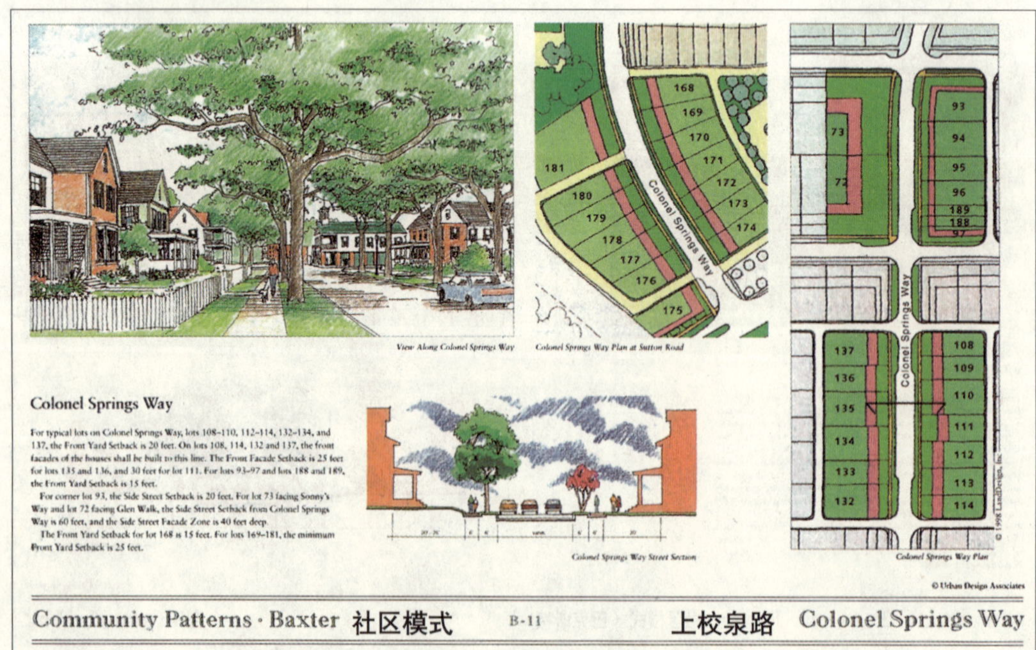
View Along Colonel Springs Way
168
169
170
171
172
173
174
175
176
177
178
179
180
181
Colonel Springs Way
Colonel Springs Way Plan at Sutton Road
93
94
95
96
189
188
73
72
137
136
135
134
133
132
108
109
110
111
112
113
114
Colonel Springs Way
Colonel Springs Way Plan
Colonel Springs Way
For typical lots on Colonel Springs Way, lots 108–110, 112–114, 132–134, and 137, the Front Yard Setback is 20 feet. On lots 108, 114, 132 and 137, the front facades of the houses shall be built to this line. The Front Facade Setback is 25 feet for lots 135 and 136, and 30 feet for lot 111. For lots 93–97 and lots 188 and 189, the Front Yard Setback is 15 feet.
For corner lot 93, the Side Street Setback is 20 feet. For lot 73 facing Sonny's Way and lot 72 facing Glen Walk, the Side Street Setback from Colonel Springs Way is 60 feet, and the Side Street Facade Zone is 40 feet deep.
The Front Yard Setback for lot 168 is 15 feet. For lots 169–181, the minimum Front Yard Setback is 25 feet.
Colonel Springs Way Street Section
© Urban Design Associates
Community Patterns · Baxter 社区模式
B-11
上校泉路 Colonel Springs Way

社区模式建成效果，巴克斯特，南卡罗来纳州福特米尔市

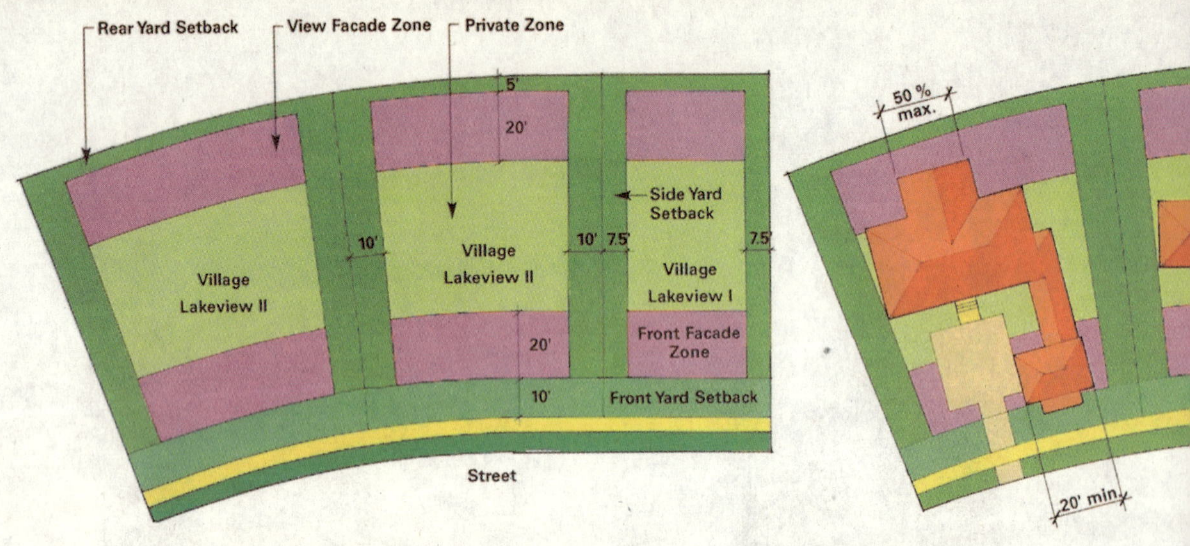

Village Lakeview Lots I and II Specifications

Lot Size
Village Lakeview I lots are approximately 50 feet wide. Village Lakeview II lots are approximately 70 feet wide.

Main Body
The width of the Main Body of the house shall be a maximum of 35 feet for Village I lots and 50 feet for Village II lots. The Main Body need not be placed in the Front Facade Zone. The house should be sited to preserve as many trees as possible.

Front Yard Setback / Front Facade Zone
The depth of the Front Yard is typically 10 feet from the front property line to the Front Yard Setback Line, unless noted otherwise in the Address section. The Front Facade Zone extends 20 feet from the Front Yard Setback Line. A minimum 20-foot-wide building mass at least one-and-one-half stories high and containing a living area shall be placed on the Front Yard Setback Line, while preserving as many trees as possible.

Side Yard Setback
Structures shall be set back a minimum of 7.5 feet from the side property line for Village Lakeview I lots and 10 feet setbacks for Village Lakeview II lots.

Side Street Setback / Side Street Facade Zone
Structures shall be set back a minimum of 5 feet from the side street property line. The Side Street Facade shall be defined by the side facades of the Main Body and any Rear Wings or Out Buildings. Where there is no building structure, the Side Street Facade shall be delineated by a fence or hedge.

Rear Yard Setback/ View Facade Zone
Structures shall be set back a minimum of 5 feet (unless noted otherwise in the Address section) from the rear property line, including porches and swimming pools. The View Facade Zone extends 20 feet from the setback line. The total width of the buildings within the View Facade Zone may be no more than 50 percent of the width of the lot at the rear property line. A full-facade, two-story porch, which may be partially or fully enclosed, is required on all Main Body or Rear Wir the View Facade

Encroachments
Only porch steps Front Yard and Si Zone.

Out Building Rec
Garages may be e ings or one-and-c masses attached w nection to the M doors may not fa houses, garages w may be placed in Zone. 'Porch coc designed to be in porch) may be pl Facade Zone.

Building Placement – Village Lakeview

沃特卡勒

佛罗里达州潘汉德尔地区

在沃特卡勒有很多不同的邻里，每一个都在与自然的关系、海岸环境和总体规划设计方面具有自己独一无二的特征。它所面临的一个挑战是住宅和地块关系的设计，地块上的每一栋住宅都既临街又滨水。这一页所介绍的是在保持了强制性的临街空间和步行导向道路的扇形地块平面上，为适宜的停车场和车库所确定的导则。立面分区有助于建立一种前院的特征，这可以缓解在公共区域中汽车所带来的影响。

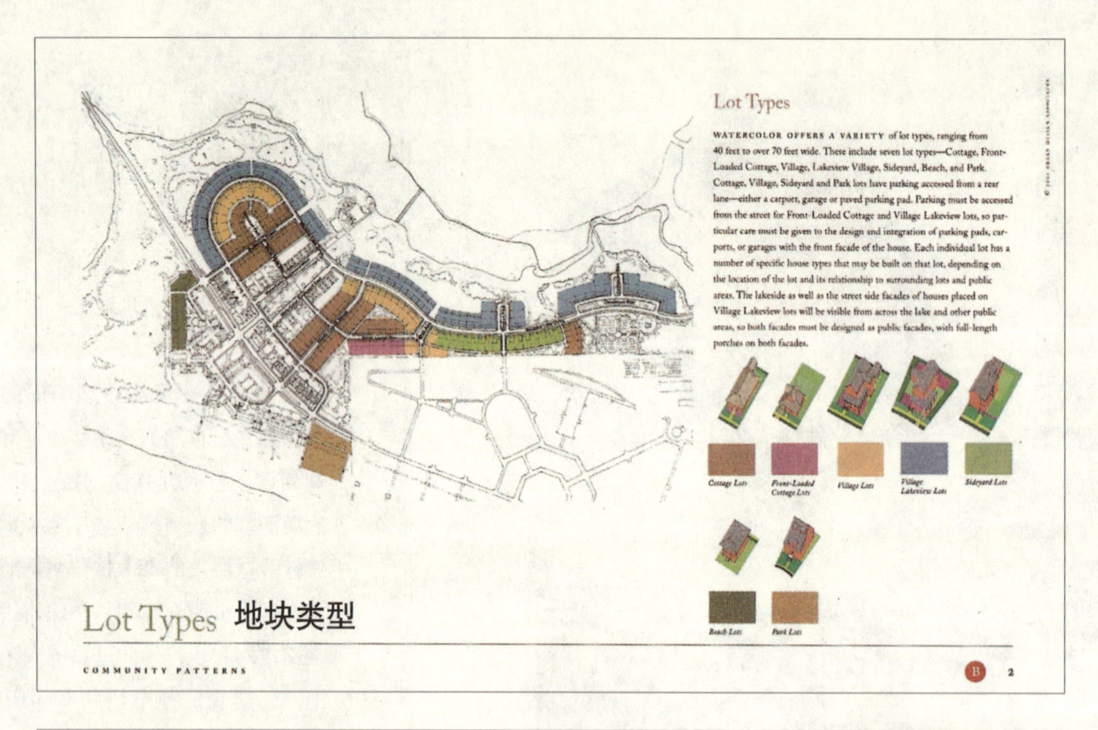

Lot Types

WATERCOLOR OFFERS A VARIETY of lot types, ranging from 40 feet to over 70 feet wide. These include seven lot types—Cottage, Front-Loaded Cottage, Village, Lakeview Village, Sideyard, Beach, and Park. Cottage, Village, Sideyard and Park lots have parking accessed from a rear lane—either a carport, garage or paved parking pad. Parking must be accessed from the street for Front-Loaded Cottage and Village Lakeview lots, so particular care must be given to the design and integration of parking pads, carports, or garages with the front facade of the house. Each individual lot has a number of specific house types that may be built on that lot, depending on the location of the lot and its relationship to surrounding lots and public areas. The lakeside as well as the street side facades of houses placed on Village Lakeview lots will be visible from across the lake and other public areas, so both facades must be designed as public facades, with full-length porches on both facades.

Lot Types 地块类型

COMMUNITY PATTERNS　B　2

Cottage Lots I and II Specifications

Lot Size
Cottage I lots are 40 feet wide. Cottage II lots are 50.

Main Body
The width of the Main Body of the house shall be a maximum of 30 feet for Cottage I and 40 feet for Cottage II lots.

Front Yard Setback / Front Facade Zone
The depth of the Front Yard is typically 10 feet from the front property line to the Front Yard Setback Line, unless noted otherwise in the Address section. The Front Facade Zone extends 20 feet from the Front Yard Setback Line. The front porch shall be placed as close to the Front Yard Setback Line while siting the house to preserve as many trees as possible.

Side Yard Setback
Minimum 5-foot setback from the side property line.

Side Street Setback / Side Street Facade Zone
Structures shall be set back a minimum of 5 feet from the side street property line, unless noted otherwise in the Address section. The Side Street Facade shall be defined by the side facades of the Main Body and any Rear Wings or Out Buildings. Wraparound porches are encouraged on the Main Body. Where there is no building structure, the Side Street Facade shall be delineated by a fence or hedge. Houses on corner lots shall have a garage, carport, or other Out Building placed on the Rear Yard Setback line within the Side Street Facade Zone.

Rear Yard Setback
All structures shall be set back a minimum of 5 feet from the rear lane right-of-way.

Encroachments
Only porch steps may extend into the Front Yard and Side Street Setback Zone.

Out Building Requirements
Garages and carports shall be set back either 5 feet from the rear property line or a minimum of 18 feet. Garages may be either detached or attached to the Main Body by a one-story rear wing. Garage doors opening onto public streets are not permitted.

Building Placement – Cottage Lots 建筑安排—别墅地段

COMMUNITY PATTERNS　B　3

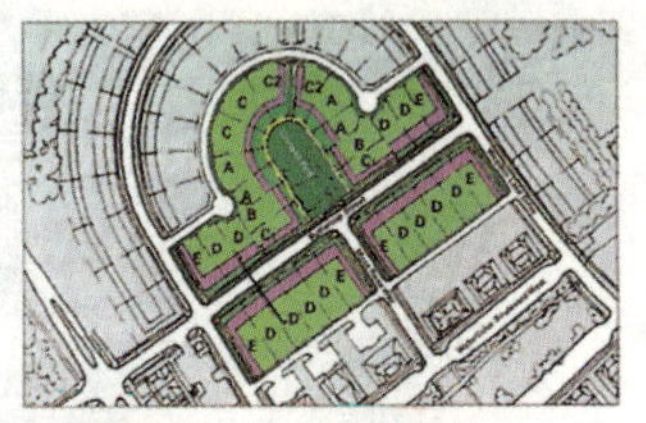

Crescent Frontage Lots (Cottage Lots)

Main Body Types
For (**A**) lots, either one- or one-and-one-half-story Single cottages; for (**B**) lots, one-and-one-half-story Single cottages; and for (**C**) lots, two-story Side Hall or T-Shaped houses.

House Placement
Front porches should be set back 20 feet from the front property line, except for the **B** and **C1** lots where the Front Yard Setback and Side Yard Setback Lines are 10 feet from the front property line. Towers and tall elements should be used to mark corners on **C1** lots and along the pedestrian path (**C2**).

Colors
Warm vibrant tones of red, pink, yellow, and orange; see Section E.

Street Frontage Lots (Cottage Lots)

Main Body Types
For (**D**) lots, either one- or one-and-one-half-story Single or Creole cottages; for (**E**) lots, two-story Side Hall or T-Shaped houses.

Placement
Front porches should be set back 10 feet from the front property line.

Colors
Buildings are a range of grays and neutral colors, with trim and special elements in bright colors. See Section E.

Rose Garden Mews
Rose Garden Mews consists of a crescent-shaped park with a walkway lined with brightly colored cottages, and a street, lined with similar houses, that leads to the water. The continuous porches and landscaping create an intimate neighborhood within the town. The ends of blocks and the corners of the Crescent are defined by two-story houses.

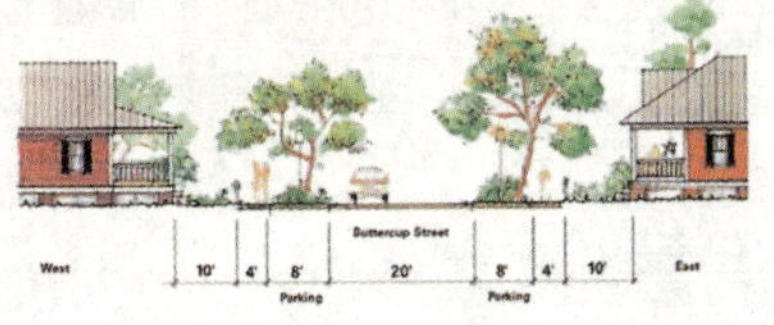

Addresses – Rose Garden Mews 地址—玫瑰花园地

Western Lake Drive
Western Lake Drive is WaterColor's main residential street which will connect the future neighborhoods of WaterColor to the Town Center. Along it are a variety of houses and neighborhood greens, each with a distinct character. Houses on the north side of the street have backyards which look across the lake or wetlands. They are designed to take advantage of these views while still providing view corridors for houses on the south side. There are a variety of house types along the street.

South Side Lots (Village Lots and Sideyard Lots)

Main Body Types
For (**C**) lots, two- or two-and-one-half-story Side Hall houses, and for (**D**) lots, two-and-one-half-story Sideyard houses.

Placement
Lots on the western block (**C**) have a 12-foot Front Yard Setback on which front porches are to be placed. The eastern blocks (**D**) have a 10-foot Front Yard Setback. The front facades of these Sideyard houses are to be placed on the Front Yard Setback Line. **C1** and **C2** houses have 10-foot front and side yard setbacks. These houses should have wraparound porches and use other elements to provide a continuous facade around the park. The front facade of **C2** should address the park.

Colors
Limited to a few muted colors per house, with white trim in white; houses on Viridian Park have accents in red and darker sandtones; see Section E.

North Side Lots (Village Lakeview Lots)

Main Body Types
For (**A**) lots, two- or two-and-one-half-story Side Hall houses for 50-foot lots, and for (**B**) lots, one-and-one-half-, two-, or two-and-one-half-story Center Hall houses for the wider lots.

Placement
Houses are to be set back a minimum of 10 feet from the front property line. **B** and **B1** lots have 20-foot rear lot setbacks. Towers and tall elements should be used on **B1** lots. Wraparound porches and other elements should be used at the corners on **B2** lots to define the public open spaces along Western Lake Drive.

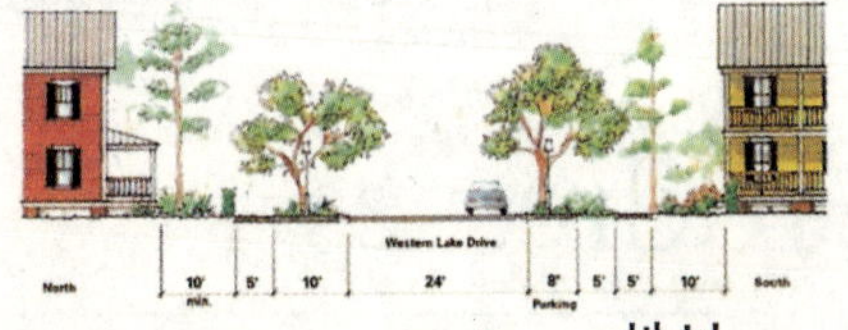

Addresses – Western Lake Drive 地址—西湖车道

Character Sketch of Pretty Lake Waterfront

Typical Section of Pretty Lake Avenue (east of Little Annapolis)

Typical Section of Pretty Lake Avenue

East Beach Addresses

COMMUNITY PATTERNS 社区模式

Pretty Lake District

THE MARINAS ALONG PRETTY LAKE AVENUE create the backdrop for the character of a mixed-use waterfront district termed "Little Annapolis". The area adjacent to Shore Drive will have a mix of shops and residential units in the upper stories with broad sidewalks and a direct connection to the marinas along the shore. Further west, Pretty Lake Avenue will have a mix of attached and detached houses that look out to the marinas.

Images from Annapolis, Maryland

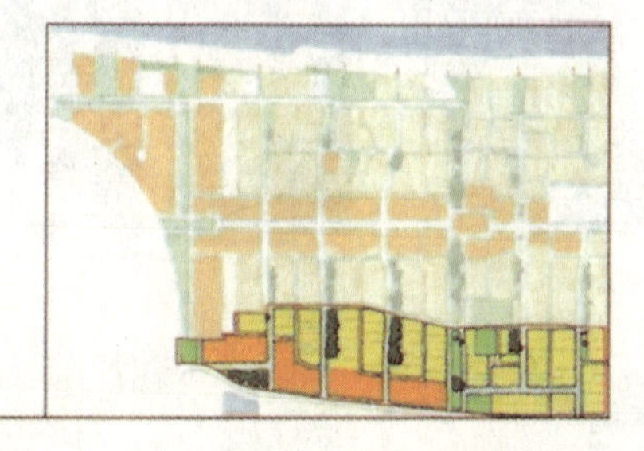

B 3

伊斯特比奇

弗吉尼亚州诺福克市

弗吉尼亚州诺福克市的伊斯特比奇，位于切萨皮克湾、内陆湖和被称为普瑞提湖的海港之间的独特场地之上。这种具有两面临街立面的特征是与相距仅三个街区的新沃特卡勒村镇相一致的。普瑞提湖的滨水空间位于场地的西部边缘地带，这里被计划在传统滨水居住区——诸如切萨皮克湾北部稍远的安纳波利斯——的基础上建设成为码头地区。通过建筑类型、建筑特征及其与码头的关系，为诺福克塑造了一处新的公共空间。

East Beach Character

THE MASTER PLAN FOR EAST BEACH creates a series of distinct addresses that will be built over time. Each park space or street has a unique sense of place defined by the character of the civic space and the houses or buildings that form the neighborhood. This section of the *Pattern Book - Community Patterns* - illustrates examples of the intended character generated by the setbacks, placement and type of buildings and the architectural vocabularies described in the *Architectural Patterns* and *Landscape Patterns* sections of the *Pattern Book*.

Site Plan Delineating Principal Addresses

Shore Drive | *The Bay Front* | *Pleasant Avenue* | *Pretty Lake*

Addresses Within The Plan 平面图中的地段

COMMUNITY PATTERNS　B　1

Pleasant Avenue District

PLEASANT AVENUE IS THE HEART of the East Beach neighborhood. The street will have a mix of large and small houses, attached and detached to create a varied, rich character along its length. There are two public squares that define addresses along the Avenue. The westernmost square at Shore Drive is flanked by mixed-use buildings and serves as a gateway into the neighborhood. On the eastern end, the Avenue splits around a civic site framed by attached and detached houses. The site provides a prominent terminus for Pleasant Avenue while creating a unique address for this district. A series of parks along the north-south streets connect Pleasant Avenue to both waterfronts – Chesapeake Bay and Pretty Lake.

Perspective view of Pleasant Avenue

Images from Annapolis, Maryland (left) and Edenton, North Carolina (right)

Typical Section of Pleasant Avenue

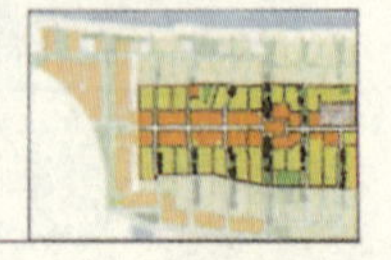

East Beach Addresses 伊斯特比奇地段

COMMUNITY PATTERNS　B　4

Perspective view of typical neighborhood street

North-South Streets

THE NORTH-SOUTH STREETS within East Beach will have a marvelous character, each street different from the next. Many of these streets will have small, informal parks with mature trees and landscape. The character of the streets will also change from north to south. Some sections will have narrow cartways defined by soft edges and meandering walkways. A mix of cottages, attached houses and a variety of lot sizes and types will help create a sense of diversity and interesting character throughout the neighborhoods. These streets connect the Cheasapeake Bay to Pretty Lake providing easy walking access to both shores. Porches will be important elements for houses lining these streets.

Neighborhood images from Easton (left) and Annapolis (right) in Maryland

Typical Section (north of Pleasant Avenue)

Typical Section (south of Pleasant Avenue)

East Beach Addresses 伊斯特比奇地段

COMMUNITY PATTERNS

5

Perspective view along Shore Drive

Shore Drive

THE ENTRANCE TO EAST BEACH is from Shore Drive, a major route linking all of Ocean View to the surrounding region. The image along this Drive is drawn from the precedent of an Admirals' Row of large houses, similar in character and scale, facing the drive across a linear park. This stately image will set the overall character of East Beach as residents and visitors cross the bridge to the bayfront.

Officers' Housing at the Presidio in San Francisco

Captains' Houses in Nantucket

Typical Section at Shore Drive

East Beach Addresses 伊斯特比奇地段

COMMUNITY PATTERNS

B 7

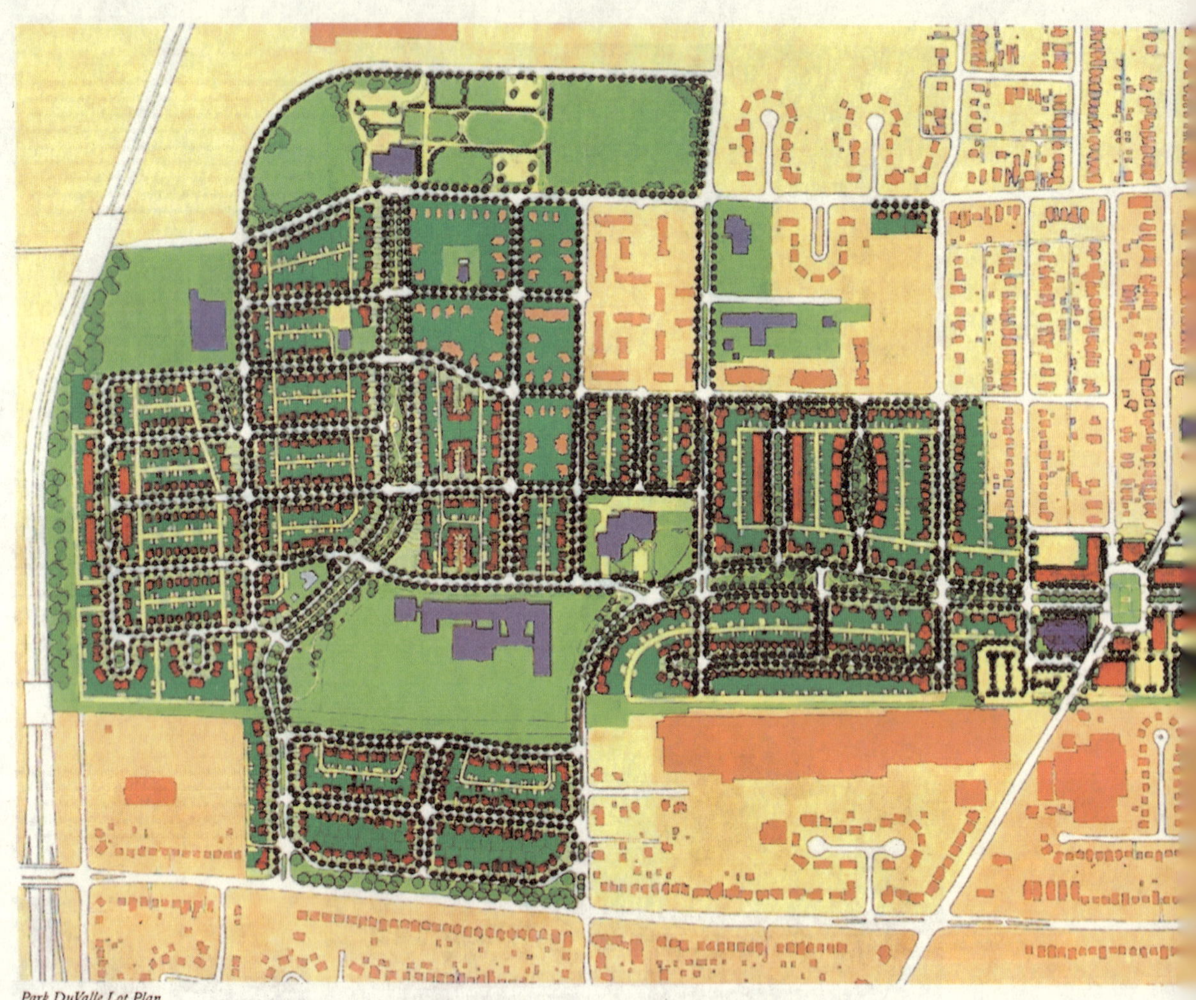
Park DuValle Lot Plan

Addresses 地段

COMMUNITY PATTERNS

帕克杜瓦拉

Creating Park DuValle Addresses

FOLLOWING LOUISVILLE'S GREAT TRADITIONS of neighborhood development, Park DuValle is created as a series of attractive addresses, each with its own character. Constant setbacks from property lines, building massing, and a mix of building types and architectural styles create the public spaces and streets of the addresses.

The addresses of Park DuValle include: Algonquin Parkway, Park Drive, typical Local Neighborhood Streets, and typical Community Streets. Architectural styles include: Victorian, Colonial Revival, and Craftsman.

B　I

帕克杜瓦拉

肯塔基州路易斯维尔市

帕克杜瓦拉是一个公私联合的开发项目，旨在将很大范围的衰退公共住房及其周边的出租住宅再开发为适宜居住的、混合收入的邻里。开发的努力是在公共程序和用来在总体设计下协调开发活动的模式图则的引导下进行的。建筑类型、街区模式、公共开放空间和建筑，是模式图则的主要构成部分。再开发的关键部分，是对整个规划中能够改变场地已往观感的独特场所的设计。在社区模式部分对每一处场所都通过透视图、平面图和剖面图来进行说明。

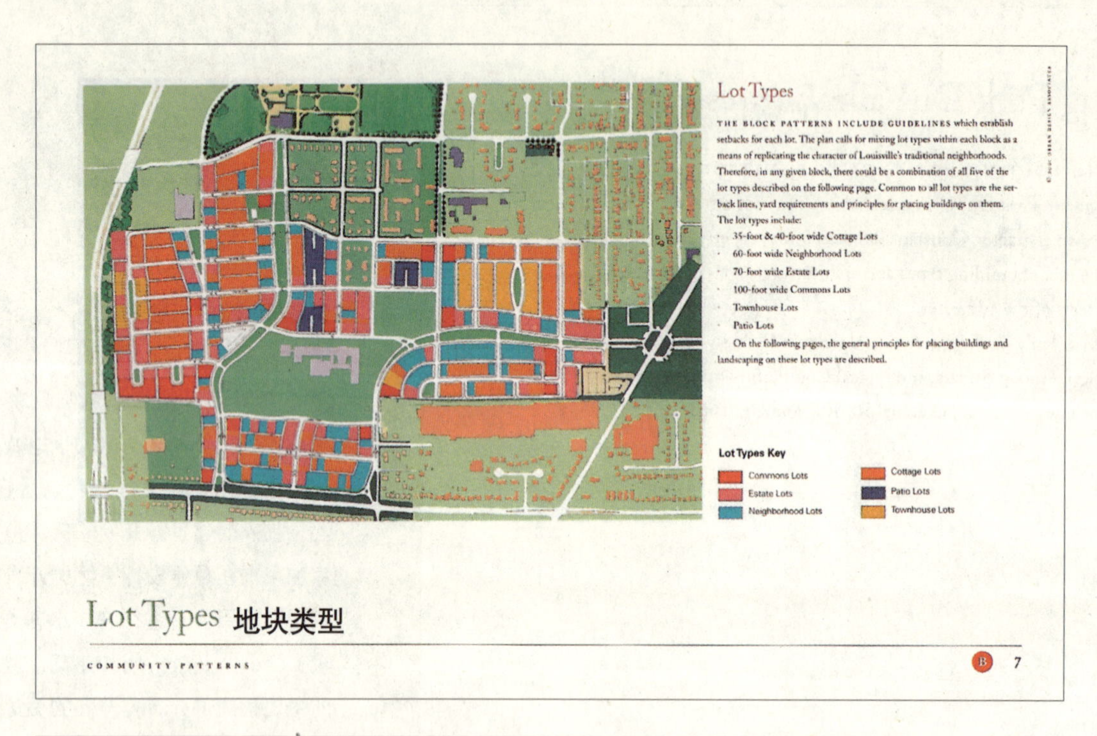

Lot Types

THE BLOCK PATTERNS INCLUDE GUIDELINES which establish setbacks for each lot. The plan calls for mixing lot types within each block as a means of replicating the character of Louisville's traditional neighborhoods. Therefore, in any given block, there could be a combination of all five of the lot types described on the following page. Common to all lot types are the setback lines, yard requirements and principles for placing buildings on them. The lot types include:

- 35-foot & 40-foot wide Cottage Lots
- 60-foot wide Neighborhood Lots
- 70-foot wide Estate Lots
- 100-foot wide Commons Lots
- Townhouse Lots
- Patio Lots

On the following pages, the general principles for placing buildings and landscaping on these lot types are described.

Lot Types 地块类型

COMMUNITY PATTERNS

B 7

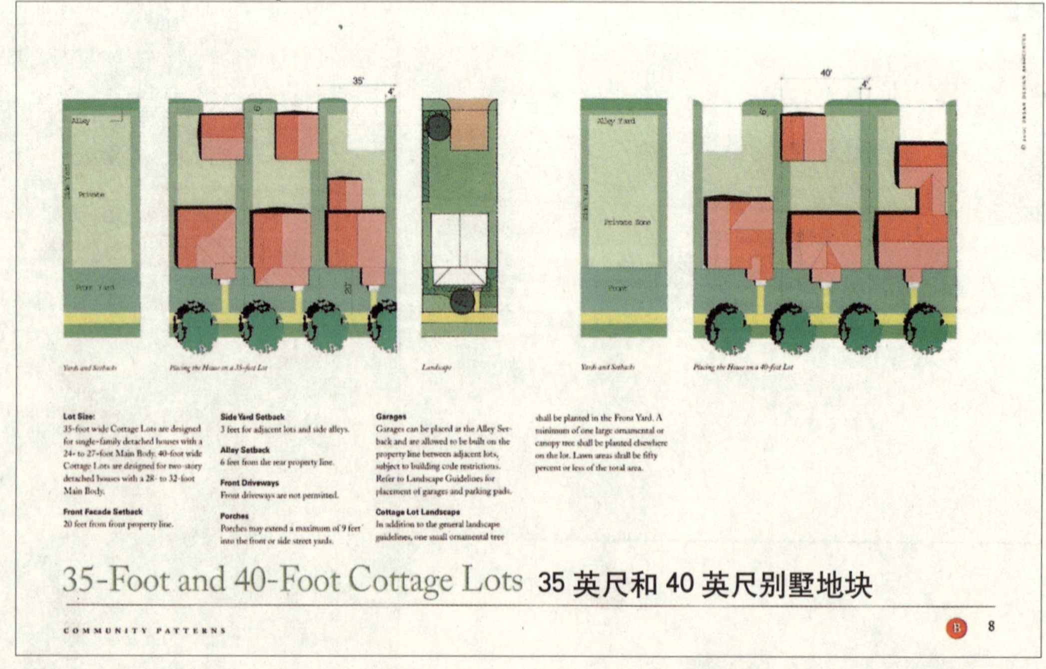

Yards and Setbacks

Placing the House on a 35-foot Lot

Landscape

Yards and Setbacks

Placing the House on a 40-foot Lot

Lot Size:
35-foot wide Cottage Lots are designed for single-family detached houses with a 24- to 27-foot Main Body. 40-foot wide Cottage Lots are designed for two-story detached houses with a 28- to 32-foot Main Body.

Front Facade Setback
20 feet from front property line.

Side Yard Setback
3 feet for adjacent lots and side alleys.

Alley Setback
6 feet from the rear property line.

Front Driveways
Front driveways are not permitted.

Porches
Porches may extend a maximum of 9 feet into the front or side street yards.

Garages
Garages can be placed at the Alley Setback and are allowed to be built on the property line between adjacent lots, subject to building code restrictions. Refer to Landscape Guidelines for placement of garages and parking pads.

Cottage Lot Landscape
In addition to the general landscape guidelines, one small ornamental tree shall be planted in the Front Yard. A minimum of one large ornamental or canopy tree shall be planted elsewhere on the lot. Lawn areas shall be fifty percent or less of the total area.

35-Foot and 40-Foot Cottage Lots 35 英尺和 40 英尺别墅地块

COMMUNITY PATTERNS

B 8

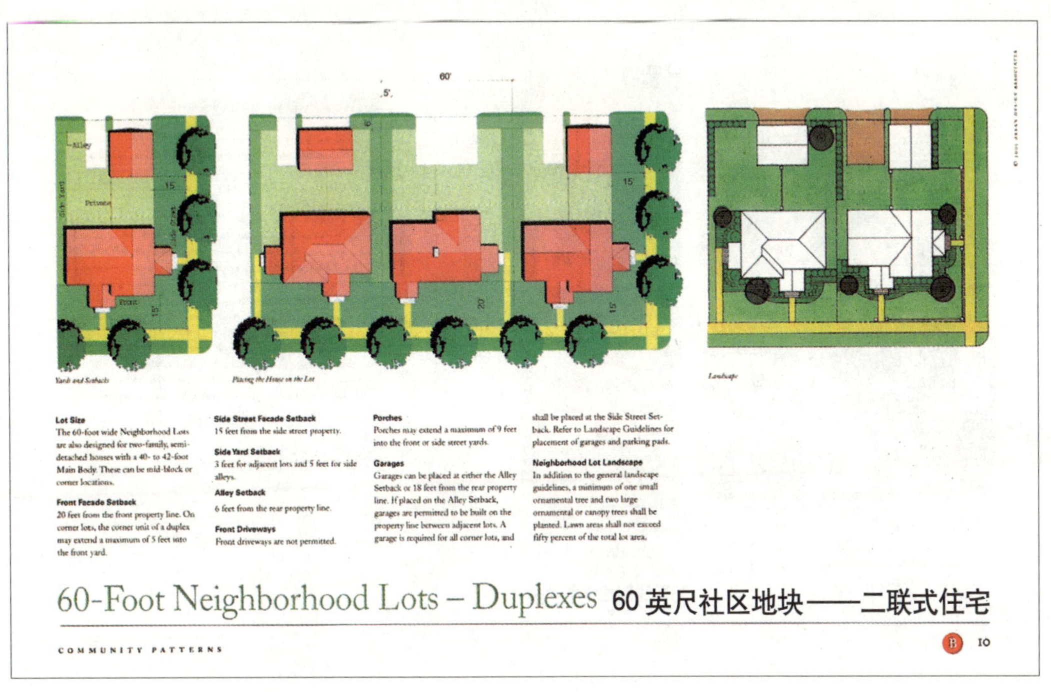

Yards and Setbacks　　Placing the House on the Lot　　Landscape

Lot Size
The 60-foot wide Neighborhood Lots are also designed for two-family, semi-detached houses with a 40- to 42-foot Main Body. These can be mid-block or corner locations.

Front Facade Setback
20 feet from the front property line. On corner lots, the corner unit of a duplex may extend a maximum of 5 feet into the front yard.

Side Street Facade Setback
15 feet from the side street property.

Side Yard Setback
3 feet for adjacent lots and 5 feet for side alleys.

Alley Setback
6 feet from the rear property line.

Front Driveways
Front driveways are not permitted.

Porches
Porches may extend a maximum of 9 feet into the front or side street yards.

Garages
Garages can be placed at either the Alley Setback or 18 feet from the rear property line. If placed on the Alley Setback, garages are permitted to be built on the property line between adjacent lots. A garage is required for all corner lots, and shall be placed at the Side Street Setback. Refer to Landscape Guidelines for placement of garages and parking pads.

Neighborhood Lot Landscape
In addition to the general landscape guidelines, a minimum of one small ornamental tree and two large ornamental or canopy trees shall be planted. Lawn areas shall not exceed fifty percent of the total lot area.

60-Foot Neighborhood Lots – Duplexes 60英尺社区地块——二联式住宅

COMMUNITY PATTERNS　　B 10

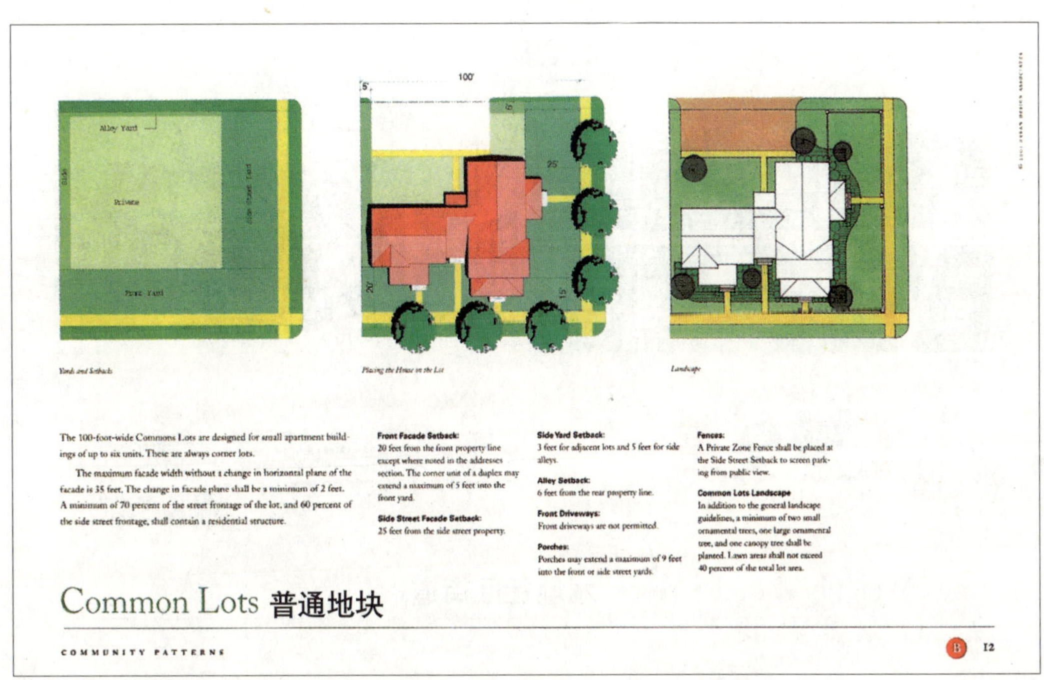

Yards and Setbacks　　Placing the House on the Lot　　Landscape

The 100-foot-wide Commons Lots are designed for small apartment buildings of up to six units. These are always corner lots.

The maximum facade width without a change in horizontal plane of the facade is 35 feet. The change in facade plane shall be a minimum of 2 feet. A minimum of 70 percent of the street frontage of the lot, and 60 percent of the side street frontage, shall contain a residential structure.

Front Facade Setback:
20 feet from the front property line except where noted in the addresses section. The corner unit of a duplex may extend a maximum of 5 feet into the front yard.

Side Street Facade Setback:
25 feet from the side street property.

Side Yard Setback:
3 feet for adjacent lots and 5 feet for side alleys.

Alley Setback:
6 feet from the rear property line.

Front Driveways:
Front driveways are not permitted.

Porches:
Porches may extend a maximum of 9 feet into the front or side street yards.

Fences:
A Private Zone Fence shall be placed at the Side Street Setback to screen parking from public view.

Common Lots Landscape
In addition to the general landscape guidelines, a minimum of two small ornamental trees, one large ornamental tree, and one canopy tree shall be planted. Lawn areas shall not exceed 40 percent of the total lot area.

Common Lots 普通地块

COMMUNITY PATTERNS　　B 12

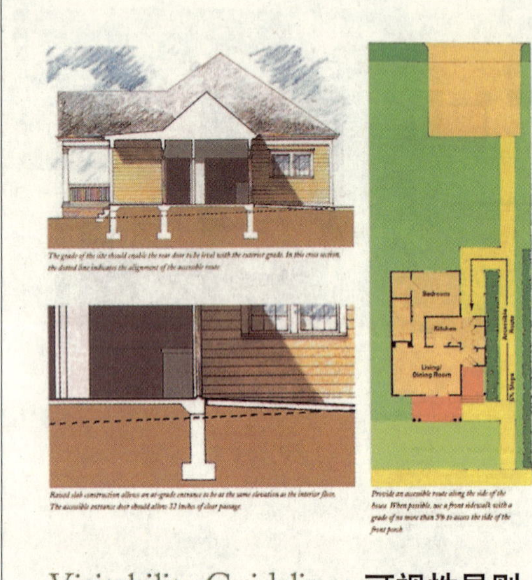

The grade of the site should enable the rear door to be level with the exterior grade. In this cross section, the dotted line indicates the alignment of the accessible route

Raised slab construction allows an at-grade entrance to be at the same elevation as the interior floor. The accessible entrance door should allow 32 inches of clear passage

Provide an accessible route along the side of the house. When possible, use a front sidewalk with a grade of no more than 5% to access the side of the front porch

Visitability Guidelines

PARK DUVALLE IS INTENDED to become a diverse, mixed-income, stable neighborhood. Contributing to its diversity will be the development of housing that is both accessible to persons with disabilities and marketable to all. 'Visitability' is loosely defined as making it possible for a disabled person to visit a friend's home and enjoy a meal, without having to be lifted up a step or be unable to get through the doorway. Therefore, house designs should provide at least one level access at grade without a step at the threshold.

Visitability can be accomplished by using raised slab foundation construction. This allows the rear entrance at grade to be at the same elevation as the interior floor. It also makes it possible to have a sloped walk along the side of the house with its change in grade taken up along the side of the foundation wall.

The house should also have a bathroom on the ground floor that can be used by a disabled person. In those cases in which it is a small powder room, this can be accomplished by simply employing an outward opening door yielding a 32-inch clear passage space and path inside.

Visitability not only permits disabled persons to visit their neighbors but can allow currently physically-able persons to age comfortably in their homes without the need for extensive renovations and the necessity for ungainly add-on ramps if disability should occur as part of the normal aging process.

Visitability Guidelines 可视性导则

COMMUNITY PATTERNS B 15

Partial Plan of Von Spiegel Avenue

View of Von Spiegel Avenue

Typical Section

Local Neighborhood Streets

THE MOST INTIMATE of neighborhood streets, local streets discourage high volumes of through-traffic and create a human, pedestrian scale. Cartways are narrow, with parking on both sides of the street, and houses are placed closer to the street to define a more enclosed space.

Local Neighborhood Streets 本地社区街道

COMMUNITY PATTERNS B 5

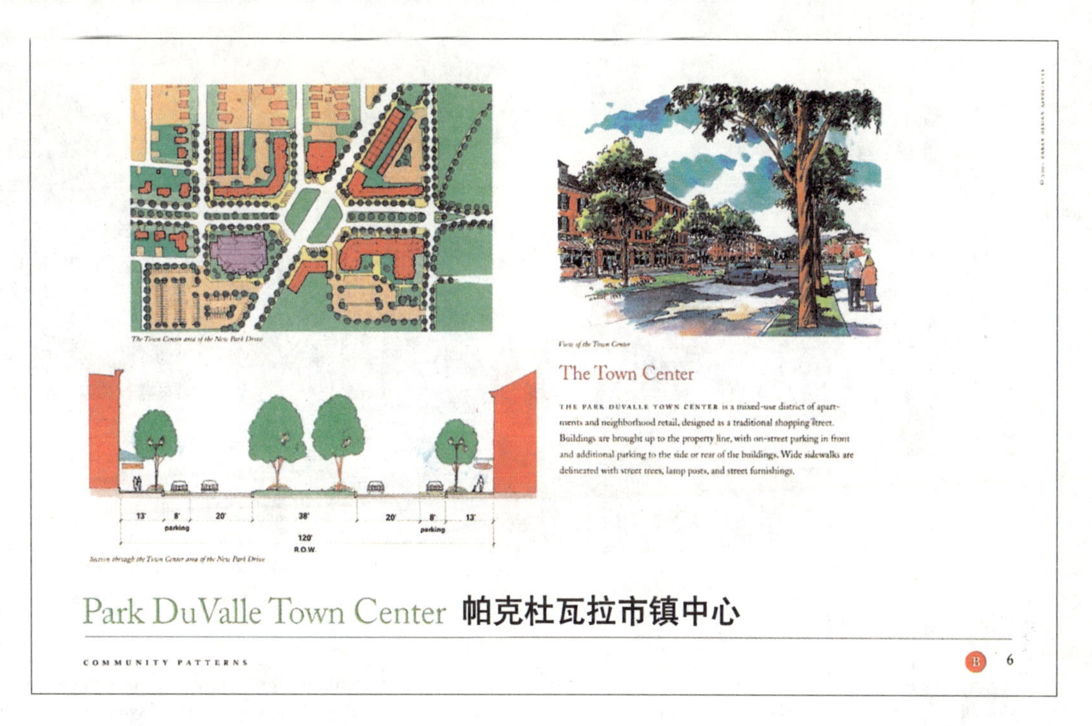

The Town Center area of the New Park Drive

Section through the Town Center area of the New Park Drive

View of the Town Center

The Town Center

THE PARK DUVALLE TOWN CENTER is a mixed-use district of apartments and neighborhood retail, designed as a traditional shopping street. Buildings are brought up to the property line, with on-street parking in front and additional parking to the side or rear of the buildings. Wide sidewalks are delineated with street trees, lamp posts, and street furnishings.

Park DuValle Town Center 帕克杜瓦拉市镇中心

COMMUNITY PATTERNS　B 6

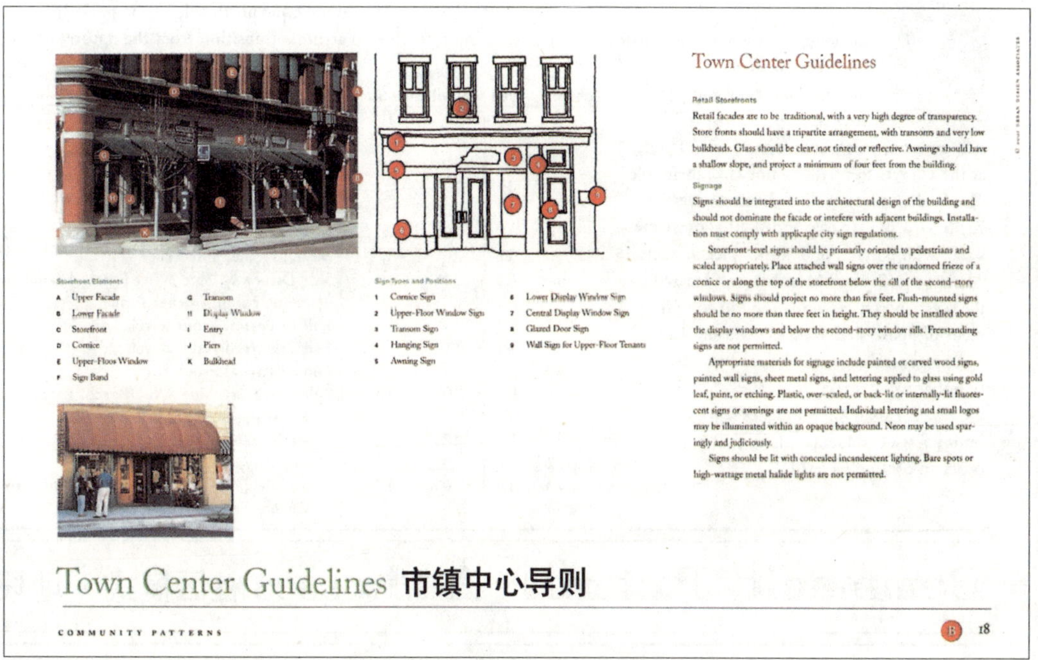

Storefront Elements

- A Upper Facade
- B Lower Facade
- C Storefront
- D Cornice
- E Upper-Floos Window
- F Sign Band
- G Transom
- H Display Window
- I Entry
- J Piers
- K Bulkhead

Sign Types and Positions

1. Cornice Sign
2. Upper-Floor Window Sign
3. Transom Sign
4. Hanging Sign
5. Awning Sign
6. Lower Display Window Sign
7. Central Display Window Sign
8. Glazed Door Sign
9. Wall Sign for Upper-Floor Tenants

Town Center Guidelines

Retail Storefronts

Retail facades are to be traditional, with a very high degree of transparency. Store fronts should have a tripartite arrangement, with transoms and very low bulkheads. Glass should be clear, not tinted or reflective. Awnings should have a shallow slope, and project a minimum of four feet from the building.

Signage

Signs should be integrated into the architectural design of the building and should not dominate the facade or intefere with adjacent buildings. Installation must comply with applicaple city sign regulations.

Storefront-level signs should be primarily oriented to pedestrians and scaled appropriately. Place attached wall signs over the unadorned frieze of a cornice or along the top of the storefront below the sill of the second-story windows. Signs should project no more than five feet. Flush-mounted signs should be no more than three feet in height. They should be installed above the display windows and below the second-story window sills. Freestanding signs are not permitted.

Appropriate materials for signage include painted or carved wood signs, painted wall signs, sheet metal signs, and lettering applied to glass using gold leaf, paint, or etching. Plastic, over-scaled, or back-lit or internally-lit fluorescent signs or awnings are not permitted. Individual lettering and small logos may be illuminated within an opaque background. Neon may be used sparingly and judiciously.

Signs should be lit with concealed incandescent lighting. Bare spots or high-wattage metal halide lights are not permitted.

Town Center Guidelines 市镇中心导则

COMMUNITY PATTERNS　B 18

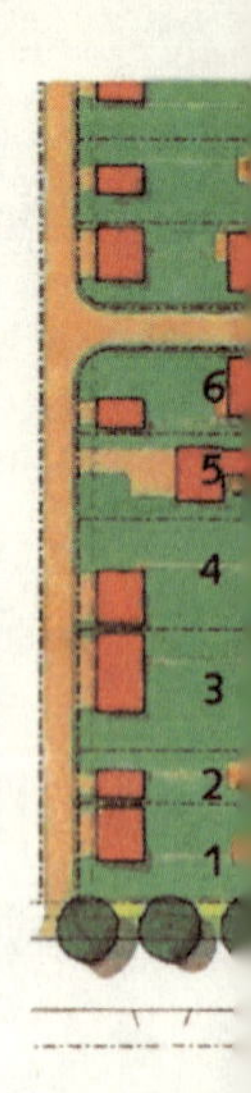
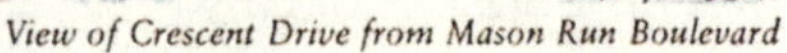

View of Crescent Drive from Mason Run Boulevard

Crescent Park

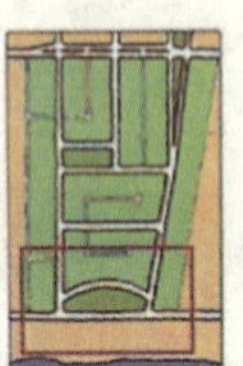

Located on Elm Street, Crescent Park sets the image and creates an address for Mason Run. As with any park, the character and appearance of surrounding homes dictate the quality of the space. Here, a symmetrical arrangement of Neighborhood and Estate houses (Lots 41-48) embrace the north side of the park. It is recommended that these homes possess full or partial front porches to create a transition from the natural to the built environment. The Estate houses that "bookend" this row (Lots 41 & 48) will have side wings or side porches and picket fences at the corners to further emphasize their role. The Duplex units on Elm Avenue (Lots 1 & 73) have a similar role and shall address the street with porches and corner picket fences. Also of note is the "Model Row," located to the west and representing the full compliment of house sizes found in Mason Run.

As a general rule, any house that sits next to either a side street or an alley (Lots 1, 6, 41, 48, 73) shall have a picket fence or a built garage connection along that edge in order to contain its yard.

Lot No.	Type	Setbacks Front	Setbacks Side Street	Special Details
1	D	20'	10'	Model Home, Full or Partial Side Porch, Corner Fence
2	D	20'		Model Home, Full Front Porch
3	E	20'		Model Home
4	N	20'		Model Home
5	V	20'		Model Home
6	C	20'		Model Home, Side Fence
41	E2	20'	10'	Side Driveway, Side Wing/Porch, Partial Front Porch, Corner Fence
42	N2	25'		Full or Partial Front Porch
43–46	N1	25'		Full or Partial Front Porch
47	N2	25'		Full or Partial Front Porch
48	E2	20'	10'	Side Driveway, Side Wing/Porch, Partial Front Porch, Corner Fence
71	E1	20'		Front Driveway, Full or Partial Front Porch
72	D	20'		Shared drive w/73
73	D	20'	10'	Shared drive w/72, Full or Partial Side Porch, Corner Fence

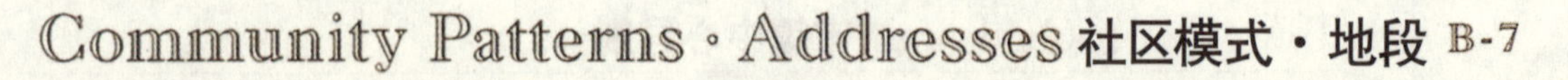

Community Patterns · Addresses 社区模式・地段 B-7

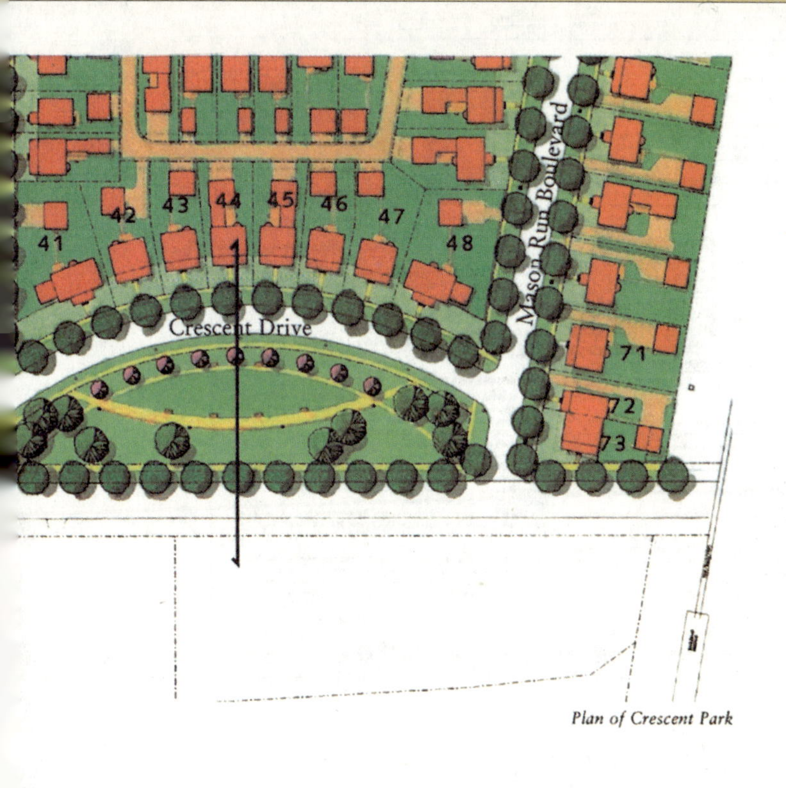

Plan of Crescent Park

Crescent Park Street Section

新月公园 Crescent Park

梅森兰

密歇根州门罗市

这一项目的社区模式部分联合使用了规划设计中重要场所推荐特征的透视图、带有地块名称的场地详细设计平面图、在街道或公园处的典型剖面图以及说明了必需建筑后退或特殊条件的列表。设计方案是作为棕地再开发策略的一部分，由私人开发商与镇区官员共同完成的。模式的提出是基于周边的邻里和在整个城镇中发现的多种多样的建筑风格和语汇。

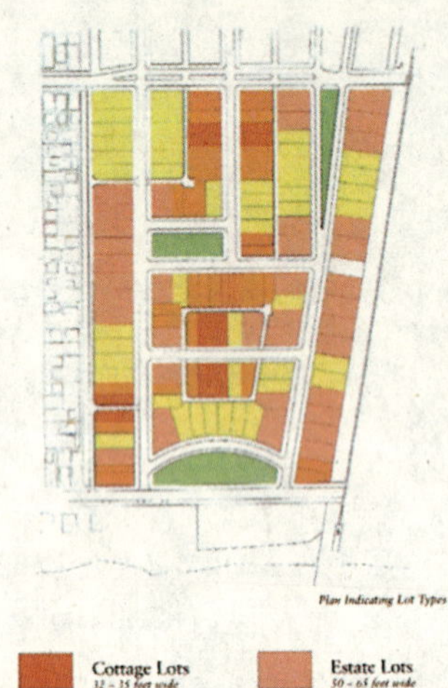

Plan Indicating Lot Types

Cottage Lots 32 – 35 feet wide		Estate Lots 50 – 65 feet wide	
Village Lots 35 feet wide		Duplex Lots Varied width	
Neighborhood Lots 40 – 50 feet wide		Townhouse Lots Varied width	

Community Patterns

Lot Types

The Mason Run site plan contains a range of differently sized lots to accommodate the wide variety of house and unit types – from twenty-foot wide, rear-loaded, Townhouse lots, to 65 foot wide, front-loaded Estate lots. Lots are typically 100 to 110 feet deep and are either served by alleys or by driveways from the street.

The mix of large and small houses provides great variety along neighborhood streets. Corner lots with generous front and side yards are particularly important to the image of the neighborhood because they are highly visible and serve as important landmarks.

Setbacks and Zones

The overall image of a street is created by the way the houses and the landscaping work together to define space. Because the location of the house is crucial to creating the overall effect, each lot has setback lines that define minimum distances from front, rear and side lot lines. In addition, there are also a number of *setback zones* which allow a greater degree of flexibility in house placement.

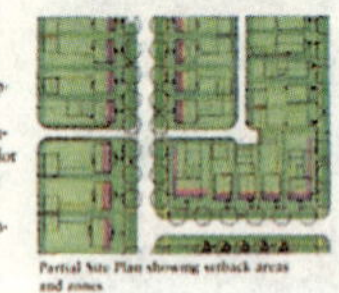

Partial Site Plan showing setback areas and zones

Front Setback Line: Measured from the front property line.

Front Facade Zone: The zone in which the Front Facade must be placed. The zone begins at the *Front Setback Line* and continues a specified distance into the property. The Zone varies for each lot type and is subject to specific guidelines set out in this section.

Front Yard: The area between the front property line and the front facade of the house. Porches, Stoops and Bay Windows may encroach into the *Front Yard.*

Side Street Setback Line: Measured from the side property line on the street side.

Side Street Facade Zone: The zone in which the Side Street Facade must be placed. The zone begins at the ***Side Street Setback Line*** and continues a specified distance into the property. The Zone varies for each lot type and is subject to specific guidelines set out in this section.

Side Setback Line: Measured from the interior side property line.

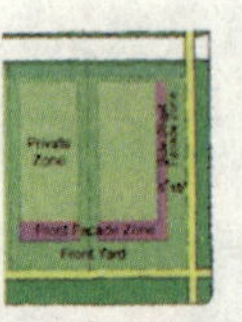

Typical Lot Diagram showing setback areas and zones

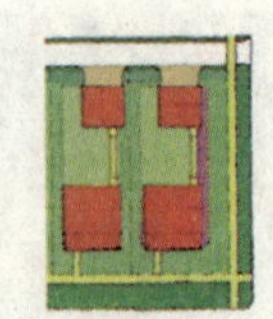

Typical Lot Diagram showing building placement on the site

Garage Setback Line: Garages facing alleys must be placed eight feet or sixteen feet or more from the rear property line, and a minimum of three feet from the interior side property line. (Sixteen feet provides enough driveway space for parking.)

Utility Easement Zone: Begins at the rear property line and extends six feet into the back yard. Only lawn, ground cover, annuals and perennials may be planted. (See Landscape Guidelines Section).

Architectural Variety

One of the things that people admire most about older, traditional neighborhoods is the architectural variety. When creating a new, traditional neighborhood it is extremely important to create and maintain variety in the streetscape. This can be done by creating a site plan that mixes different house types throughout the plan, and by establishing rules requiring diversity among facade elements. For example, if one house is brick with a full front porch and double windows, the adjacent house could be clad in siding with a partial porch and bay windows.

The *Architectural Patterns* section describes the basic principles of Mason Run residential architectural styles and provides the key elements for those parts of the house that face a public street or area.

In Mason Run there are four basic elements that contribute to the character and feel of a house facade: 1) Materials, 2) Color Palette, 3) Porch Design, and 4) Window Type

The Rule: In order to achieve architectural variety along the street, no one elevation may be repeated within a row of four houses, nor may it be repeated within the three elevations across the street. Within the row of four houses, each consecutive house must also differ in at least two of the four facade elements.

Community Patterns · Lots 社区模式 · 地块 B-1 地块类型，后退和划区 Lot Types, Setbacks & Zones

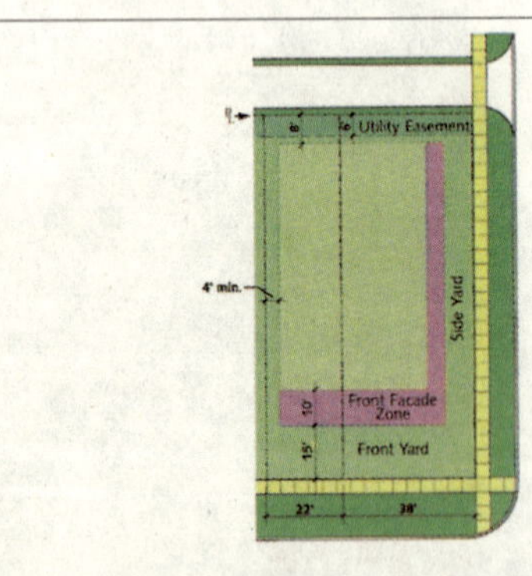

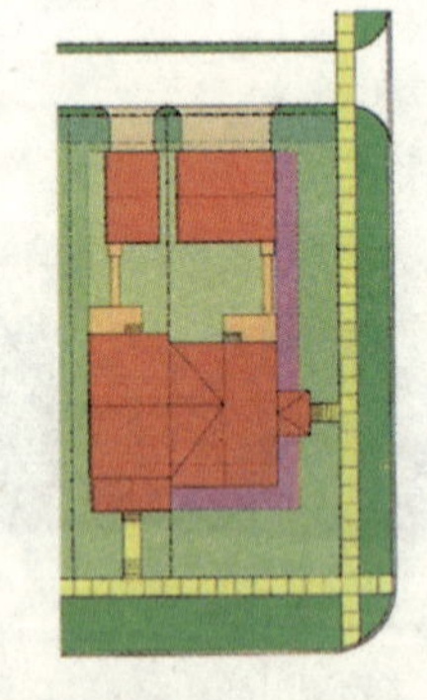

Duplex Lots

Duplex Lots are typically 22 and 38 feet wide by 100 to 110 feet deep. They may vary in size from lot to lot depending on location and type.

House Width: 18 and 24 feet typical.

Front Yard Setback: 15 feet (See Lot Information Table for specific setback information).

Front Facade Zone: 10 feet.

Side Yard Setback: 4 feet minimum (exterior side), 0 (zero) feet (interior side).

Side Street Setback: 10 feet minimum (See Lot Information Table for specific setback information).

Side Street Facade Zone: 5 feet.

Garage Side Setback: 3 feet.

Garage Size: One- and two-car garages permitted.

Side Street Fences: Picket fences are required on all corner lots. See Landscape Guidelines for specific rules about type and location.

Porches: May extend into the front yard and side yard a maximum of 8 feet.

Fireplaces: Must be located a minimum of 8 feet behind the front facade. Fireplaces on side street facade must have a chimney.

Rear (Alley) Access Lots

Utility Easement: 6 feet.

Rear Fence Line: 6 feet from the rear property line.

Garage Rear Setback: 8 feet, or 16 feet or more from the rear property line.

Garage Orientation: Garage doors must face the alley.

Front (Driveway) Access Lots

Garages: Garages on front access lots shall be positioned at least 16 feet behind the front facade of the house and no less than 3 feet from the rear property line.

Driveway: A maximum width of 8 feet maximum between curb and rear of house.

Rear Fence: Permitted along the rear property line.

Community Patterns · Mason Run 社区模式 · 梅森兰 B-6 二联式住宅地段 Duplex Lots

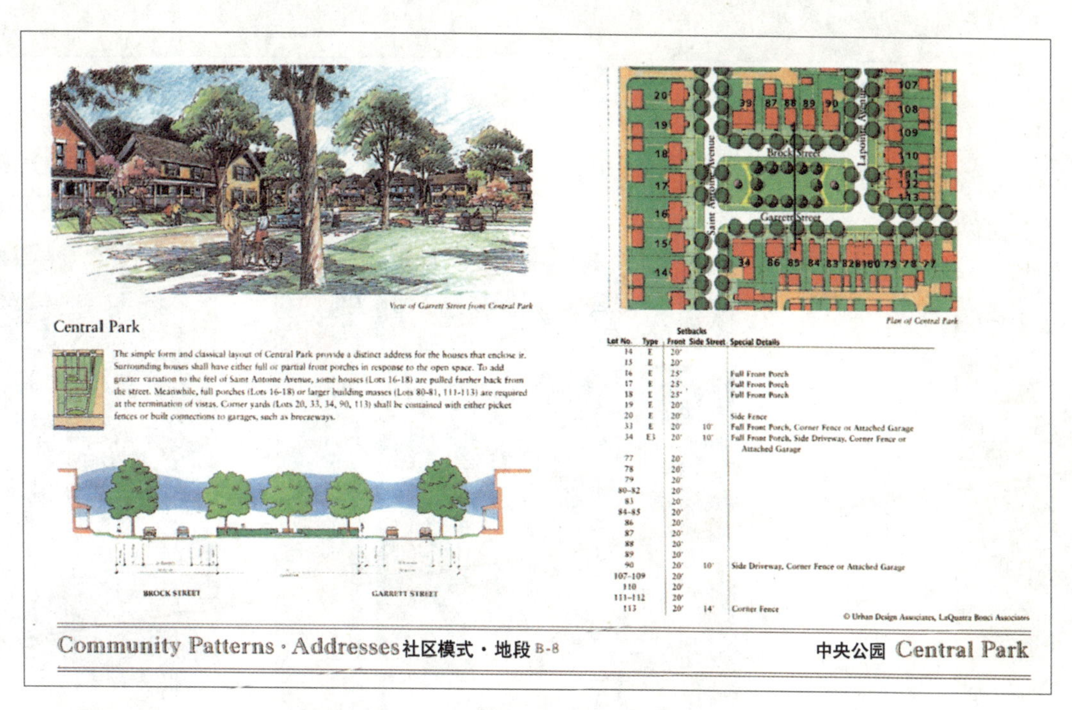

View of Garrett Street from Central Park

Central Park

The simple form and classical layout of Central Park provide a distinct address for the houses that enclose it. Surrounding houses shall have either full or partial front porches in response to the open space. To add greater variation to the feel of Saint Antoine Avenue, some houses (Lots 16-18) are pulled farther back from the street. Meanwhile, full porches (Lots 16-18) or larger building masses (Lots 80-81, 111-113) are required at the termination of vistas. Corner yards (Lots 20, 33, 34, 90, 113) shall be contained with either picket fences or built connections to garages, such as breezeways.

Plan of Central Park

Lot No.	Type	Setbacks Front	Setbacks Side Street	Special Details
14	E	20'		
15	E	20'		
16	E	25'		Full Front Porch
17	E	25'		Full Front Porch
18	E	25'		Full Front Porch
19	E	20'		
20	E	20'		Side Fence
33	E	20'	10'	Full Front Porch, Corner Fence or Attached Garage
34	E3	20'	10'	Full Front Porch, Side Driveway, Corner Fence or Attached Garage
77		20'		
78		20'		
79		20'		
80-82		20'		
83		20'		
84-85		20'		
86		20'		
87		20'		
88		20'		
89		20'		
90		20'	10'	Side Driveway, Corner Fence or Attached Garage
107-109		20'		
110		20'		
111-112		20'		
113		20'	14'	Corner Fence

© Urban Design Associates, LaQuatra Bonci Associates

Community Patterns · Addresses 社区模式 · 地段 B-8　　中央公园 Central Park

Houses with gracious front porches overlook Mason Run Boulevard

Mason Run Boulevard

Mason Run Boulevard begins as a simple, neighborhood street and broadens into a lush, landscaped boulevard as it continues north to E. Noble Avenue. Of special note along Mason Run Boulevard are the houses on the western side, south of Garrett Street (Lots 49-51, 64, 74-76). Instead of mimicking the angle of the road, they remain orthogonal to the grid and "step" toward the linear park to the north, framing the view corridor. Further north along the park, some houses (Lots 116-119, 127-130) are moved closer to the street to contain the park area and emphasize that address. Finally, houses with corner lots (Lots 51, 64, 76, 114, 123, 124) shall have either a picket fence or a built connection (such as a breezeway) to the garage in order to contain the backyards.

Lot No.	Type	Setbacks Front	Setbacks Side Street	Special Details
49	E1	20'		Parallel to Fontaine St.
50	N1	20'		Parallel to Fontaine St.
51	N2	20'	10'	Parallel to Fontaine St., Corner Fence or Attached Garage
64	E1	20'	10'	Parallel to Fontaine St.,Corner Fence or Attached Garage
65	E2	25'		Front Driveway, Full Front Porch
66	N2	25'		Front Driveway, Full Front Porch
67	N2	20'		Front Driveway
68-70	E1	20'		Front Driveway
74		20'		Parallel to Fontaine St.
75		20'		Parallel to Fontaine St.
76		20'	10'	Parallel to Fontaine St., Side Driveway, Corner Fence or Attached Garage
114		25'	10'	Corner Fence or Attached Garage
115		25		
116-119		20'		
120-122		25'		
123		25'	10'	Corner Fence or Attached Garage
124		25'	10'	Side Driveway, Corner Fence or Attached Garage
125-126		25'		Front Driveway
127-130		20'		Front Driveway
131		25'		Front Driveway
132		25'		Front Driveway, Side Fence
133		20'		Front Driveway, Side Fence
134-135		20'		Front Driveway

Plan of Mason Run Boulevard

© Urban Design Associates, LaQuatra Bonci Associates

Community Patterns · Addresses 社区模式 · 地段 B-9　　梅森兰林荫大道 Mason Run Boulevard

GÉNITOY EST: BUSSY SAINT GEORGES, FRANCE

Typical Block Plan

典型的街区平面图

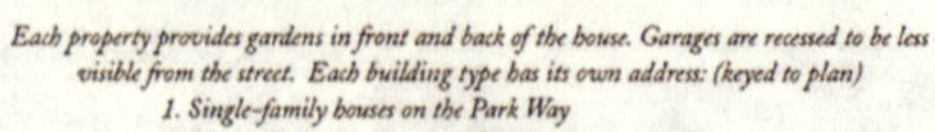

Each property provides gardens in front and back of the house. Garages are recessed to be less visible from the street. Each building type has its own address: (keyed to plan)

1. *Single-family houses on the Park Way*
2. *Single-family houses on a residential lane*
3. *Small Villas (two attached houses) on a Park Way*
4. *Small Villas and houses on a neighborhood street*
5. *Large Villas (four or more attached units) on corners*

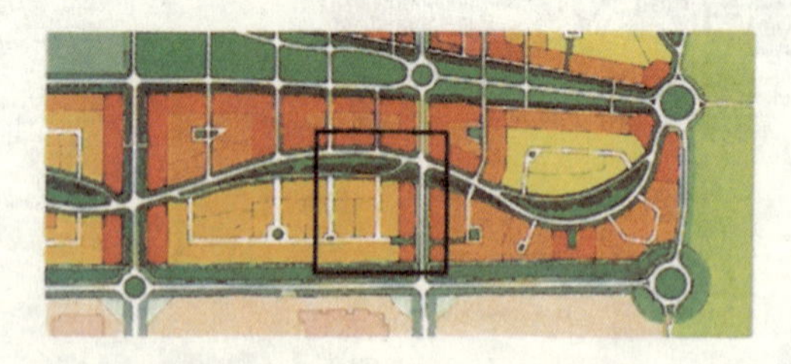

JUNE 2002

URBAN DESIGN ASSOCIATE

ERASME ÉTUDES URBAINES

Single-family houses and small villas face the Park Way and a neighborhood lane. They are compatible with the large villas which are placed on the corner of the block.

Each house has a clearly visible front door. The landscape elements and fencing define the yard without blocking views of the house. Garages are recessed.

热尼特伊

法国比西圣乔治

热尼特伊的总体规划设计是对巴黎郊外现有新城的拓展。对于这项规划而言，东部的延伸意味着在建筑类型和建筑设计方面对更为传统的法国模式的回归。模式图则是对城市设计进行的简明安排，而构成一系列街区的建筑模式和地块模式则详细说明了建筑布局策略、建筑类型和建筑风格导则。在规划设计中联合使用了不同的方法来重建场所感和空间品质，这也是这一地区法国传统居住区的组成部分。

GÉNITOY EST: BUSSY SAINT GEORGES, FRANCE
Key Concepts of the Plan: Les Parcs du Génitoy
平面图的主要概念
URBAN DESIGN ASSOCIATES
ERASME ÉTUDES URBAINES
JUNE 2002

GÉNITOY EST: BUSSY SAINT GEORGES, FRANCE
Urban Design Plan 城市设计平面图
URBAN DESIGN ASSOCIATES
ERASME ÉTUDES URBAINES
JUNE 2002

热尼特伊

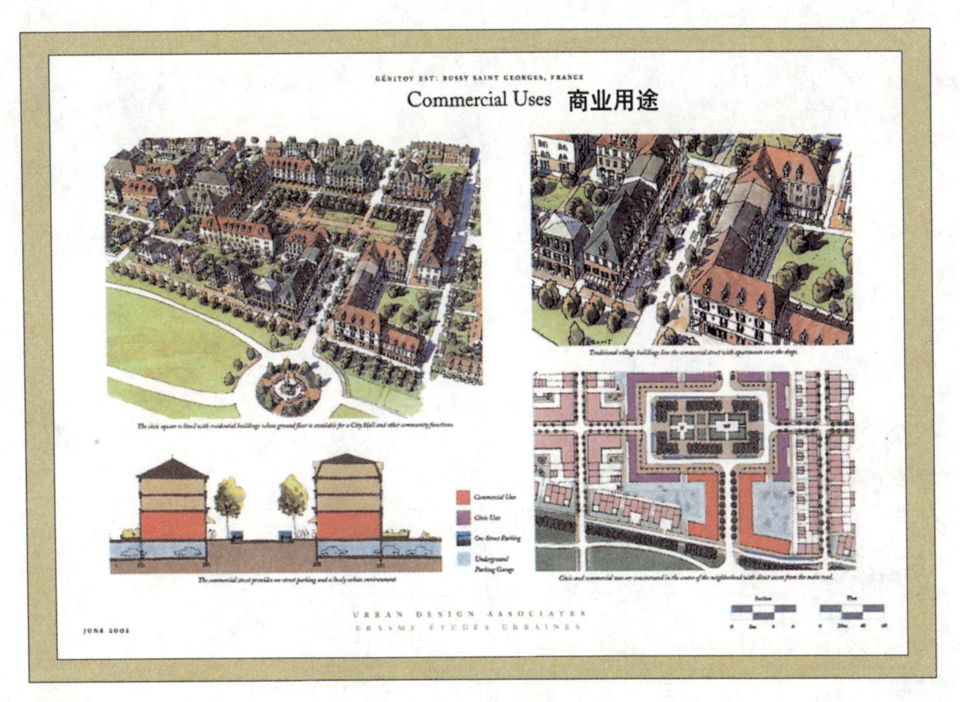
GÉNITOY EST: BUSSY SAINT GEORGES, FRANCE
Commercial Uses 商业用途
URBAN DESIGN ASSOCIATES

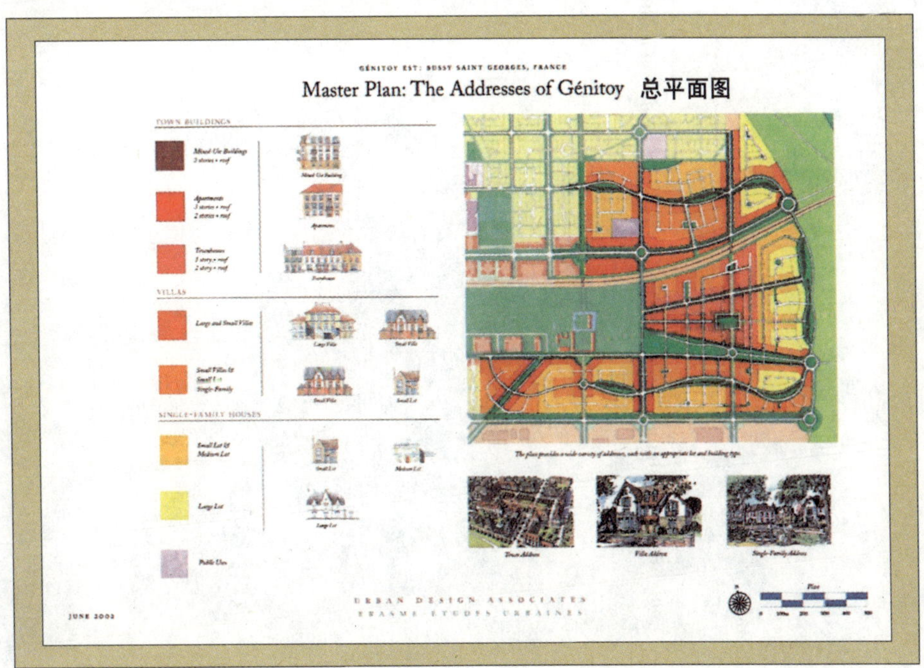
GÉNITOY EST: BUSSY SAINT GEORGES, FRANCE
Master Plan: The Addresses of Génitoy 总平面图
TOWN BUILDINGS
VILLAS
URBAN DESIGN ASSOCIATES

22′-28′
24′-32′
28′-35′
24′-28′
32′-40′
24′-28′

第八章

建筑模式册页

建筑模式部分描述了很多被认为是适合于计划开发项目、并对营造区域性独特场所来说至关重要的特定建筑语汇或建筑风格。正如在第五章中所讨论的那样，对主要建筑模式的识别确定，源自记录区域内的范例邻里和从作为相关范例的特定场所的核心特征中提炼主要建筑语汇的严谨过程。一旦这些特征风格被确定下来，经过拍照和测量，工作小组就可以开始从每一语汇中确定关键部件的过程了，这包括以下方面：体块类型，屋檐细部，材料，标准的窗和门，特殊的或独特的部件、方法和细部，以及门廊、柱子和屋顶。

通常每一种语汇都具有地方性的独特属性或特征。在出现这种情况的地方，重要的是要记录下这些细部和方法并纳入相关的建筑风格部分。在大多数案例中，模式图则会在建筑模式部分介绍和说明 3 至 6 种不同的风格。为每一种风格或语汇准备图纸和相关内容，需要进行仔细的分析和相关研究来完善这些模式。这并不需要精确地复制历史形式。作为向建筑师和建造者提供的一种资源，要对建筑细部和比例关系、典型的部件和组合加以说明。除了每种语汇的标志性元素之外，在本章所展示的范例还强调了将传统语汇进行转化的重要性，要使之能应用于当前建筑实践并适合于规划的地块尺度与类型。要对比例关系进行调整使之适合于当前开发实践中的建筑净高，并能够从生产商那里买到窗、门、柱子和装饰等合用的部件。模式图则限定住宅的体块类型，而将建筑平面图和室内的细部设计留给不同的使用者、建造者或建筑师去决定。对建筑平面的布置必须要在体块类型和策略的限定范围内进行，这是为了形成与某种特定语汇协调一致的建筑特征。一种超越大多数传统建筑而反复出现的主题，就是在每一种语汇中所找到的相对简单的体量。

建筑模式部分可以组织成 5 － 6 页的章节，来说明某种特定类型或风格的基本要素。在本章所介绍的范例中，建筑模式部分的第 1 页首先借助范例说明和照片来描述某种建筑风格的历史和特征；第 2 页记录 4 或 5 种主要的体块类型，描述其基本形式（建筑主体部分）和附属形式（侧翼部分）；第 3 页安排住宅的主要形式和重要的屋檐与檐口选择。这就限定了建筑净高及其与基地首层平面的关系；第 4 页图解说明门和窗的形式；第 5 页用来介绍门廊类型；最后一页则确定典型的建筑材料，并使用模式图则所详细描述的元素来说明各种可能的立面形式。

本书剩下的部分用来介绍 UDA 模式图则的建筑模式册页范例。它们按照建筑风格来进行分类，这样你就能够看到即使是在一种特定建筑语汇中，不同项目所进行的设计也是根据其发展目标和特定区域中的主流范例而各不相同的。本章介绍 5 种不同的风格：殖民复兴风格、维多利亚风格、工艺美术风格、古典风格和欧洲浪漫主义风格。

莱奇斯是使用在周围亚拉巴马州亨茨维尔市附近的历史性邻里和村庄中找到的关键性特色元素进行设计的新建邻里。该场地是坎伯兰高原顶部能够俯瞰周围山谷的著名的良好地段。每一处街道、公园和建设范围都是经过精心设计的，以体现场地的内在模式和对历史性聚居模式的继承。为开发小组制定的一份模式图则，旨在作为对地方建筑语汇的补充，在住宅布局和设计方面为建筑师和建造者提供指引。

殖民复兴风格

比较分析

殖民复兴风格的变体是大多数具有代表性建筑的美国邻里的基本风格，特别是那些在 19 世纪后期和 20 世纪早期建成的邻里。尽管在美国全国早期开发阶段，有很多不同的“殖民”复兴风格，通常大部分在模式图则中记录的还是属于较晚的时期，从 1900 年至 1930 年代。罗伊尔·巴里·威尔斯是协助确定了这种特定复兴风格要素的美国早期的建筑师之一。这种建筑风格成为具有代表性的宽容文化的标志，其涵盖范围包括了从最朴素的到最奢华的住宅。住宅可能是非常简朴的，带有宽宽的优雅的窗子和简洁的门廊或前门外的围篱。较为富丽的住宅则在屋檐和檐口处，以及门套和窗洞处使用带有异国情调的、厚重的装饰线脚和梁托。

在 20 世纪早期，城镇和城市开始向乡村蔓延，殖民复兴风格也常常成为住宅的选择。很多设计都是在西尔斯、阿拉丁和标准住宅等公司的销售目录的基础上形成的。通常侧重宽度的较大窗子的比例关系，双窗式的构图，有角度的凸窗或者成组的双悬窗，对诸如帕拉第奥窗或弧形上窗套的窗等特殊的窗和要素的大量使用，富有表现力的中楣带状装饰细部，以及在简洁的两坡屋顶住宅形式中使用的檐口连接，在这一风格语汇中都是常见的特点。通常，传统的本地住宅形式已经过改造从而使其适合于采用殖民复兴风格元素。

殖民复兴风格的范例册页，是从伊斯特比奇（弗吉尼亚州诺福克市）、利伯蒂（加利福尼亚州艾尔希诺湖）和弗吉尼亚州诺福克市的模式图则中摘录的。

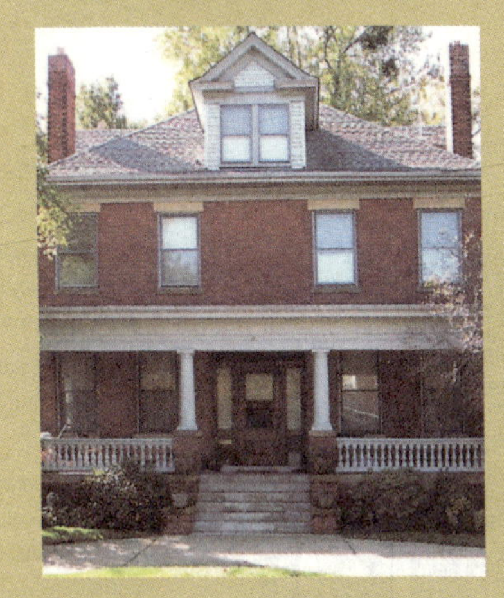

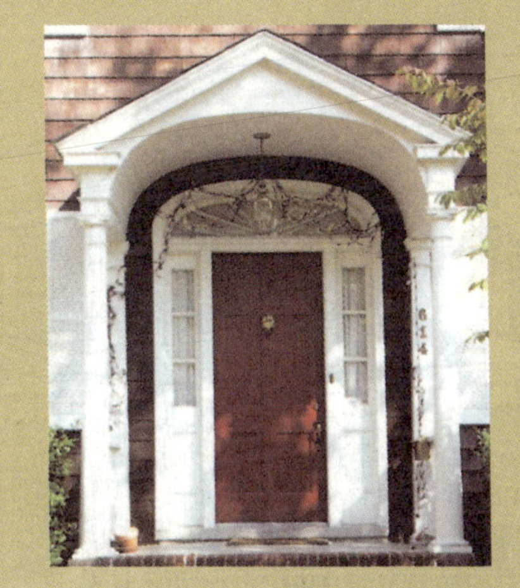

范例图片

Elizabeth City, North Carolina

Portsmouth, Virginia

Edenton, North Carolina

Essential Elements of the Tidewater Colonial Revival

1 Simple, straightforward volumes wi side wings and porches added to make more complex shapes.

2 An orderly, symmetrical relationshi between windows, doors and buildi mass.

3 Simplified versions of Classical details and columns, occasionally with Classical orders used at the entry.

4 Multi-pane windows.

Tidewater Colonial Revival

ARCHITECTURAL PATTERNS

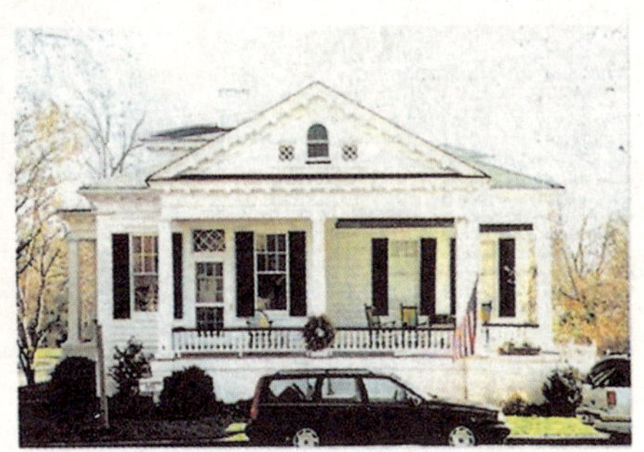

Edenton, North Carolina

History and Character

THE TIDEWATER COLONIAL REVIVAL is based on Colonial Revival styles that were prevalent throughout the country in the late nineteenth and early twentieth centuries. During this era, elements from Classical and Colonial houses were combined with and modified to produce a new vocabulary that became popular in the latter part of the nineteenth century. This mixing of influences produced a wide variety of expression and form in the Colonial Revival house. Many of these houses have more elaborate entrances, cornice treatments and window compositions. Dutch Colonial Gambrel forms are also very typical. Windows tend to be tall and narrow in proportion and more free in composition than the original Classical houses. Many of the houses in Mid-Atlantic coastal villages and neighborhoods incorporate deep front porches, running the entire face of the front façade. Accent windows are often used over the central door location.

泰德沃特殖民复兴风格

C　I

伊斯特比奇

弗吉尼亚州诺福克市

伊斯特比奇殖民复兴风格，是这种风格适用于遍及弗吉尼亚州泰德沃特地区、北卡罗来纳州和马里兰州的海滨住宅形式的一个很好的实例。典型情况下，殖民复兴风格的住宅的门廊是作为建筑主要体块的侧翼的，然而许多泰德沃特住宅传统的“正方形”平面却是四方形四坡屋顶的。大型的、充满整个建筑立面的门廊通常是附加用来保护首层房间避开夏日骄阳的，并使住户家庭能够充分利用门廊将其作为住宅布置的核心部分。符合古典主义比例的立柱、门廊栏杆和柱顶盘细部通常都是很粗壮的。最朴素的住宅较少使用装饰元素。很多小型邻里住宅只在窗框上部使用窗套装饰。屋檐和细部处理较为简单，但会采用各种各样的立柱和精致门廊栏杆。通常，这种简单的方盒子建筑会加上两层高的前翼和侧翼来构成完整的体块，并允许建筑转变为与最初的四方形形式不同的形态和特征。很多四坡顶建筑发展成为对存在于东海岸的很多非常古老的殖民风格住宅的延拓。

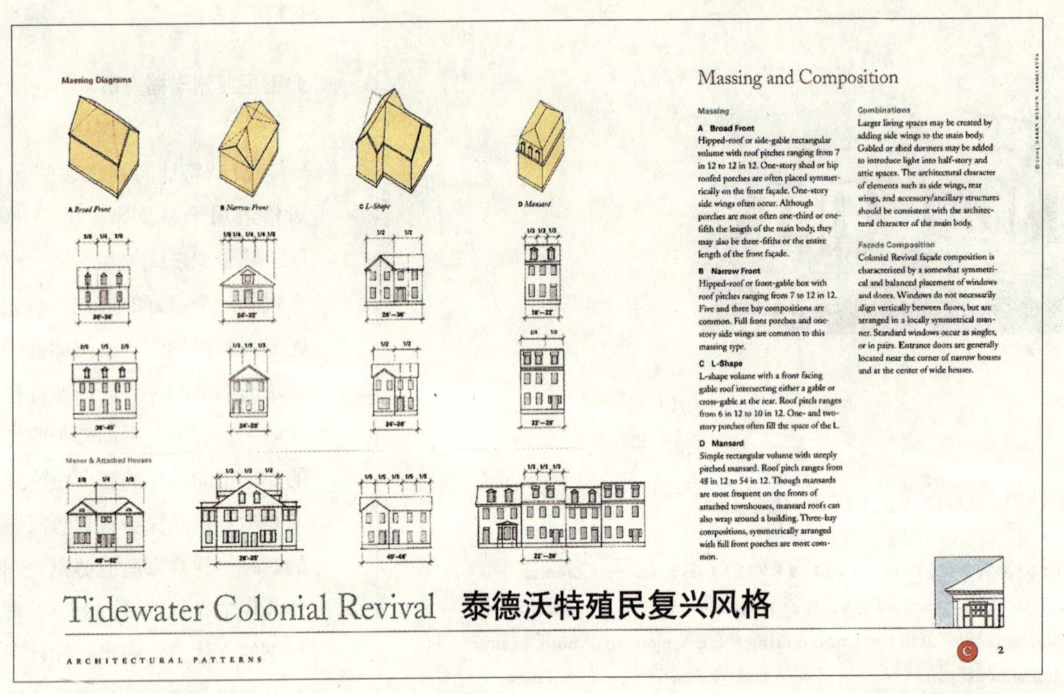

Massing Diagrams

A Broad Front

B Narrow Front

C L-Shape

D Mansard

Manor & Attached Houses

Massing and Composition

Massing

A Broad Front
Hipped-roof or side-gable rectangular volume with roof pitches ranging from 7 in 12 to 12 in 12. One-story shed or hip roofed porches are often placed symmetrically on the front façade. One-story side wings often occur. Although porches are most often one-third or one-fifth the length of the main body, they may also be three-fifths or the entire length of the front façade.

B Narrow Front
Hipped-roof or front-gable box with roof pitches ranging from 7 to 12 in 12. Five and three bay compositions are common. Full front porches and one-story side wings are common to this massing type.

C L-Shape
L-shape volume with a front facing gable roof intersecting either a gable or cross-gable at the rear. Roof pitch ranges from 6 in 12 to 10 in 12. One- and two-story porches often fill the space of the L.

D Mansard
Simple rectangular volume with steeply pitched mansard. Roof pitch ranges from 48 in 12 to 54 in 12. Though mansards are most frequent on the fronts of attached townhouses, mansard roofs can also wrap around a building. Three-bay compositions, symmetrically arranged with full front porches are most common.

Combinations
Larger living spaces may be created by adding side wings to the main body. Gabled or shed dormers may be added to introduce light into half-story and attic spaces. The architectural character of elements such as side wings, rear wings, and accessory/ancillary structures should be consistent with the architectural character of the main body.

Façade Composition
Colonial Revival façade composition is characterized by a somewhat symmetrical and balanced placement of windows and doors. Windows do not necessarily align vertically between floors, but are arranged in a locally symmetrical manner. Standard windows occur as singles, or in pairs. Entrance doors are generally located near the corner of narrow houses and at the center of wide houses.

Tidewater Colonial Revival 泰德沃特殖民复兴风格

ARCHITECTURAL PATTERNS

C 2

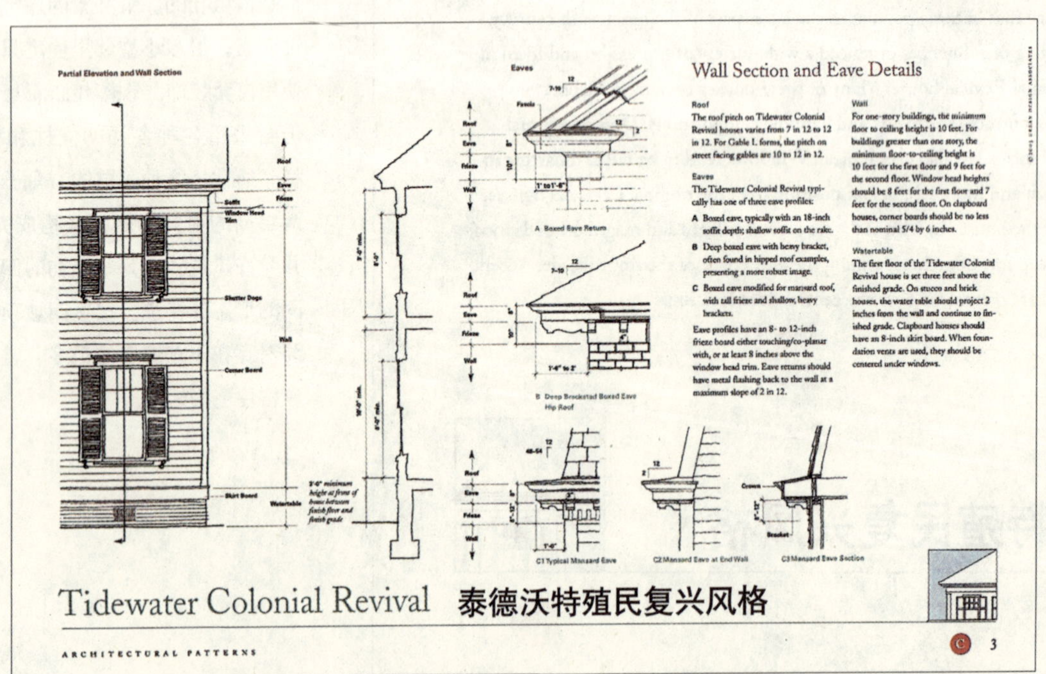

Partial Elevation and Wall Section

Eaves

A Boxed Eave Return

B Deep Bracketed Boxed Eave Hip Roof

C1 Typical Mansard Eave

C2 Mansard Eave at End Wall

C3 Mansard Eave Section

Wall Section and Eave Details

Roof
The roof pitch on Tidewater Colonial Revival houses varies from 7 in 12 to 12 in 12. For Gable L forms, the pitch on street facing gables are 10 to 12 in 12.

Eaves
The Tidewater Colonial Revival typically has one of three eave profiles:

- **A** Boxed eave, typically with an 18-inch soffit depth; shallow soffit on the rake.
- **B** Deep boxed eave with heavy bracket, often found in hipped roof examples, presenting a more robust image.
- **C** Boxed eave modified for mansard roof, with tall frieze and shallow, heavy brackets.

Eave profiles have an 8- to 12-inch frieze board either touching/co-planar with, or at least 8 inches above the window head trim. Eave returns should have metal flashing back to the wall at a maximum slope of 2 in 12.

Wall
For one-story buildings, the minimum floor to ceiling height is 10 feet. For buildings greater than one story, the minimum floor-to-ceiling height is 10 feet for the first floor and 9 feet for the second floor. Window head heights should be 8 feet for the first floor and 7 feet for the second floor. On clapboard houses, corner boards should be no less than nominal 5/4 by 6 inches.

Watertable
The first floor of the Tidewater Colonial Revival house is set three feet above the finished grade. On stucco and brick houses, the water table should project 2 inches from the wall and continue to finished grade. Clapboard houses should have an 8-inch skirt board. When foundation vents are used, they should be centered under windows.

Tidewater Colonial Revival 泰德沃特殖民复兴风格

ARCHITECTURAL PATTERNS

C 3

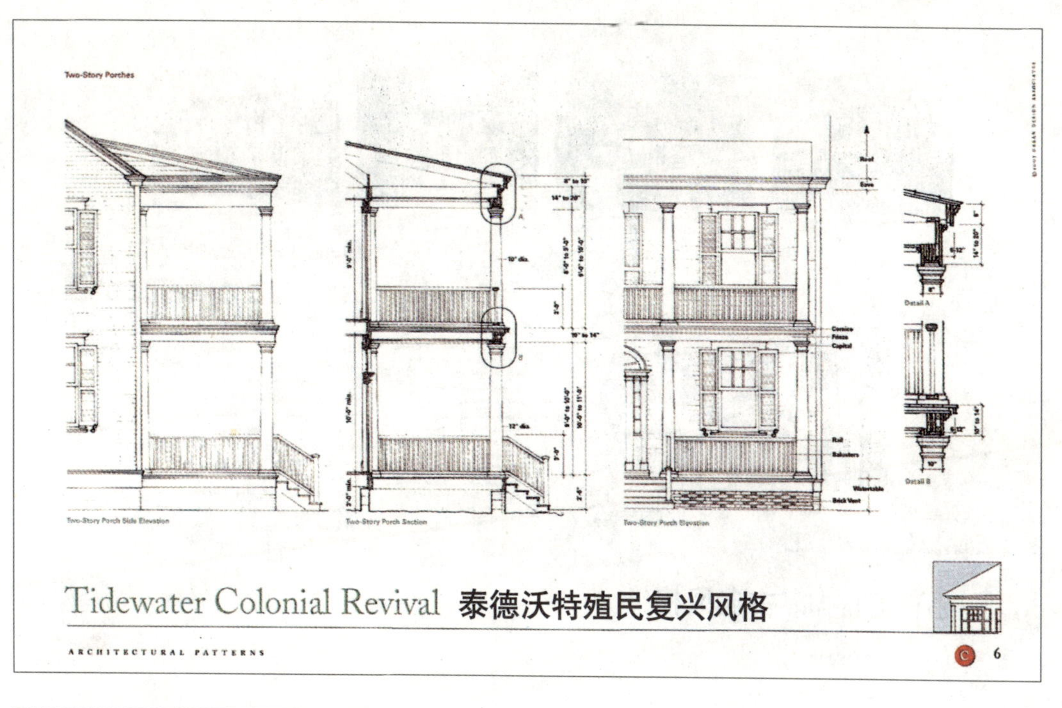

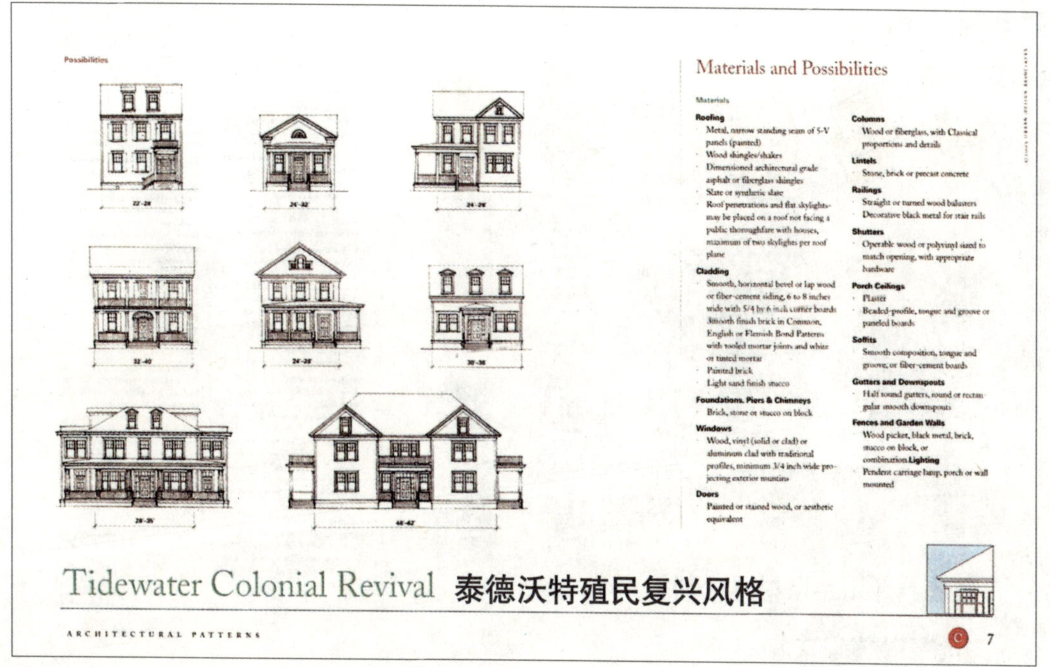
Possibilities

Materials and Possibilities

Materials

Roofing
- Metal, narrow standing seam of 5-V panels (painted)
- Wood shingles/shakes
- Dimensioned architectural grade asphalt or fiberglass shingles
- Slate or synthetic slate
- Roof penetrations and flat skylights-may be placed on a roof not facing a public thoroughfare with houses, maximum of two skylights per roof plane

Cladding
- Smooth, horizontal bevel or lap wood or fiber-cement siding, 6 to 8 inches wide with 5/4 by 6 inch corner boards Smooth finish brick in Common, English or Flemish Bond Patterns with tooled mortar joints and white or tinted mortar
- Painted brick
- Light sand finish stucco

Foundations, Piers & Chimneys
- Brick, stone or stucco on block

Windows
- Wood, vinyl (solid or clad) or aluminum clad with traditional profiles, minimum 3/4 inch wide projecting exterior muntins

Doors
- Painted or stained wood, or aesthetic equivalent

Columns
- Wood or fiberglass, with Classical proportions and details

Lintels
- Stone, brick or precast concrete

Railings
- Straight or turned wood balusters
- Decorative black metal for stair rails

Shutters
- Operable wood or polyvinyl sized to match opening, with appropriate hardware

Porch Ceilings
- Plaster
- Beaded-profile, tongue and groove or paneled boards

Soffits
- Smooth composition, tongue and groove, or fiber-cement boards

Gutters and Downspouts
- Half round gutters, round or rectangular smooth downspouts

Fences and Garden Walls
- Wood picket, black metal, brick, stucco on block, or combination

Lighting
- Pendent carriage lamp, porch or wall mounted

Tidewater Colonial Revival 泰德沃特殖民复兴风格

ARCHITECTURAL PATTERNS

7

A simple symmetrical house in Orange with a full front porch

The symmetrical arrangement of windows and dormers creates a classical composition for this Classical house in Pasadena

This house in Pasadena is a good example of horizontal proportions and the use of special windows

Early twentieth century Colonial Revival house

Colonial Revival, Madison Heights neighborhood, Pasadena, California

Essential Elements of the Liberty Classic

1. Simple, straightforward volumes with one-story side wings and porches added to make more complex shapes.
2. Symmetrical composition of doors and windows.
3. Simplified versions of Classical details and columns, often with robust and exotic Classical orders such as Ionic and Corinthian used in the porch element.
4. Multi-pane windows that are wide in proportion usually with 6 over 6 or 6 over 1 pane patterns.

History and Character

THE LIBERTY CLASSIC is based on Colonial Revival styles that were prevalent throughout the country in the early 1900s. The Colonial Revival style is evident in many California towns and cities. Interesting renditions can be found in the Los Angeles area, including Madison Heights in Pasadena. Valley communities, such as Riverside and Orange, also have a diverse collection of period houses designed in the Colonial Revival style.

The Colonial Revival style is based on Classical design principles followed in the Colonial period in this country. The interpretations, however, are often regional in character. The Southern California examples include many houses with full front porches, Dutch Colonial renditions and unusual color use on the facade and architectural elements. Cornices are often deep, with extended overhangs and brackets along the soffit on the main body and the porches.

The houses are composed of simple forms with well-proportioned windows and door surrounds. These are often more horizontal in appearance with special windows appearing in the center of the house over the front door.

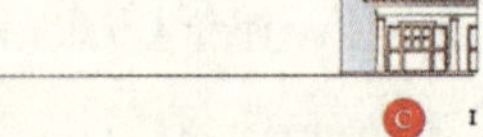

Liberty Classic 古典风格

ARCHITECTURAL PATTERNS

C 1

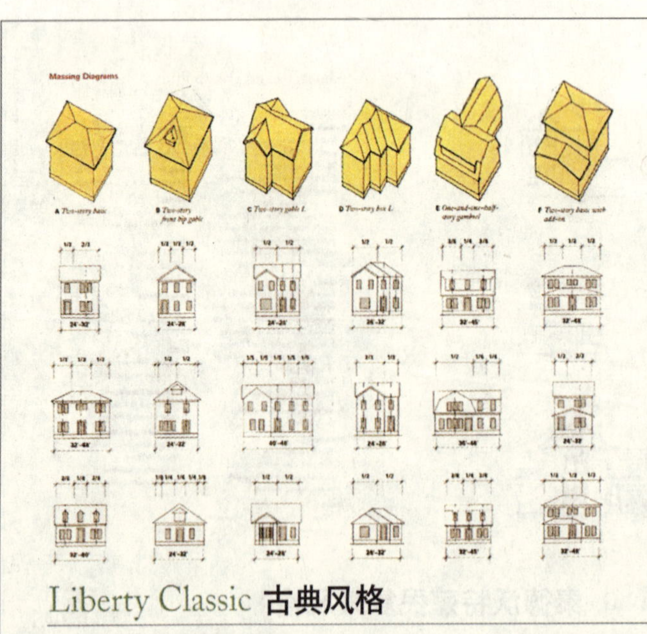

Massing and Composition

Massing

A Two-story basic
Hipped or side-gabled rectangular volume. Hip roof pitch is typically 5 in 12 and gable roof pitch is 7 in 12. One-story temple front or hip front porches, placed symmetrically on the front facade. Two-story center porches are also permitted. Although porches are most often one-fifth the length of the main body, they may also be three-fifths or the entire length of the front facade.

B Two-story front hip gable
Hipped or front-gabled rectangular volume. Hip roof pitch is typically 7 in 12 and gable roof pitch is 8 in 12. As with the basic massing, symmetrically placed gabled and/or hipped front porches are common. Porches may be either one- or two-story. Two-story porches are integral and are reserved for front-gabled houses between 24 and 30 feet wide.

C Two-story gable L
Cross-gabled volume with an 8 in 12 gable facing the street. Cross gable has a lower slope. The width of the gable facing the street is typically half that of the main body for houses up to 36 feet wide and two-fifths that of the main body for houses 36 feet and over. This massing typically accommodates a continuous porch with shed roof located between the legs of the L.

D Two-story box L
Hipped roof box with a two-story bay added to both the front and side. The bays shall have either a gabled roof with a 7 in 12 pitch or more commonly a hipped roof with a 5 in 12 pitch. Ideal for corner houses, this massing invites an L-shaped corner porch between the two bays.

E One-and-one-half-story gambrel
Rectangular volume with a gambrel roof parallel to the street. Roof pitch is nearly vertical on the lower slope and 4 to 6 in 12 at the top. One-story temple front porches centered on the front facade are typical. Porches extending three-fifths or the entire length of the front facade are also permitted.

F Two-story basic with add-on
Two-story main body with a gable or hip roof parallel to the street. An 'add-on' is usually a full- or partial-width one-story forward-projecting wing with similar roof form.

Facade Composition

Classic facade composition is characterized by a symmetrical and balanced placement of doors and windows. Standard windows most often occur as singles, or in pairs. Entrance doors are generally located in the corner of narrow houses and center of wide houses.

Liberty Classic 古典风格

ARCHITECTURAL PATTERNS

C 2

利伯蒂

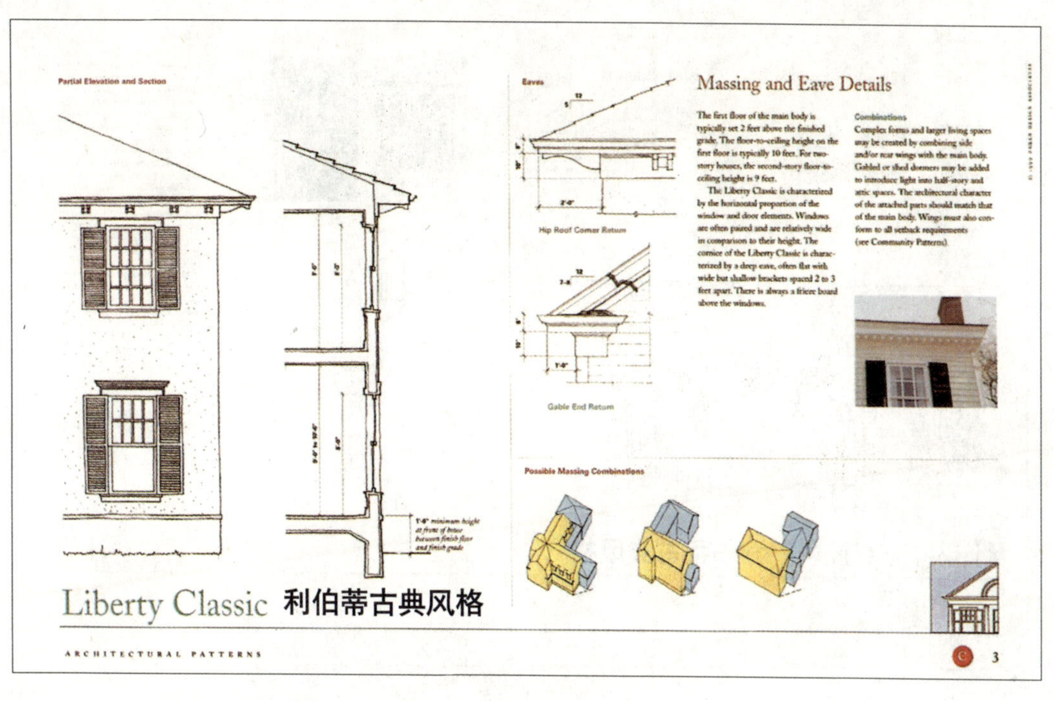

Partial Elevation and Section

Eaves

Hip Roof Corner Return

Gable End Return

Massing and Eave Details

The first floor of the main body is typically set 2 feet above the finished grade. The floor-to-ceiling height on the first floor is typically 10 feet. For two-story houses, the second-story floor-to-ceiling height is 9 feet.

The Liberty Classic is characterized by the horizontal proportion of the window and door elements. Windows are often paired and are relatively wide in comparison to their height. The cornice of the Liberty Classic is characterized by a deep eave, often flat with wide but shallow brackets spaced 2 to 3 feet apart. There is always a frieze board above the windows.

Combinations

Complex forms and larger living spaces may be created by combining side and/or rear wings with the main body. Gabled or shed dormers may be added to introduce light into half-story and attic spaces. The architectural character of the attached parts should match that of the main body. Wings must also conform to all setback requirements (see Community Patterns).

Possible Massing Combinations

Liberty Classic 利伯蒂古典风格

ARCHITECTURAL PATTERNS

3

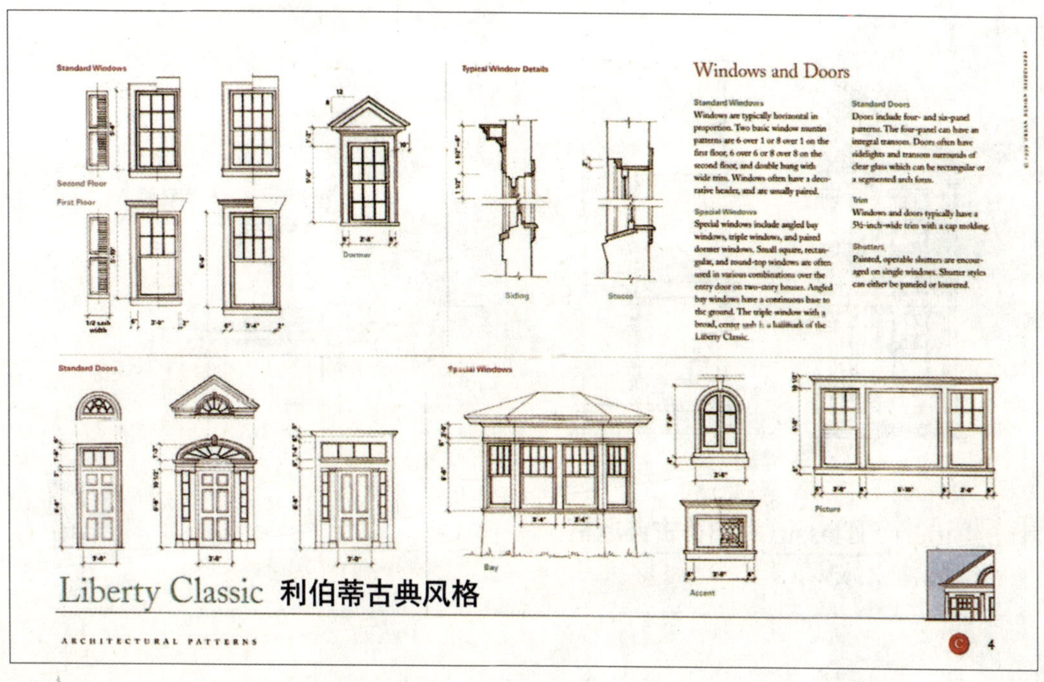

Standard Windows

Second Floor

First Floor

Dormer

Typical Window Details

Siding

Stucco

Windows and Doors

Standard Windows

Windows are typically horizontal in proportion. Two basic window muntin patterns are 6 over 1 or 8 over 1 on the first floor, 6 over 6 or 8 over 8 on the second floor, and double hung with wide trim. Windows often have a decorative header, and are usually paired.

Special Windows

Special windows include angled bay windows, triple windows, and paired dormer windows. Small square, rectangular, and round-top windows are often used in various combinations over the entry door on two-story houses. Angled bay windows have a continuous base to the ground. The triple window with a broad, center sash is a hallmark of the Liberty Classic.

Standard Doors

Doors include four- and six-panel patterns. The four-panel can have an integral transom. Doors often have sidelights and transom surrounds of clear glass which can be rectangular or a segmented arch form.

Trim

Windows and doors typically have a 5½-inch-wide trim with a cap molding.

Shutters

Painted, operable shutters are encouraged on single windows. Shutter styles can either be paneled or louvered.

Standard Doors

Special Windows

Bay

Accent

Picture

Liberty Classic 利伯蒂古典风格

ARCHITECTURAL PATTERNS

4

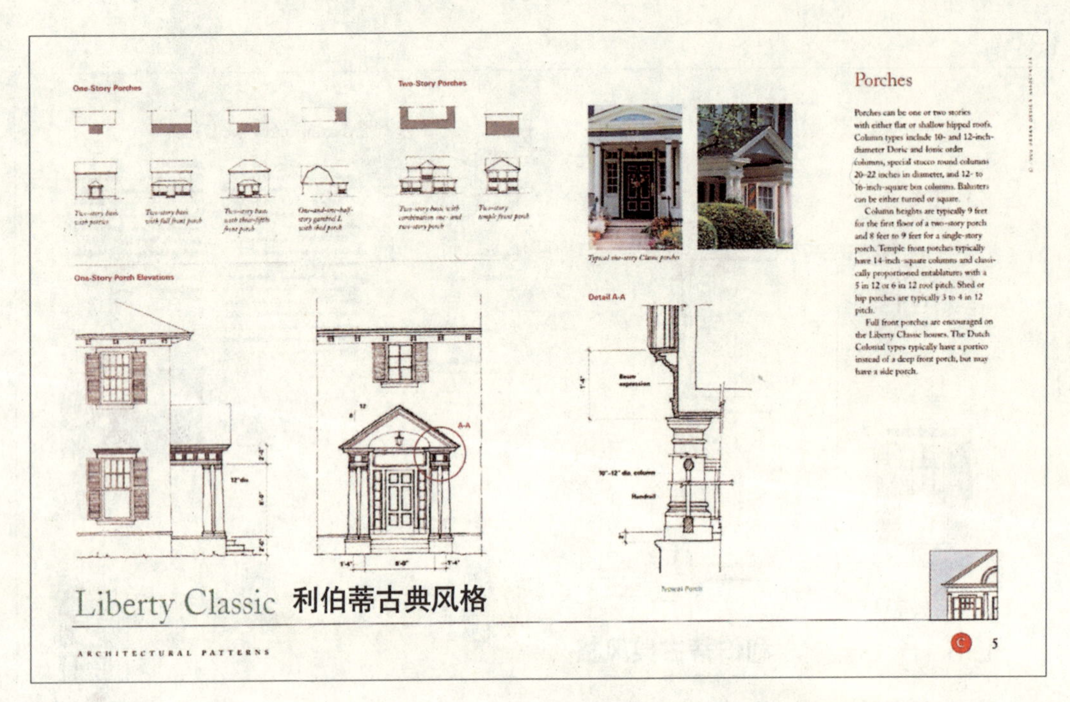
One-Story Porches
Two-Story Porches
One-Story Porch Elevations
Typical one-story Classic porches
Detail A-A
A-A
Porches
Porches can be one or two stories with either flat or shallow hipped roofs. Column types include 10- and 12-inch-diameter Doric and Ionic order columns, special stucco round columns 20-22 inches in diameter, and 12- to 16-inch-square box columns. Balusters can be either turned or square.
Column heights are typically 9 feet for the first floor of a two-story porch and 8 feet to 9 feet for a single-story porch. Temple front porches typically have 14-inch-square columns and classically proportioned entablatures with a 5 in 12 or 6 in 12 roof pitch. Shed or hip porches are typically 3 to 4 in 12 pitch.
Full front porches are encouraged on the Liberty Classic houses. The Dutch Colonial types typically have a portico instead of a deep front porch, but may have a side porch.
Liberty Classic 利伯蒂古典风格
ARCHITECTURAL PATTERNS
5

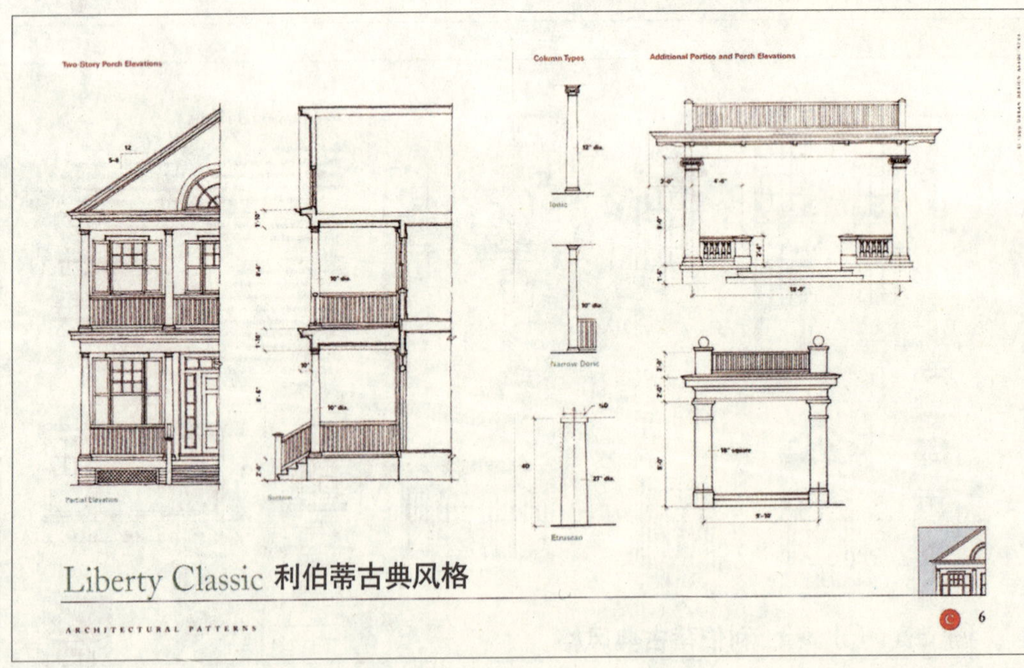
Two-Story Porch Elevations
Column Types
Additional Portico and Porch Elevations
Ionic
Narrow Doric
Etruscan
Partial Elevation
Section
Liberty Classic 利伯蒂古典风格
ARCHITECTURAL PATTERNS
6

利伯蒂和诺福克

Essential Elements of the Norfolk Colonial Revival Style

1. Simple, straightforward volumes with side wings and porches added to make more complex shapes.
2. Orderly, symmetrical relationship between windows, doors and building mass.
3. Simplified versions of Classical details and columns, occasionally with Classical orders used at the entry.
4. Multi-pane windows.

NORFOLK COLONIAL REVIVAL
诺福克殖民复兴风格

The Norfolk Colonial Revival is based on the Colonial Revival styles prevalent throughout the country in the late nineteenth and early twentieth centuries. During this era, elements from Classical and Colonial houses were combined and modified to produce a new vocabulary that became popular in the latter part of the nineteenth century. This mixing of influences produced a wide variety of expression and form in the Colonial Revival house.

Norfolk's Colonial Revival houses tend to have tall, narrow windows, elaborate entrances and cornice treatments, and deep front porches that run the entire face of the front facade. The relaxed rules of composition, frequent use of paired windows, and the occasional gambrel roof form, give these houses a comfortable quality which places them in stark contrast to the regulated order of more Classical styles.

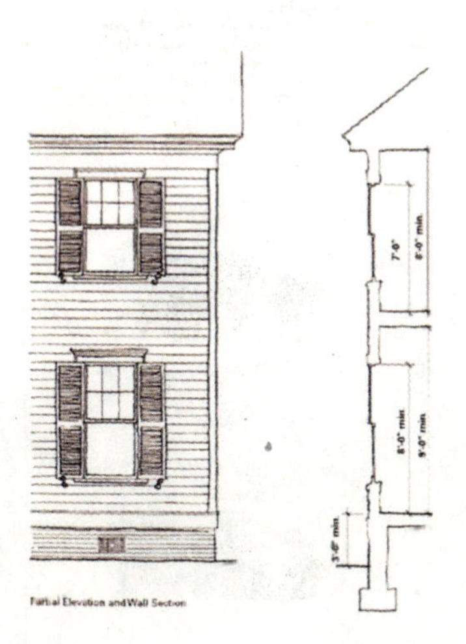

Partial Elevation and Wall Section

Massing & Composition 集中和构成

MASSING DIAGRAMS

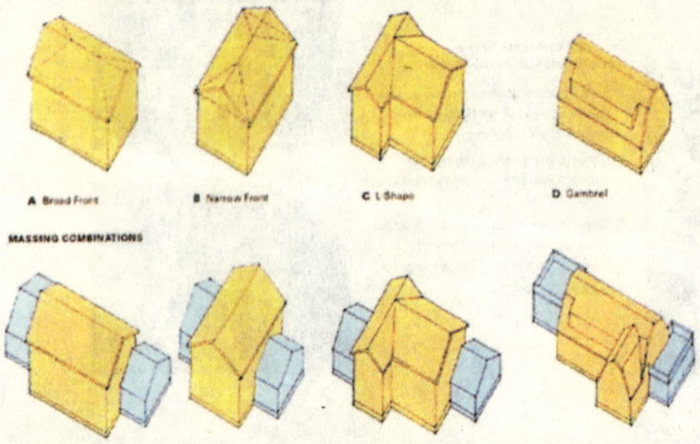

A Broad Front　B Narrow Front　C L-Shape　D Gambrel

MASSING COMBINATIONS

FACADE COMPOSITION DIAGRAMS

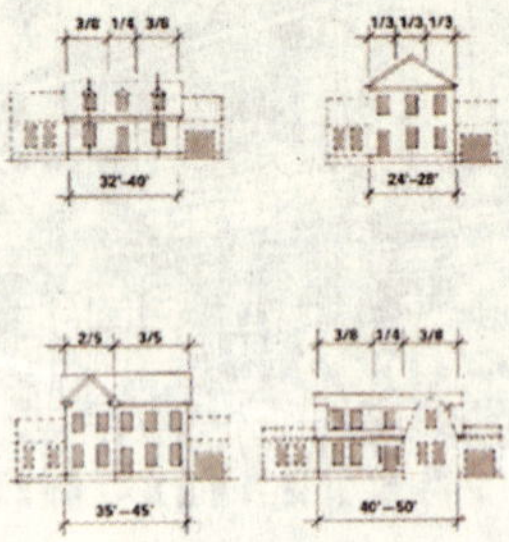

Massing

A BROAD FRONT

Hipped-roof or side-gable rectangular volume with roof pitches ranging from 7 in 12 to 12 in 12. One-story shed or hip roofed porches are often placed symmetrically on the front facade. One-story side wings often occur. Although porches are most often one-third or one-fifth the length of the main body, they may also be three-fifths or the entire length of the front facade.

B NARROW FRONT

Hipped-roof or front-gable box with roof pitches ranging from 7 to 12 in 12. Five- and three-bay compositions are common. Full front porches and one-story side-wings are common to this massing type.

C L-SHAPE

L-shape volume with a front-facing gable roof intersecting either a gable or cross-gable at the rear. Roof pitch ranges from 6 in 12 to 10 in 12. One- and two-story porches often fill the space of the L.

D GAMBREL

Rectangular volume with a gambrel roof containing a second or third story. Gambrel roofs have two roof pitches, 20 in 12 to 36 in 12 at the eave, and 6 in 12 to 10 in 12 above the pitch break. Shed dormers are common. Porches may be inset in street-facing gambrels.

Combinations

Larger living spaces may be created by adding side wings to the main body. Gabled or shed dormers may be added to introduce light into half-story and attic spaces. The architectural character of elements such as side wings, rear wings and accessory/ancillary structures should be consistent with the architectural character of the main body.

Facade Composition

Colonial Revival facade composition is characterized by a symmetrical and balanced placement of windows and doors. Standard windows occur as singles, or in pairs. Entrance doors are generally located near the corner of narrow houses and at the center of wide houses.

Roof

The roof pitch on Norfolk Colonial Revival houses varies from 6 in 12 to 12 in 12. For L-shape forms, the pitch on street-facing gables is 10 to 12 in 12.

Eaves

The Colonial Revival house typically has one of three eave profiles:

1 Boxed eave, typically with an 18-inch soffit depth; shallow soffit on the rake.

2 Deep boxed eave with heavy bracket, often found in hipped-roof examples, presenting a more robust image.

3 Boxed eave modified for gambrel roof, with tall frieze and shallow, heavy brackets.

Eave profiles have an 8- to 12-inch frieze board at least 8 inches above the window head trim. Eave returns should have metal flashing back to the wall at a maximum slope of 2 in 12.

Wall Section & Eave Details

The first floor of the main body is typically set three feet above the finished grade. The floor-to-ceiling height on the first floor is typically 10 feet. For two-story houses, the second story floor-to-ceiling height is 9 feet. Window head heights should be 8 feet for the first floor and 7 feet for the second floor.

On clapboard houses, corner boards should be no less than nominal 5/4 by 6 inches. On stucco and brick houses, the watertable should project 2 inches from the wall. Clapboard houses should have an 8-inch skirt board. When foundation vents are used, they should be centered under windows.

TYPICAL EAVE DETAILS

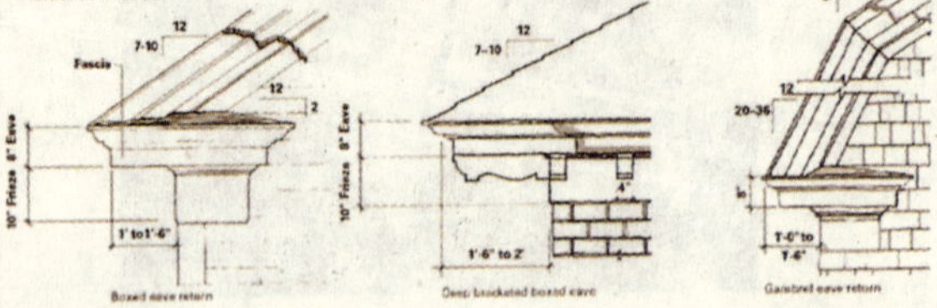

Boxed eave return　Deep bracketed boxed eave　Gambrel eave return

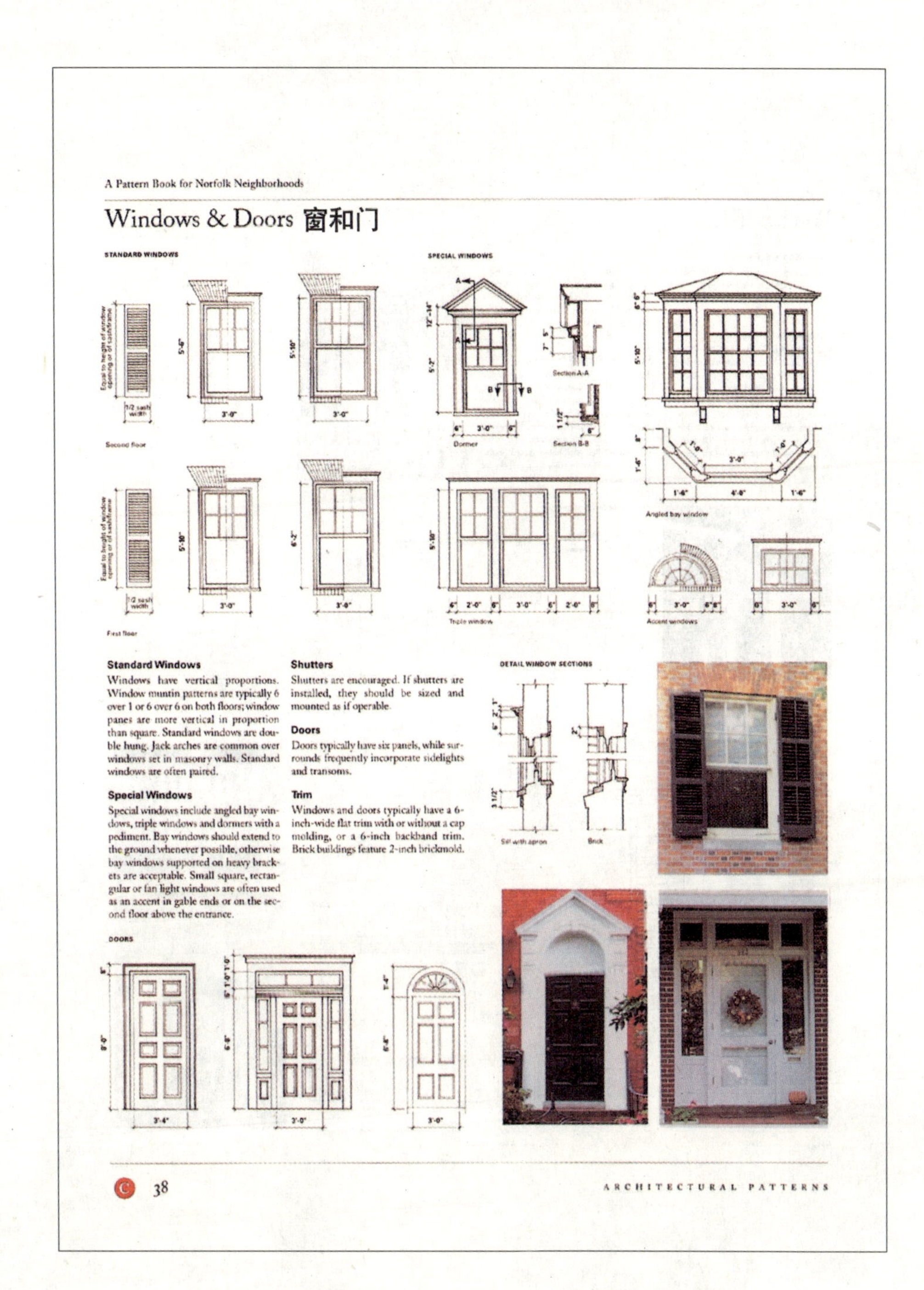

Windows & Doors 窗和门

Standard Windows

Windows have vertical proportions. Window muntin patterns are typically 6 over 1 or 6 over 6 on both floors; window panes are more vertical in proportion than square. Standard windows are double hung. Jack arches are common over windows set in masonry walls. Standard windows are often paired.

Special Windows

Special windows include angled bay windows, triple windows and dormers with a pediment. Bay windows should extend to the ground whenever possible, otherwise bay windows supported on heavy brackets are acceptable. Small square, rectangular or fan light windows are often used as an accent in gable ends or on the second floor above the entrance.

Shutters

Shutters are encouraged. If shutters are installed, they should be sized and mounted as if operable.

Doors

Doors typically have six panels, while surrounds frequently incorporate sidelights and transoms.

Trim

Windows and doors typically have a 6-inch-wide flat trim with or without a cap molding, or a 6-inch backband trim. Brick buildings feature 2-inch brickmold.

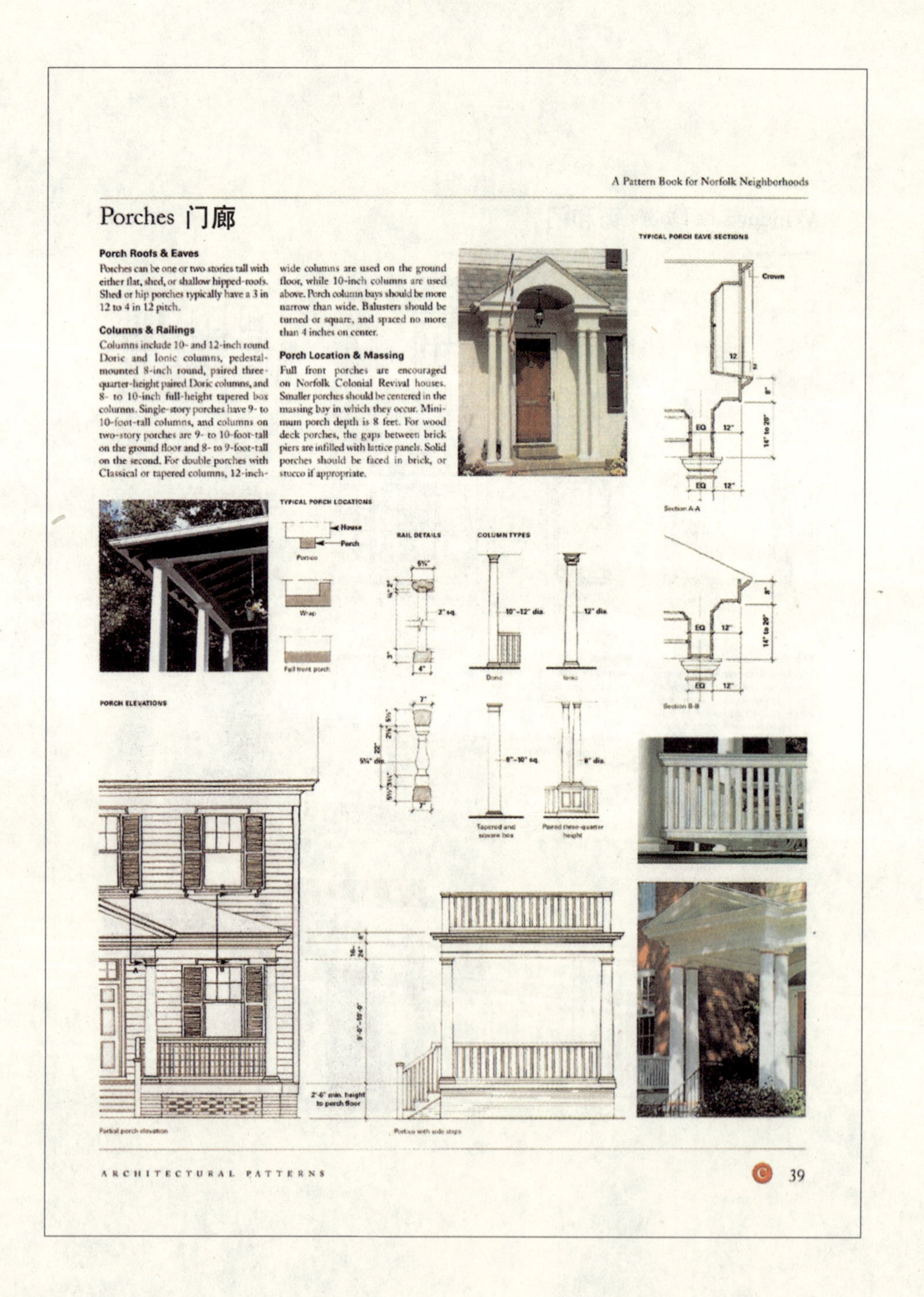

A Pattern Book for Norfolk Neighborhoods

Porches 门廊

Porch Roofs & Eaves

Porches can be one or two stories tall with either flat, shed, or shallow hipped-roofs. Shed or hip porches typically have a 3 in 12 to 4 in 12 pitch.

Columns & Railings

Columns include 10- and 12-inch round Doric and Ionic columns, pedestal-mounted 8-inch round, paired three-quarter-height paired Doric columns, and 8- to 10-inch full-height tapered box columns. Single-story porches have 9- to 10-foot-tall columns, and columns on two-story porches are 9- to 10-foot-tall on the ground floor and 8- to 9-foot-tall on the second. For double porches with Classical or tapered columns, 12-inch-wide columns are used on the ground floor, while 10-inch columns are used above. Porch column bays should be more narrow than wide. Balusters should be turned or square, and spaced no more than 4 inches on center.

Porch Location & Massing

Full front porches are encouraged on Norfolk Colonial Revival houses. Smaller porches should be centered in the massing bay in which they occur. Minimum porch depth is 8 feet. For wood deck porches, the gaps between brick piers are infilled with lattice panels. Solid porches should be faced in brick, or stucco if appropriate.

ARCHITECTURAL PATTERNS

C 39

A Pattern Book for Norfolk Neighborhoods

Materials & Applications 材料和运用

Roofing

- Slate (including manufactured slate products), laminated asphalt or composition shingles with a slate pattern, or flat clay tile

Soffits

- Smooth-finish composition board, tongue-and-groove wood boards, or fiber-cement panels

Gutters & Downspouts

- Half-round or ogee profile gutters with round or rectangular down-spouts in copper, painted or prefinished metal

Windows

- Painted wood or solid cellular PVC, or clad wood or vinyl with brick veneer only; true divided light or simulated divided light (SDL) sash with traditional exterior muntin profile (⅞ inch wide)

Doors

- Wood, fiberglass or steel with traditional stile-and-rail proportions and raised panel profiles, painted or stained

Shutters

- Wood or composite, sized to match window sash and mounted with hardware to appear operable

Cladding

- Smooth-finish wood or fiber-cement lap siding, 6- to 8-inch exposure, or random-width cut shingles
- Sand-molded or smooth-finish brick in Common, English or Flemish bond patterns
- Light sand-finish stucco

Trim

- Wood, composite, cellular PVC or polyurethane millwork; stucco, stone or cast stone

Foundations & Chimneys

- Brick, stucco or stone veneer

Columns

- Architecturally correct Classical proportions and details in wood, fiberglass, or composite material

Railings

- Milled wood top and bottom rails with square or turned balusters
- Wrought iron or solid bar stock square metal picket

Porch Ceilings

- Plaster, tongue-and-groove wood or composite boards, or beaded-profile plywood

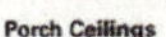

Front Yard Fences

- Wood picket, or wood, wrought iron or solid bar stock metal picket with brick or stucco finish masonry piers

Lighting

- Porch pendant or wall-mounted carriage lantern

40　ARCHITECTURAL PATTERNS

Gallery of Examples 楼座范例

弗吉尼亚州诺福克市，门

维多利亚风格

比较分析

这一类型实际上是指建筑和设计的一个时代，其涵盖了广泛类型和风格处理方式，其中包括了哥特木构风格、安妮女王风格、斯蒂克风格、意大利风格、第二帝国风格和理查森罗马风格。总体而言，这些是从 1840 年至 1900 年间在美国产生的浪漫主义风格。不同的地区常常受到其中某个时期的强烈影响。这一时限是与大部分地区的经济发展变化相一致的；到 19 世纪后期工业发展出现了集聚增长，建筑部件也开始进行批量生产。维多利亚风格很多变体的轻型形式，例如安妮女王风格和斯蒂克风格，以及哥特木构风格的住宅，都从轻骨构造而不是重型木构架的建造方式中获益。住宅的形式可以变得更为有机，使建造者能够自由地组合外部附加体块、屋顶悬挑和更为复杂的体块类型。

很多早期的住宅，都是按照采用带有陡峭坡屋顶或双坡附加体块的简单形式的哥特复兴风格或哥特木构风格建设的。通常情况下，进行建设所参考的书籍包括亚历山大·杰克逊·戴维斯在 1837 年出版的《乡村住宅》，以及应用得更为广泛一些出版物，例如安德鲁·杰克逊·唐宁的著作《农舍住宅》和《乡村住宅建筑》。其中包括在 18 世纪末期和 19 世纪初期英格兰流行的主要郊区住宅、小型别墅和乡村住宅模式。占主导地位的建筑材料是木制壁板和装饰，或定型木构立柱、定型椽子和装饰性贴面。城市中的变体则倾向于采用第二帝国风格和意大利风格——更为正式，并且既能适用于联排住宅又能适用于较大型的住宅。在大多数邻里中这些变体是独特的“香料”，并与在小城镇和很多城市的独户分离式住宅邻里中存在的更为稳重的殖民主义风格、古典主义风格和后来的殖民复兴风格的住宅形成对比。

维多利亚风格的范例册页，是从利伯蒂（加利福尼亚州艾尔希诺湖）、伊斯特比奇（弗吉尼亚州诺福克市）和伊斯特加里森（加利福尼亚州蒙特雷县）的模式图则中摘录的。

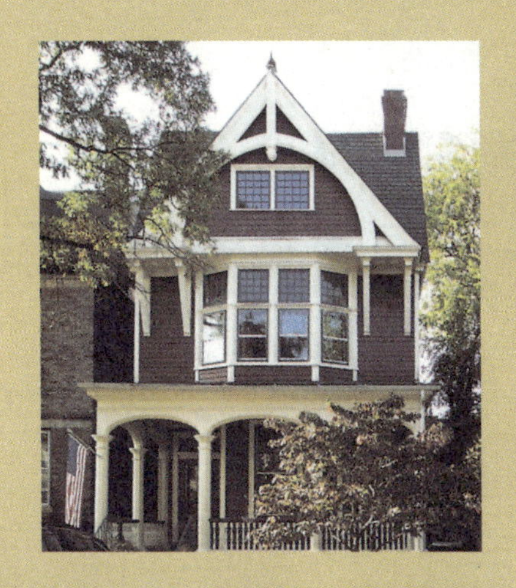

范例图片

Victorian-era farmhouse in historic Orange, California

Blair House, Mendocino, California, 1888

Tinele House, Bridgeport, California, 1878

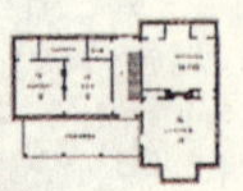

Bracketed American farmhouse and historic floor plan from The Architecture of Country Houses *by A.J.Downing*

Essential Elements of Liberty Victorian

1. Steeply pitched gable roofs.
2. Cut wood ornament, often with natural forms such as leaves and vines.
3. Wood clapboard siding.
4. Vertical proportions for windows and doors.

History and Character

THE LIBERTY VICTORIAN STYLE builds on the early 'Carpenter Gothic' cottages imported into the Western Region from the East. While the style became fashionable in the 1800s in the Bay area, as its popularity grew, it spread north and south from San Francisco. Much of the source for the early house designs came from Pattern Books published by Andrew Jackson Downing. Publications such as *The Horticulturist* influenced the preferences of the public and provided an especially dramatic contrast to the inherited Spanish and adobe building types prevalent throughout Southern California. Early resorts and larger country estates began to adopt the style with more and more exotic variations that included Eastlake, Queen Anne, and Italianate detailing.

The Liberty Victorian is centered on the simple, elegant forms that were adapted to houses in the smaller towns and the rural farmhouse. The massing forms are quite simple, and the ornament is restrained and typically limited to the porch and the cornice. The Victorian is a spice style, sprinkled in among other more predominant styles.

Liberty Victorian 维多利亚风格

ARCHITECTURAL PATTERNS

15

Massing Diagrams

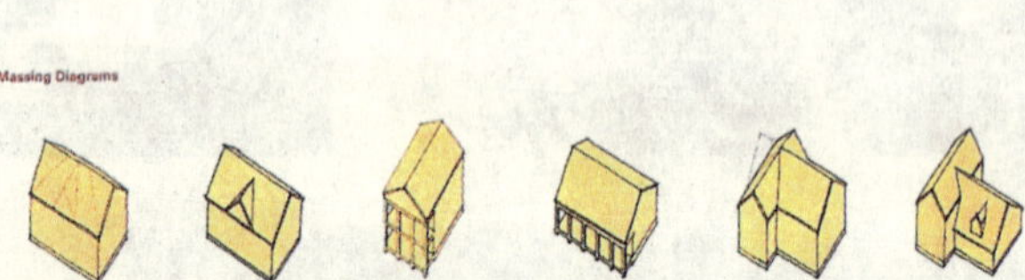

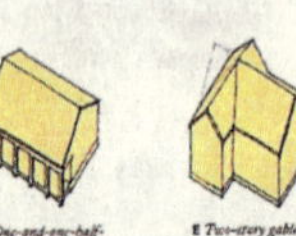

Massing and Composition

Massing

A Two-Story Basic

Hipped or side-gabled rectangular volume, often with a dormer flush to the front facade. Roof pitch is typically 8 to 10 in 12. One-story shed or hip front porches from one-fifth to the full length of the main body. Two-story full front porches are also permitted.

B One-and-One-Half-Story Basic

Side-gabled rectangular volume, often with a dormer flush to the front facade. Roof pitch is typically 10 in 12 for the main body and 12 in 12 for the dormer. One-story shed or hip front porches from one-fifth to the full length of the main body.

C One- and Two-Story Front Gable

Rectangular volume with 8 in 12 roof pitch and gable facing the street. One-story partial, full, or wrapping front porch with 3 in 12 hip roof is common. Integral full front porches are also permitted on one- and two-story main bodies.

D One-and-One-Half-Story Integral Porch

Square volume with 8 in 12 side-gabled roof. Integral front porch along the full length of the front facade. Symmetrically placed gabled or shed dormers with 6 in 12 or 12 in 12 roof pitch.

E Two-Story Gable L

Cross-gabled volume with a 12 in 12 gable facing the street. The width of the gable facing the street is typically half that of the main body for houses up to 36 feet wide and two-fifths that of the main body for houses 36 feet and over. This massing typically accommodates a continuous porch with shed roof located between the legs of the L. Corner house porches should wrap the corner.

F One-and-One-Half-Story Gable L

Cross-gabled volume with a 12 in 12 gable facing the street, often with a dormer flush to the front facade. The width of the gable facing the street is typically one-third that of the main body for houses up to 36 feet wide and two-fifths that of the main body for houses 36 feet and over. Full front porches are typical between the legs of the L.

Composition

Victorian facade composition is characterized by a symmetrical and balanced placement of doors and windows. Individual double-hung windows are the most common type. Entrance doors are generally located in the corner of narrow houses and the center of wide houses. Bay windows are typically used on the ground floor. Paired windows are often used in the forward gable of the gable L massing types E and F.

Liberty Victorian 维多利亚风格

ARCHITECTURAL PATTERNS

16

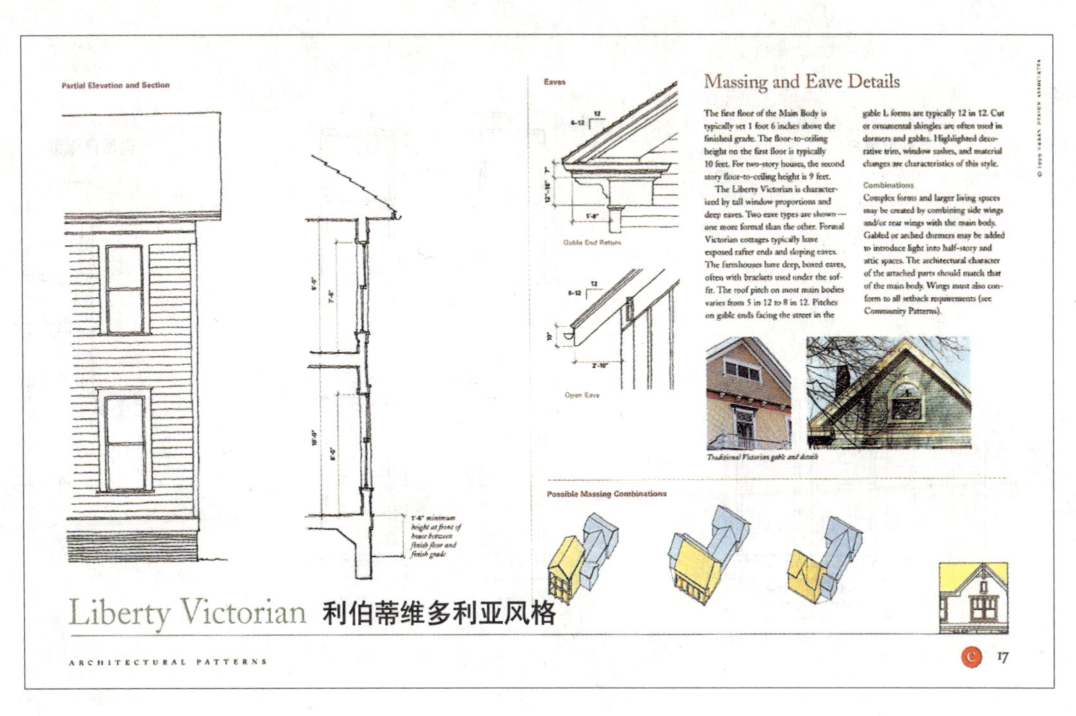
Partial Elevation and Section

Eaves

Massing and Eave Details

The first floor of the Main Body is typically set 1 foot 6 inches above the finished grade. The floor-to-ceiling height on the first floor is typically 10 feet. For two-story houses, the second story floor-to-ceiling height is 9 feet.

The Liberty Victorian is characterized by tall window proportions and deep eaves. Two eave types are shown — one more formal than the other. Formal Victorian cottages typically have exposed rafter ends and sloping eaves. The farmhouses have deep, boxed eaves, often with brackets used under the soffit. The roof pitch on most main bodies varies from 5 in 12 to 8 in 12. Pitches on gable ends facing the street in the gable L forms are typically 12 in 12. Cut or ornamental shingles are often used in dormers and gables. Highlighted decorative trim, window sashes, and material changes are characteristics of this style.

Combinations

Complex forms and larger living spaces may be created by combining side wings and/or rear wings with the main body. Gabled or arched dormers may be added to introduce light into half-story and attic spaces. The architectural character of the attached parts should match that of the main body. Wings must also conform to all setback requirements (see Community Patterns).

Traditional Victorian gable and detail

Possible Massing Combinations

Liberty Victorian　**利伯蒂维多利亚风格**

ARCHITECTURAL PATTERNS　17

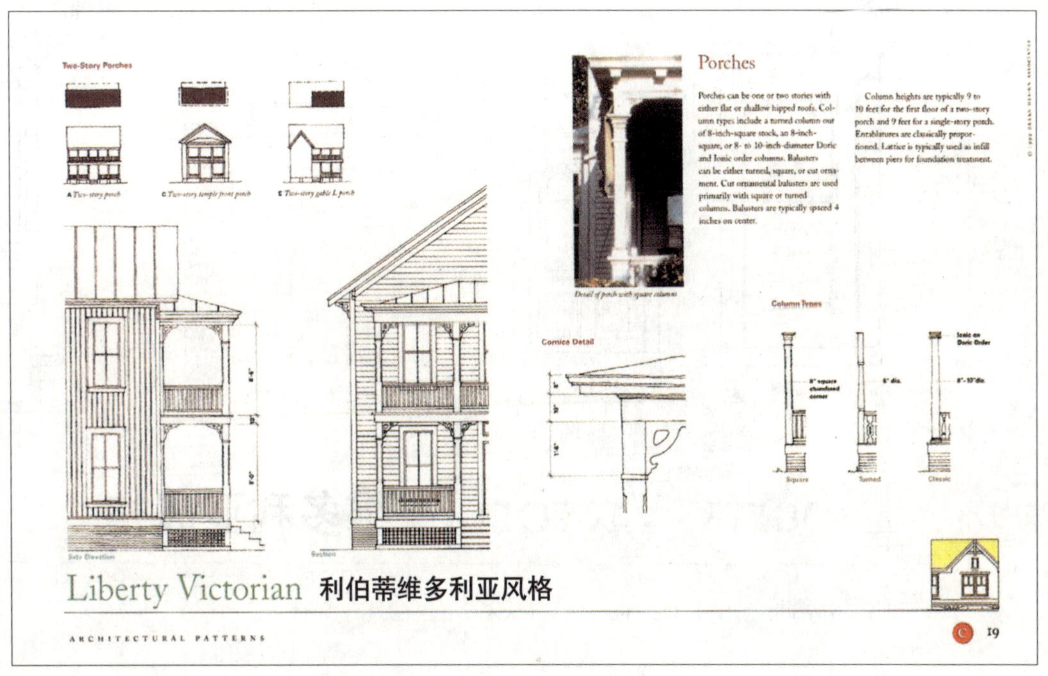
Two-Story Porches

Porches

Porches can be one or two stories with either flat or shallow hipped roofs. Column types include a turned column out of 8-inch-square stock, an 8-inch-square, or 8- to 10-inch-diameter Doric and Ionic order columns. Balusters can be either turned, square, or cut ornament. Cut ornamental balusters are used primarily with square or turned columns. Balusters are typically spaced 4 inches on center.

Column heights are typically 9 to 10 feet for the first floor of a two-story porch and 9 feet for a single-story porch. Entablatures are classically proportioned. Lattice is typically used as infill between piers for foundation treatment.

Detail of porch with square columns

Column Types

Cornice Detail

Liberty Victorian　**利伯蒂维多利亚风格**

ARCHITECTURAL PATTERNS　19

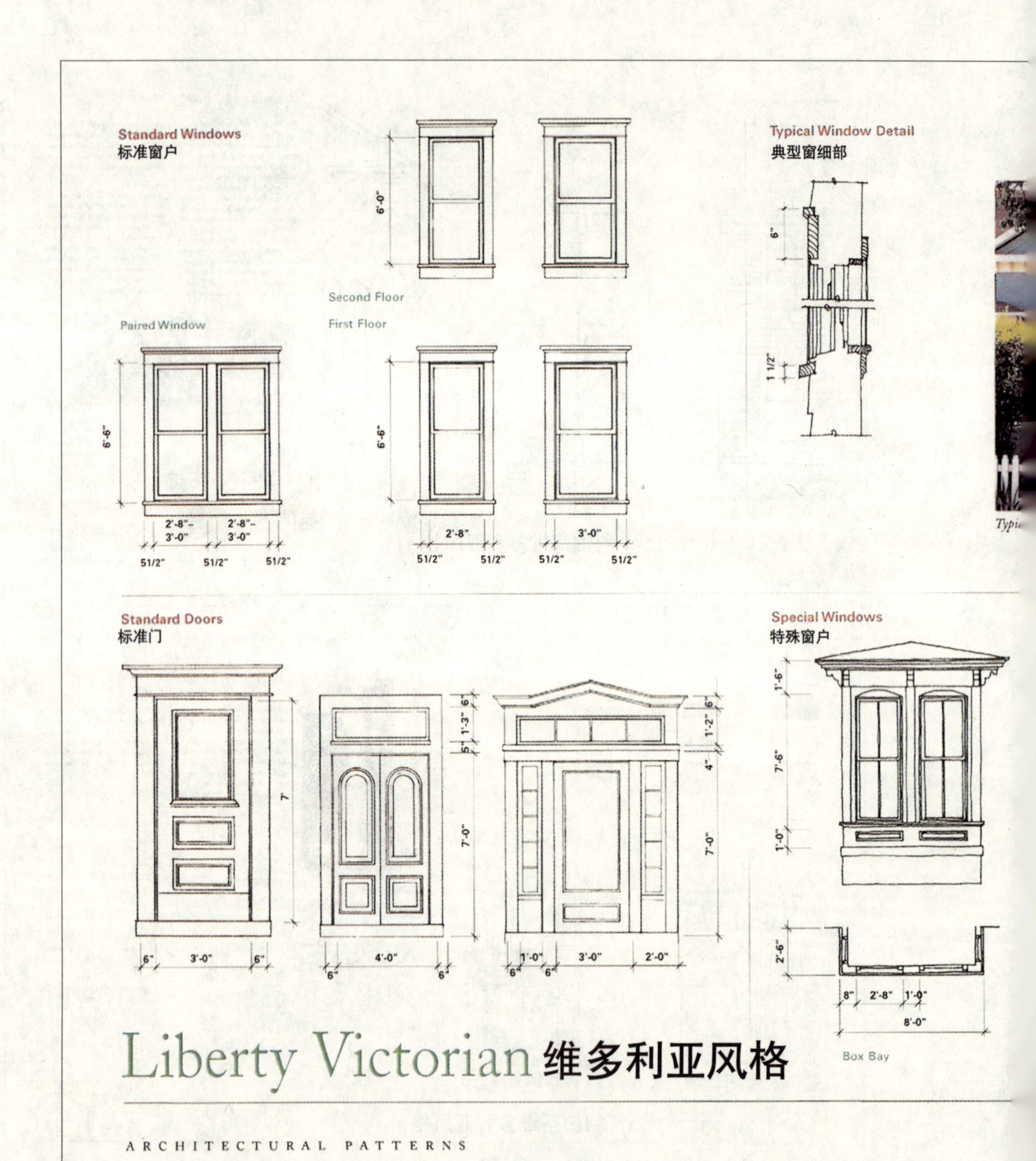

Liberty Victorian 维多利亚风格

ARCHITECTURAL PATTERNS

利伯蒂

Windows and Doors 窗和门

Standard Windows

Windows are typically vertical in proportion. Two basic window patterns are 1 over 1 and 2 over 2, double-hung with wide trim. Paired windows are often used in gable L houses, or as accents where bay windows might also be used. The window often has a decorative header. Some houses may have windows with rounded upper sashes and ornate trim for window hoods.

Standard Doors

Doors are centered in their bays and are either paneled or glazed wood doors. Double doors are often used, as well as single doors with sidelights and transoms.

Special Windows

Special windows include box bay and angled bay windows, paired dormer windows, and round top windows. Box bay and angled bay windows have a continuous base to the ground.

Trim

Windows and doors typically have a 5½-inch-wide trim with a cap molding.

Shutters

Painted, operable shutters are encouraged on single windows. Shutter styles can either be paneled or louvered.

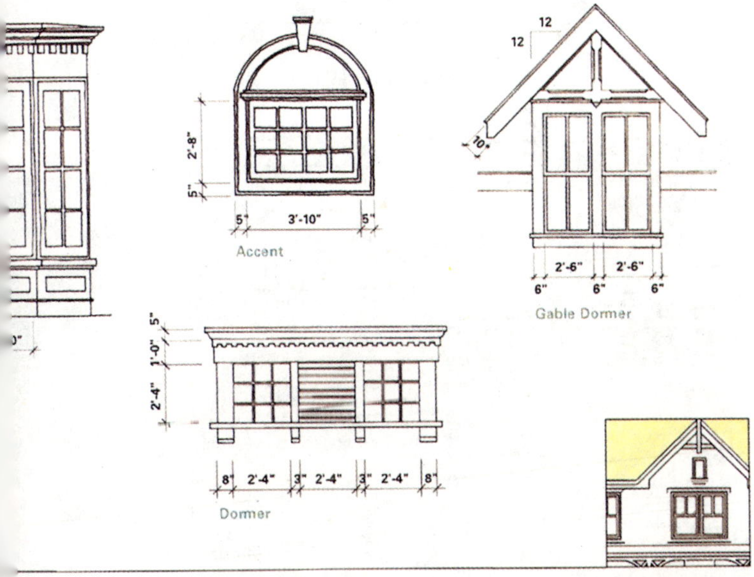

Accent

Gable Dormer

Dormer

C 18

利伯蒂

加利福尼亚州艾尔希诺湖市

利伯蒂的维多利亚风格住宅是一种有趣的混合体，它借鉴了古典主义和殖民复兴风格的形式与建筑细部，并受到了工艺美术运动的影响。利伯蒂维多利亚风格是通过对在南加利福尼亚的滨海和内陆的城镇与邻里，例如奥伦奇、里弗赛德、拉古纳比奇、长滩、纽波特比奇和帕萨迪纳等存在的前例的广泛研究而确定的。不同风格要素的混合形成了一种有趣的特征，并与其他风格融为一体。主要可参考的形式是具有简单体块类型的民间维多利亚风格的村舍。屋檐和门廊等附属部分成为使体块转化为不同变体的主要要素。其中一种应用方式包括带有托架的封闭式屋檐，很像带有门廊、使用爱奥尼柱式和旋转栏杆的殖民复兴风格住宅。哥特木构风格的方法则使用了带有转角柱或削角柱以及装饰性切边饰面的坡屋顶。

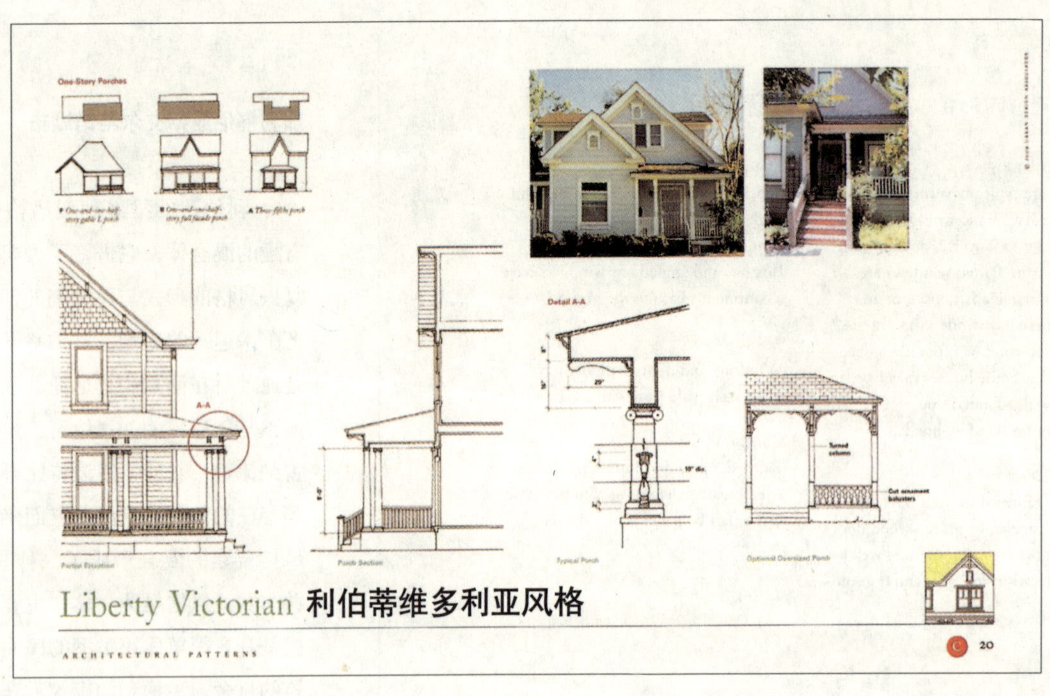

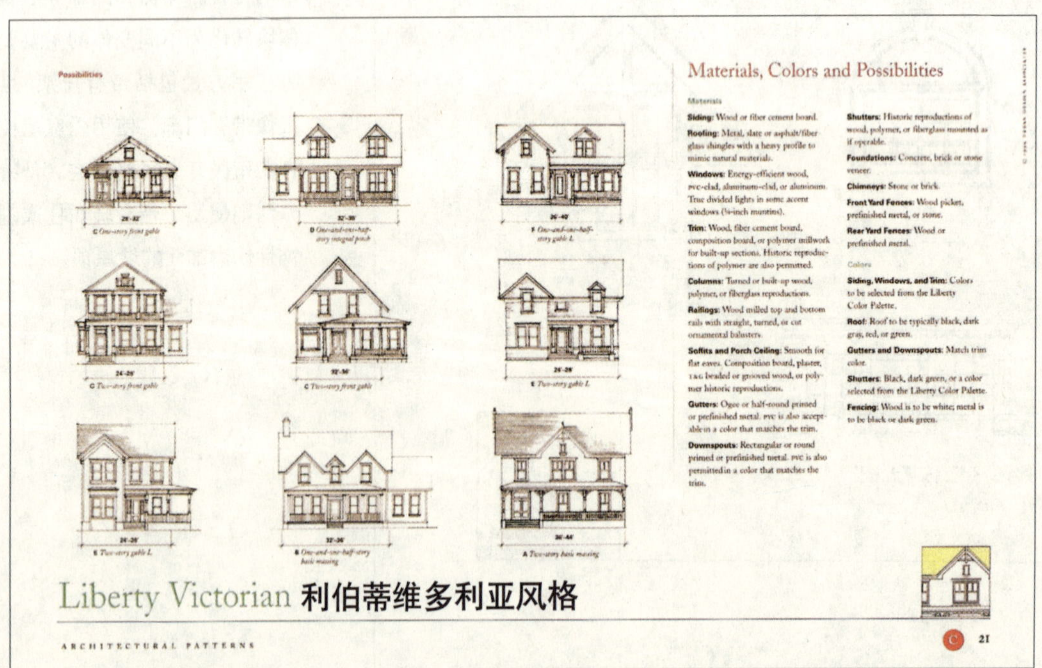

Materials, Colors and Possibilities

Materials

Siding: Wood or fiber cement board.

Roofing: Metal, slate or asphalt/fiberglass shingles with a heavy profile to mimic natural materials.

Windows: Energy-efficient wood, PVC-clad, aluminum-clad, or aluminum. True divided lights in some accent windows (⅞-inch muntins).

Trim: Wood, fiber cement board, composition board, or polymer millwork for built-up sections. Historic reproductions of polymer are also permitted.

Columns: Turned or built-up wood, polymer, or fiberglass reproductions.

Railings: Wood milled top and bottom rails with straight, turned, or cut ornamental balusters.

Soffits and Porch Ceiling: Smooth for flat eaves. Composition board, plaster, T&G beaded or grooved wood, or polymer historic reproductions.

Gutters: Ogee or half-round primed or prefinished metal. PVC is also acceptable in a color that matches the trim.

Downspouts: Rectangular or round primed or prefinished metal. PVC is also permitted in a color that matches the trim.

Shutters: Historic reproductions of wood, polymer, or fiberglass mounted as if operable.

Foundations: Concrete, brick or stone veneer.

Chimneys: Stone or brick.

Front Yard Fences: Wood picket, prefinished metal, or stone.

Rear Yard Fences: Wood or prefinished metal.

Colors

Siding, Windows, and Trim: Colors to be selected from the Liberty Color Palette.

Roof: Roof to be typically black, dark gray, red, or green.

Gutters and Downspouts: Match trim color.

Shutters: Black, dark green, or a color selected from the Liberty Color Palette.

Fencing: Wood is to be white; metal is to be black or dark green.

利伯蒂和伊斯特比奇

Edenton, North Carolina

Edenton, North Carolina

Annapolis, Maryland

Essential Elements of Tidewater Victorian

1 Steeply pitched gable roofs.
2 Cut wood ornament, often with natural forms such as leaves and vines, or simple shape cutouts and arched forms.
3 Clapboard siding, with siding, shingles, or beadboard in gable ends.
4 Vertical proportions for windows and doors, windows with two- and four-pane sashes.

History and Character

THE TIDEWATER VICTORIAN STYLE builds on the 'Carpenter Gothic' cottages abundant in early rail-served coastal resorts. Pattern Books published by Andrew Jackson Downing and others were the source for many of these early house designs. These books made it easier for the builders of early resorts, country estates, and even modest dwellings to adopt the style. Although exotic Victorian houses incorporating Eastlake, Queen Anne, and Italianate details grew in popularity, folk-based Victorian homes survive in this region.

The Tidewater Victorian is based on the simple, elegant forms adapted to houses in small towns and rural farmhouses. The massing forms are simple, while ornament is typically restrained and limited to the porch and the building's cornice.

Tidewater Victorian 泰德沃特维多利亚风格

ARCHITECTURAL PATTERNS

15

Massing Diagrams

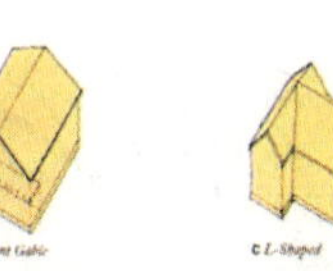

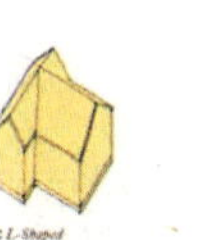

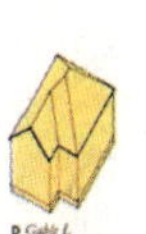

A Side Gable　B Front Gable　C L-Shaped　D Gable L

Mansion and Attached Houses

Massing and Composition

Massing

A Side Gable

Side-gabled rectangular volume, often with a steeply pitched, gabled dormer flush to the front façade. Roof pitch is typically 8 in 12 to 10 in 12, and one- or two-story front porches typically extend across the full front of the house.

B Front Gable

Front-gabled rectangular volume with a roof pitch ranging from 8 in 12 to 12 in 12 for the main body. One-story shed or hip front porches from one-third to the full width of the main body are common. Often, two-story porches are integrated under the main roof form.

C L-Shaped

Cross-gabled volume with a 9 in 12 gable facing the street. The width of the gable facing the street is typically two-fifths that of the main body. This massing typically accommodates a one- or two-story continuous porch with a shed or hipped roof that dies into the side of the projecting wing.

D Gable L

Square volume with hipped roof from which a front-facing gabled wing extends. Roof pitches range from 8 in 12 to 12 in 12. Front porch extends the full length of the front façade, or is occasionally a single-bay, hipped porch at the main body.

Combinations

Complex forms and larger living spaces may be created by combining side wings and/or rear wings with the main body. Gabled or arched dormers may be added to introduce light into half-story and attic spaces. The architectural character of the attached parts should match that of the main body.

Façade Composition

Victorian façade composition is characterized by a symmetrical and balanced placement of doors and windows. Individual double-hung windows are the most common type. Front doors are generally located in the corner of narrow houses and at the center of wide houses. Paired or bay windows are often used in the forward gable of the gable L massing types. Bay windows may be one or two stories tall.

Tidewater Victorian 泰德沃特维多利亚风格

ARCHITECTURAL PATTERNS

16

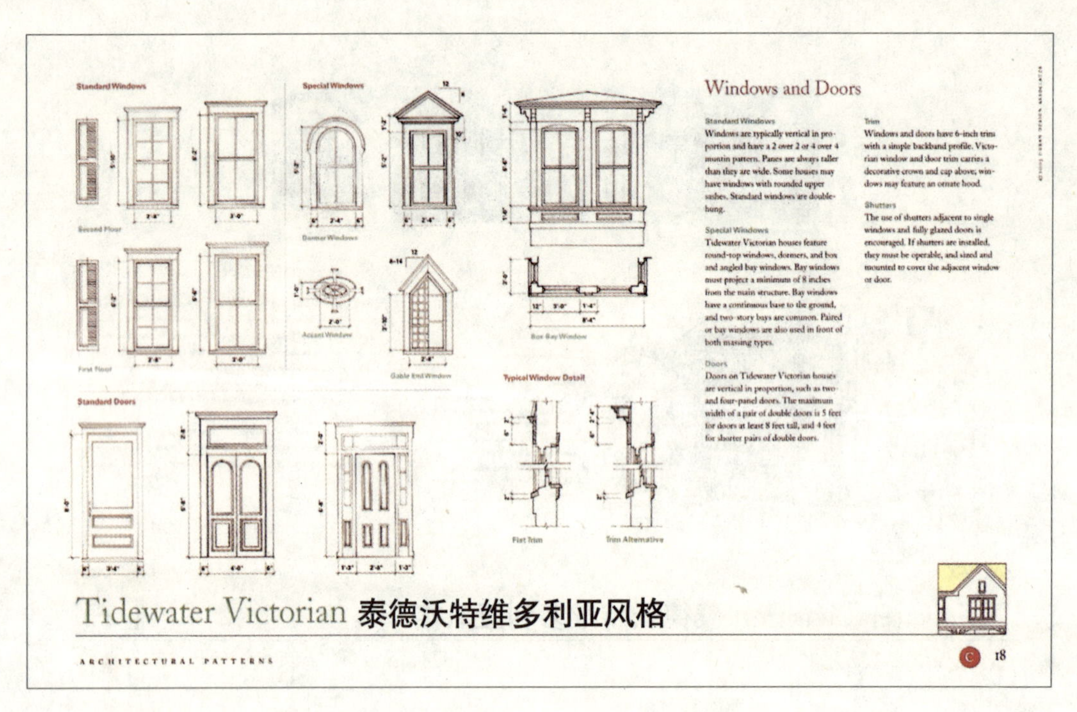

Windows and Doors

Standard Windows
Windows are typically vertical in proportion and have a 2 over 2 or 4 over 4 muntin pattern. Panes are always taller than they are wide. Some houses may have windows with rounded upper sashes. Standard windows are double-hung.

Special Windows
Tidewater Victorian houses feature round-top windows, dormers, and box and angled bay windows. Bay windows must project a minimum of 8 inches from the main structure. Bay windows have a continuous base to the ground, and two-story bays are common. Paired or bay windows are also used in front of both massing types.

Doors
Doors on Tidewater Victorian houses are vertical in proportion, such as two- and four-panel doors. The maximum width of a pair of double doors is 5 feet for doors at least 8 feet tall, and 4 feet for shorter pairs of double doors.

Trim
Windows and doors have 6-inch trim with a simple backband profile. Victorian window and door trim carries a decorative crown and cap above; windows may feature an ornate hood.

Shutters
The use of shutters adjacent to single windows and fully glazed doors is encouraged. If shutters are installed, they must be operable, and sized and mounted to cover the adjacent window or door.

Tidewater Victorian 泰德沃特维多利亚风格

ARCHITECTURAL PATTERNS

C 18

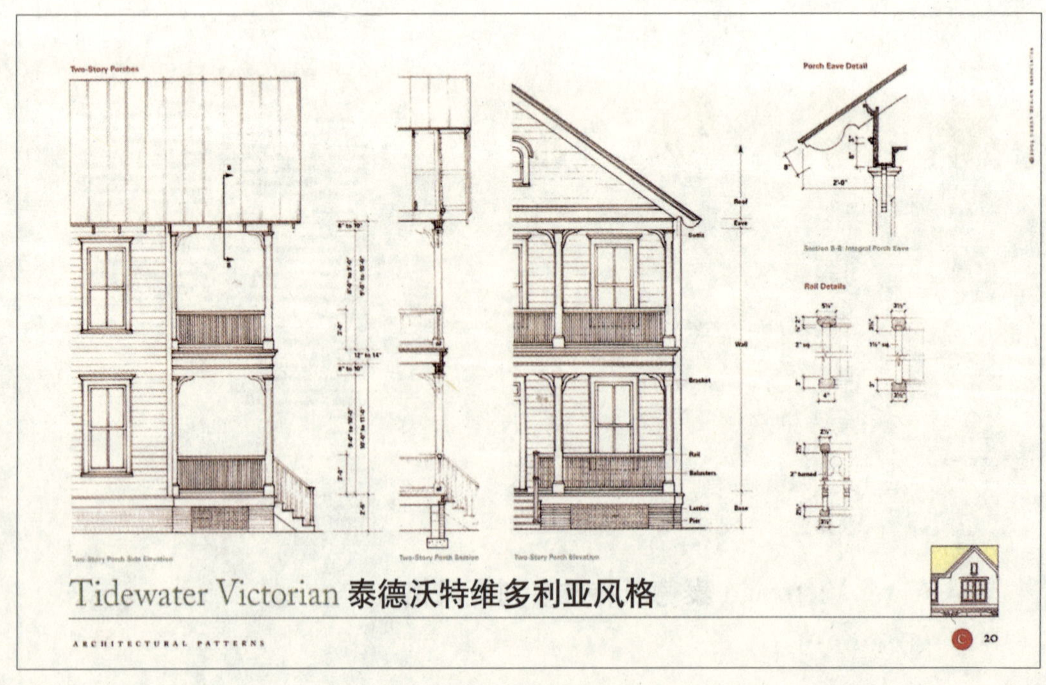

Tidewater Victorian 泰德沃特维多利亚风格

ARCHITECTURAL PATTERNS

C 20

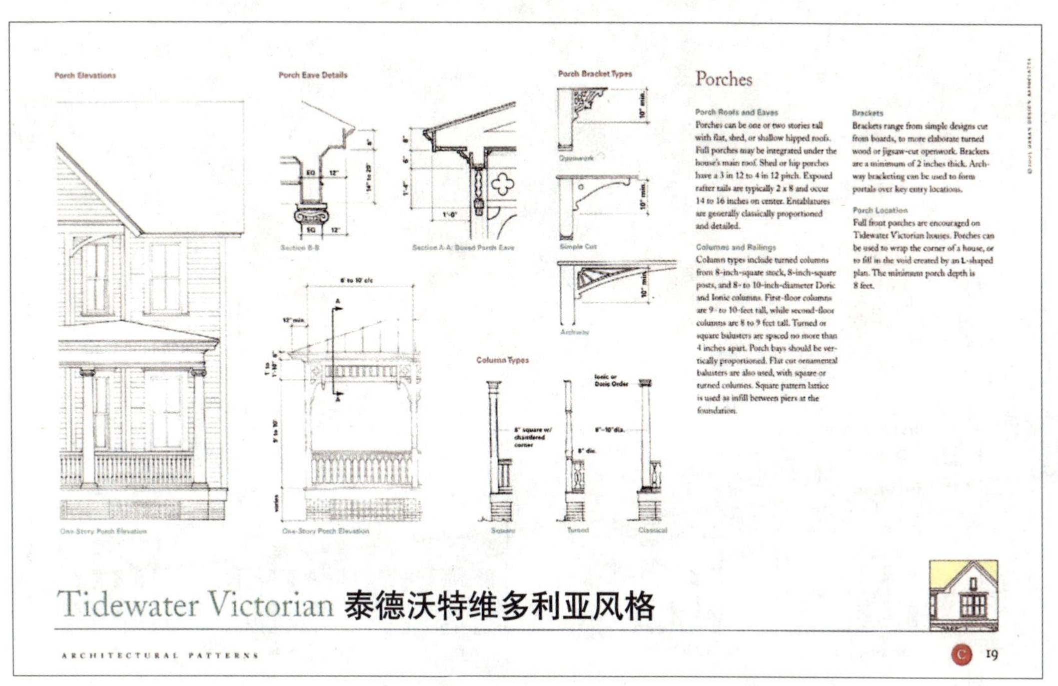

Porch Elevations

Porch Eave Details

Porch Bracket Types

Section B-B

Section A-A: Boxed Porch Eave

Openwork

Simple Cut

Archway

Column Types

One-Story Porch Elevation

One-Story Porch Elevation

Square

Turned

Classical

Porches

Porch Roofs and Eaves

Porches can be one or two stories tall with flat, shed, or shallow hipped roofs. Full porches may be integrated under the house's main roof. Shed or hip porches have a 3 in 12 to 4 in 12 pitch. Exposed rafter tails are typically 2 x 8 and occur 14 to 16 inches on center. Entablatures are generally classically proportioned and detailed.

Columns and Railings

Column types include turned columns from 8-inch-square stock, 8-inch-square posts, and 8- to 10-inch-diameter Doric and Ionic columns. First-floor columns are 9- to 10-feet tall, while second-floor columns are 8 to 9 feet tall. Turned or square balusters are spaced no more than 4 inches apart. Porch bays should be vertically proportioned. Flat cut ornamental balusters are also used, with square or turned columns. Square pattern lattice is used as infill between piers at the foundation.

Brackets

Brackets range from simple designs cut from boards, to more elaborate turned wood or jigsaw-cut openwork. Brackets are a minimum of 2 inches thick. Archway bracketing can be used to form portals over key entry locations.

Porch Location

Full front porches are encouraged on Tidewater Victorian houses. Porches can be used to wrap the corner of a house, or to fill in the void created by an L-shaped plan. The minimum porch depth is 8 feet.

Tidewater Victorian 泰德沃特维多利亚风格

ARCHITECTURAL PATTERNS　C 19

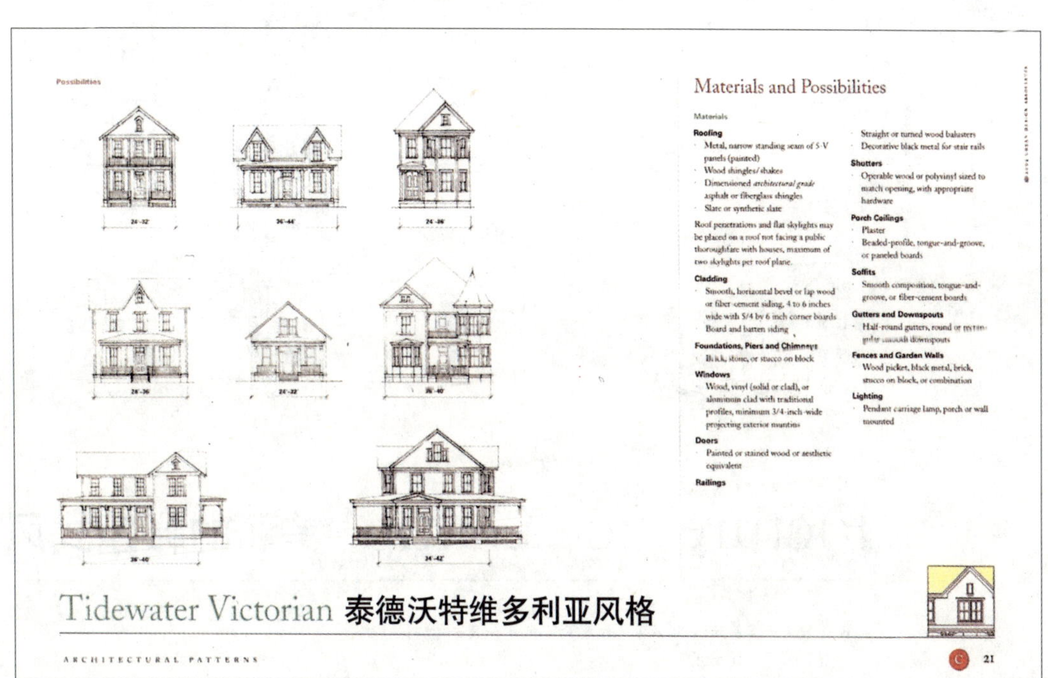

Possibilities

Materials and Possibilities

Materials

Roofing

- Metal, narrow standing seam of 5-V panels (painted)
- Wood shingles/shakes
- Dimensioned *architectural grade* asphalt or fiberglass shingles
- Slate or synthetic slate

Roof penetrations and flat skylights may be placed on a roof not facing a public thoroughfare with houses, maximum of two skylights per roof plane.

Cladding

- Smooth, horizontal bevel or lap wood or fiber-cement siding, 4 to 6 inches wide with 5/4 by 6 inch corner boards
- Board and batten siding

Foundations, Piers and Chimneys

- Brick, stone, or stucco on block

Windows

- Wood, vinyl (solid or clad), or aluminum clad with traditional profiles, minimum 3/4-inch-wide projecting exterior muntins

Doors

- Painted or stained wood or aesthetic equivalent

Railings

- Straight or turned wood balusters
- Decorative black metal for stair rails

Shutters

- Operable wood or polyvinyl sized to match opening, with appropriate hardware

Porch Ceilings

- Plaster
- Beaded-profile, tongue-and-groove, or paneled boards

Soffits

- Smooth composition, tongue-and-groove, or fiber-cement boards

Gutters and Downspouts

- Half-round gutters, round or rectangular smooth downspouts

Fences and Garden Walls

- Wood picket, black metal, brick, stucco on block, or combination

Lighting

- Pendant carriage lamp, porch or wall mounted

Tidewater Victorian 泰德沃特维多利亚风格

ARCHITECTURAL PATTERNS　C 21

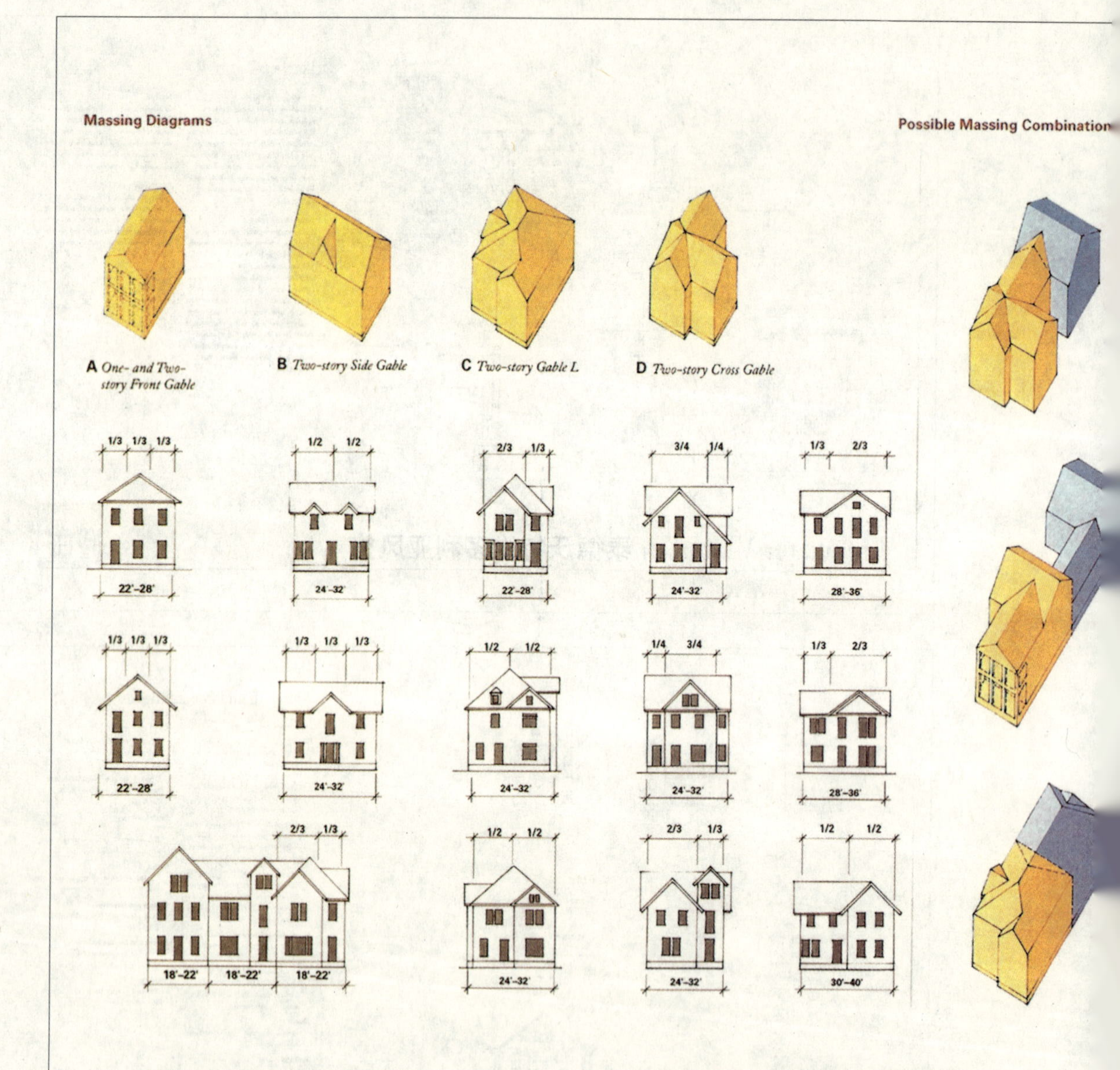

Picturesque Camp 美丽的帐篷式风格

ARCHITECTURAL PATTERNS

Massing and Composition 集中和构成

Massing

A　Two-story Front Gable
Rectangular volume with 10 in 12 roof pitch and gable facing the street. One-story partial, full, or wrapping front porch with shed or hip roof is common. Integral full front porches are also typical.

B　Two-story Side Gable
Side-gabled rectangular volume, often with a steeply-pitched, gabled dormer flush to the front façade. Front gable roof pitch is typically 10 in 12 to 12 in 12, and the side gable is less steeply pitched. One- or two-story front porches typically extend across the full front of the house.

C　Two-story Gable-L
Two-story rectangular volume with hipped roof and front gable. Front bay and gable roof can encroach into the porch zone a maximum of 3 feet and is limited to 14 feet in width.The roof pitch is typically 10 in 12. One- or two-story front wraparound porch with shed or hipped roof is most common.

D　Two-story Cross Gable
Two-story rectangular volume, with centrally-intersecting gable roofs. Gable roof pitch is typically 10 in 12. One- or two-story, full-length or wraparound front porch with shed or hipped roof.

Facade Composition

The facade composition is characterized by a symmetrical and balanced placement of doors and windows in regularly spaced bays that reflect the bays of the porch and projecting wings.

Combinations

Complex forms and larger living spaces may be created by combining side wings and/or rear wings with the main body. Gabled dormers may be added to introduce light into half-story and attic spaces. The architectural character of the attached parts should match that of the main body. Wings must also conform to all setback requirements (see *Community Patterns*).

9

伊斯特加里森

加利福尼亚州蒙特雷县

别致露营风格的体块类型，产生于19世纪的从临时性帐篷向更为永久性的住宅发展的早期教区露营住宅。这些乡村住宅最初是季节性的，并常常被设计为哥特复兴风格或斯蒂克风格住宅的微型版本。这些浪漫主义的处理方式造就了永久性的住宅，而且在露营之后就不再使用了。这样的房地产常常被并入附近的城镇。很多这类的营地，例如加利福尼亚州的“宁静树林”营地或楠塔基特的“橡林山崖”营地，都还完整地保留着这些住宅类型的重要实例。

Victorian era Picturesque Camp house in Napa Valley, California

Thin trim details and three quarter scale are characteristics of the Picturesque Camp Style.

Vertical openings and a full, expressive porch create a "Best" Picturesque Camp house.

Bracketed American farmhouse from The Architecture of Country Houses by A.J.Downing

History and Character

THE PICTURESQUE CAMP STYLE builds on the early 'Carpenter Gothic' cottages built in the western region of the United States from early Pattern Books. While the style became fashionable in the 1800s in the Bay Area, its popularity grew and it spread north and south from San Francisco. The source of many early examples were the Pattern Books published by Andrew Jackson Downing. Publications such as The Horticulturist influenced the preferences of the public and provided an especially dramatic contrast to the inherited Spanish and adobe building types prevalent at the time. Early religious camp settlements, such as Pacific Grove in Monterey and Chautauqua in New York, adopted the style with more and more exotic variations that included Eastlake, Queen Anne, and Italianate detailing.

The Picturesque Camp style is centered on the simple, elegant forms that were adapted to the more informal houses on small lots, much like the early camp housing. This style emphasizes the connection to nature and has expressive porch elements, decorative trim elements, and vertical windows and doors.

Essential Elements of Picturesque Camp

1. Steeply pitched, front facing gable roofs.
2. Cut wood ornament, often with natural forms such as leaves and vines.
3. Clapboard or cut shingle siding.
4. Vertical proportions for windows and doors.
5. Box bay and cutaway bay windows.

Picturesque Camp 美丽的帐篷式风格

ARCHITECTURAL PATTERNS

C 8

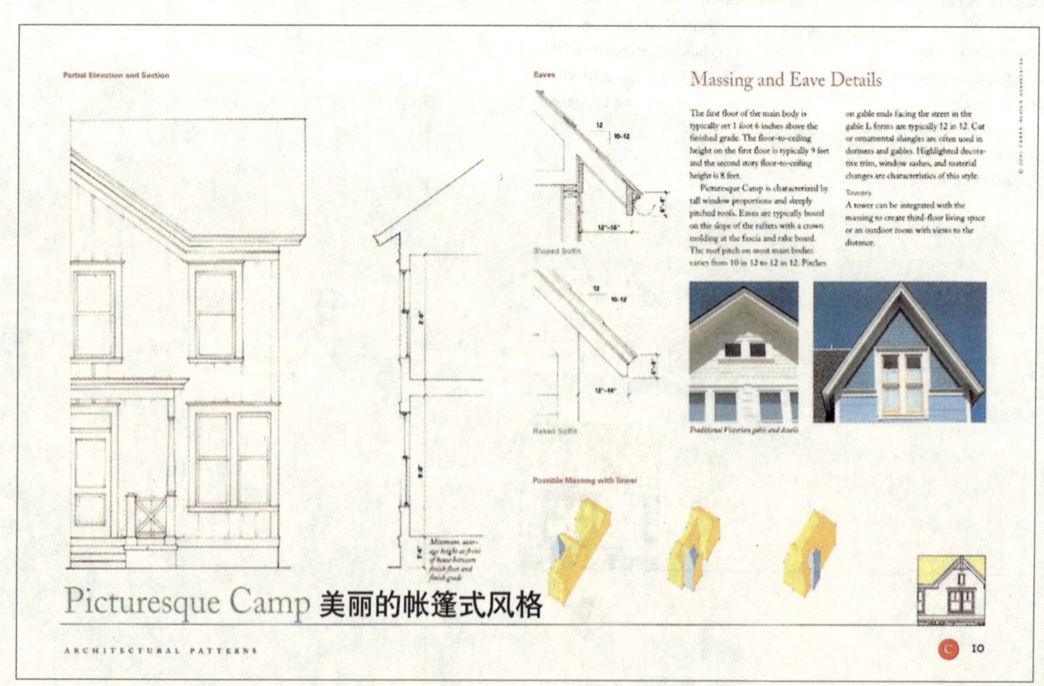

Massing and Eave Details

The first floor of the main body is typically set 1 foot 6 inches above the finished grade. The floor-to-ceiling height on the first floor is typically 9 feet and the second story floor-to-ceiling height is 8 feet.

Picturesque Camp is characterized by tall window proportions and steeply pitched roofs. Eaves are typically boxed on the slope of the rafters with a crown molding at the fascia and rake board. The roof pitch on most main bodies varies from 10 in 12 to 12 in 12. Pitches on gable ends facing the street in the gable L forms are typically 12 in 12. Cut or ornamental shingles are often used in dormers and gables. Highlighted decorative trim, window sashes, and material changes are characteristics of this style.

Towers

A tower can be integrated with the massing to create third-floor living space or an outdoor room with views to the distance.

Traditional Victorian gable and details

Picturesque Camp 美丽的帐篷式风格

ARCHITECTURAL PATTERNS

C 10

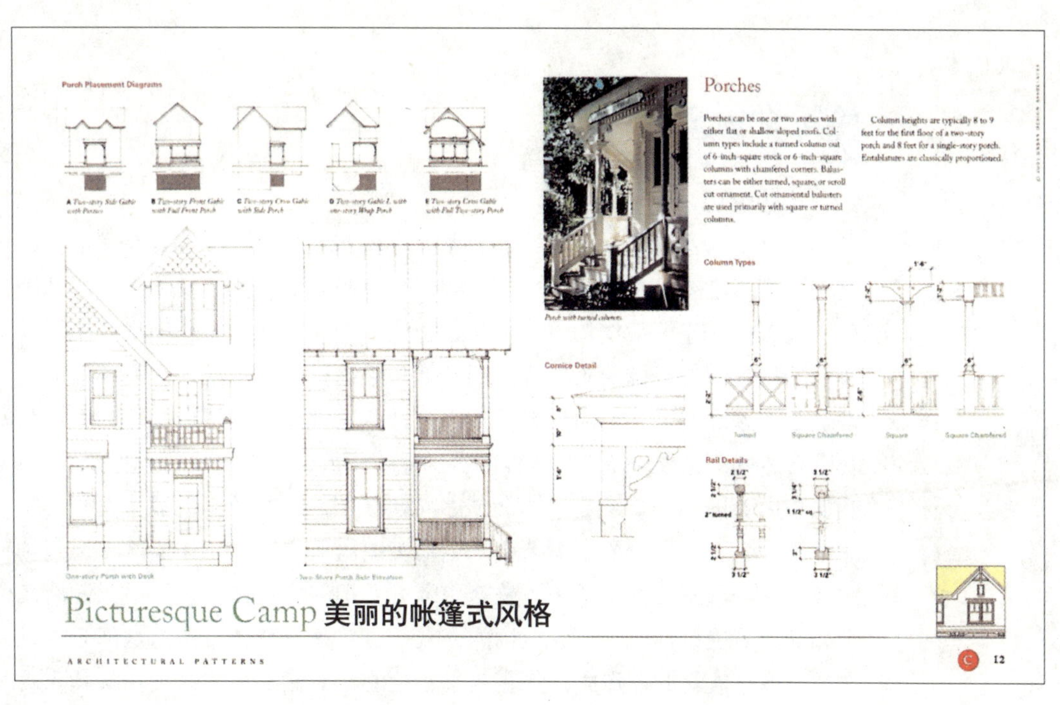
Porch Placement Diagrams
Porches
Porches can be one or two stories with either flat or shallow sloped roofs. Column types include a turned column out of 6-inch-square stock or 6-inch-square columns with chamfered corners. Balusters can be either turned, square, or scroll cut ornament. Cut ornamental balusters are used primarily with square or turned columns.
Column heights are typically 8 to 9 feet for the first floor of a two-story porch and 8 feet for a single-story porch. Entablatures are classically proportioned.
Column Types
Cornice Detail
Rail Details
Picturesque Camp 美丽的帐篷式风格
ARCHITECTURAL PATTERNS
12

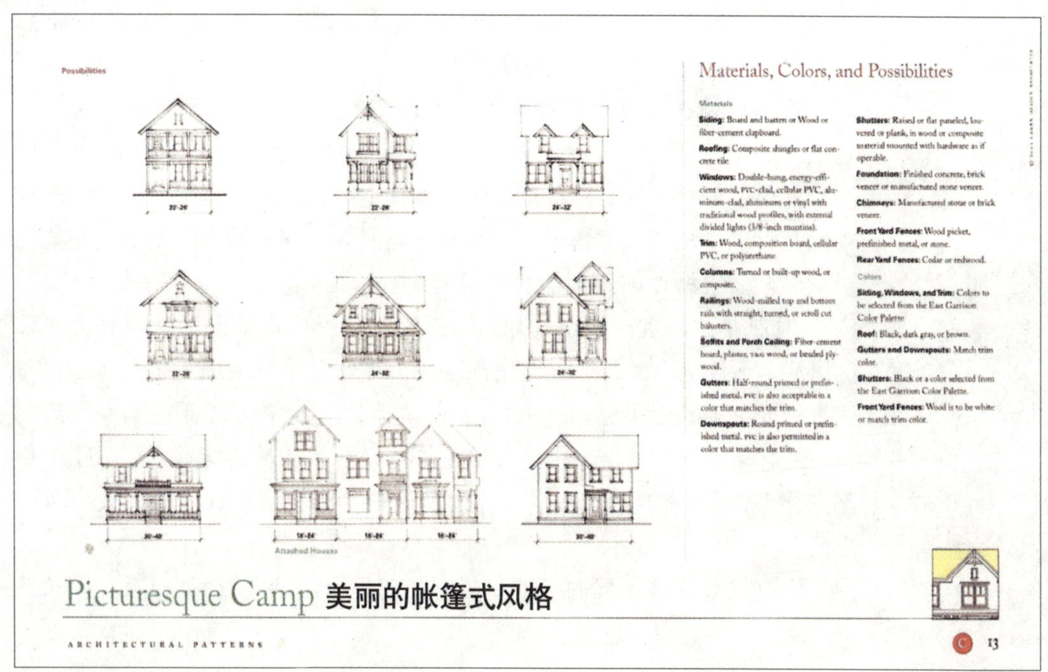
Possibilities
Materials, Colors, and Possibilities
Attached Houses
Picturesque Camp 美丽的帐篷式风格
ARCHITECTURAL PATTERNS
13

工艺美术风格

比较分析

工艺美术运动是一个国际性的现象，20 世纪初期在美国得到了显著的发展。英国建筑师威廉·莫里斯（1834 － 1896 年）是工艺美术运动思想著作出版工作和建设工作的主要推动者。这一运动是针对正在出现的制品和材料的机器制造、批量生产体系而发展起来的。建筑被视为是一项已经变得贫乏而冰冷的工艺，而失去了精心设计的感觉、经过精良制作的陈设品和以自然材料建造的住宅。其他的英国建筑师的作品，例如 C·F·A· 沃伊齐，刊发在由英国工艺美术展示协会创办的期刊上，这些期刊也被分发给美国的建筑师们。这也刺激了工艺美术运动美国语汇的发展。作为这项运动构成部分的一项核心社会任务，是努力以可负担的价格去将各种手工制造艺术和高品质的设计带给更大范围的公众。虽然这一社会目标只是部分得以实现，但有大量的创造团体和富有思想性的作品在相关的建筑和家具陈设艺术的很多领域中涌现出来，其中包括灯具、织物、耐火黏土和瓷砖制造、家具设计和建造、建筑、室内设计和平面图形设计。

这种风格在美国的早期实例，可以在新英格兰州沿海地区和加利福尼亚州的滨海社区中找到，例如帕萨迪纳市和圣巴巴拉市。在 19 世纪后半叶，很多东北部的——特别是纽约和波士顿的——建筑师开始尝试一种安妮女王风格和殖民复兴形式的变体，这种风格被文森特·斯卡利在其关于该风格发展演化的开创性研究中定义为木瓦风格。麦金、米德和怀特、H·H· 理查森和很多其他建筑师，为木瓦风格的住宅发展出一套非常优雅的语汇，无论是室外部分还是室内部分。住宅形式发展成为包裹在木瓦和壁板外壳之下的形态更为有机的构图，它们常常建造在厚重的石头基础之上，看起来就像是从土地中生长出来的一样。这一时期受到了英国建筑师理查德·诺曼·肖的极大影响，他借助这种语汇尝试设计了很多大型乡村庄园。

木瓦风格建筑发展的后期与英国工艺美术运动在时间上是相重合的，

实际上，它们共有很多相同的主题和要素。在西海岸地区，古斯塔夫斯·斯蒂克雷、格林兄弟(亨利和查尔斯)、朱莉娅·摩根和伯纳德·梅贝克等建筑师，都对工艺美术运动进行了独特的诠释。小型的、可支付的艺匠风格别墅的发展源自加利福尼亚地区的作品，并最终促成了带回廊的小住宅和小型艺匠风格别墅的大量生产。这些设计在很大程度上是由于1920年代住宅的编目销售而发展起来的。其显著的特征包括具有创造性和高效率的内置式家具，带有宽阔窗子的很大的开敞房间，不对称的门窗构图，对结构性要素的表现，自然材料的混合使用，以及手工制作的感觉和住宅内部与外部的手工细部。

工艺美术风格的范例册页，是从达克尔山（北卡罗来纳州阿什维尔附近）、伊斯特比奇（弗吉尼亚州诺福克市）和伊斯特加里森（加利福尼亚州蒙特雷县）的模式图则中摘录的。

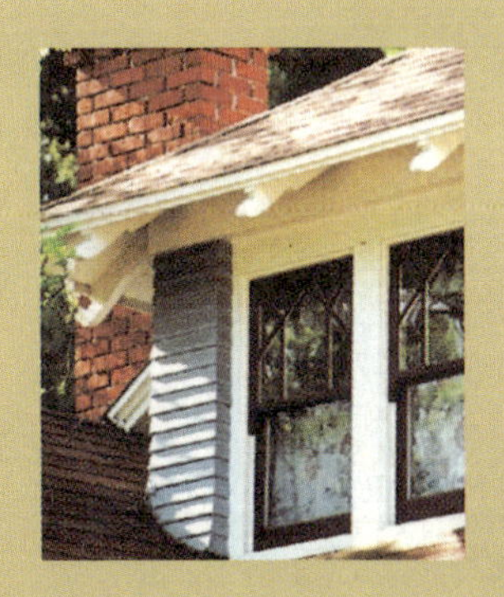
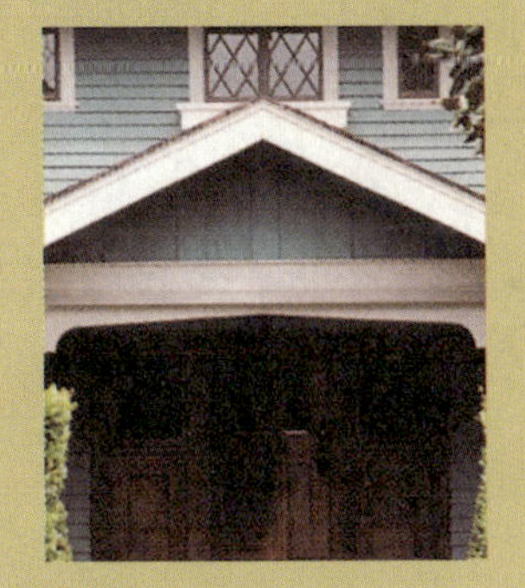

范例图片

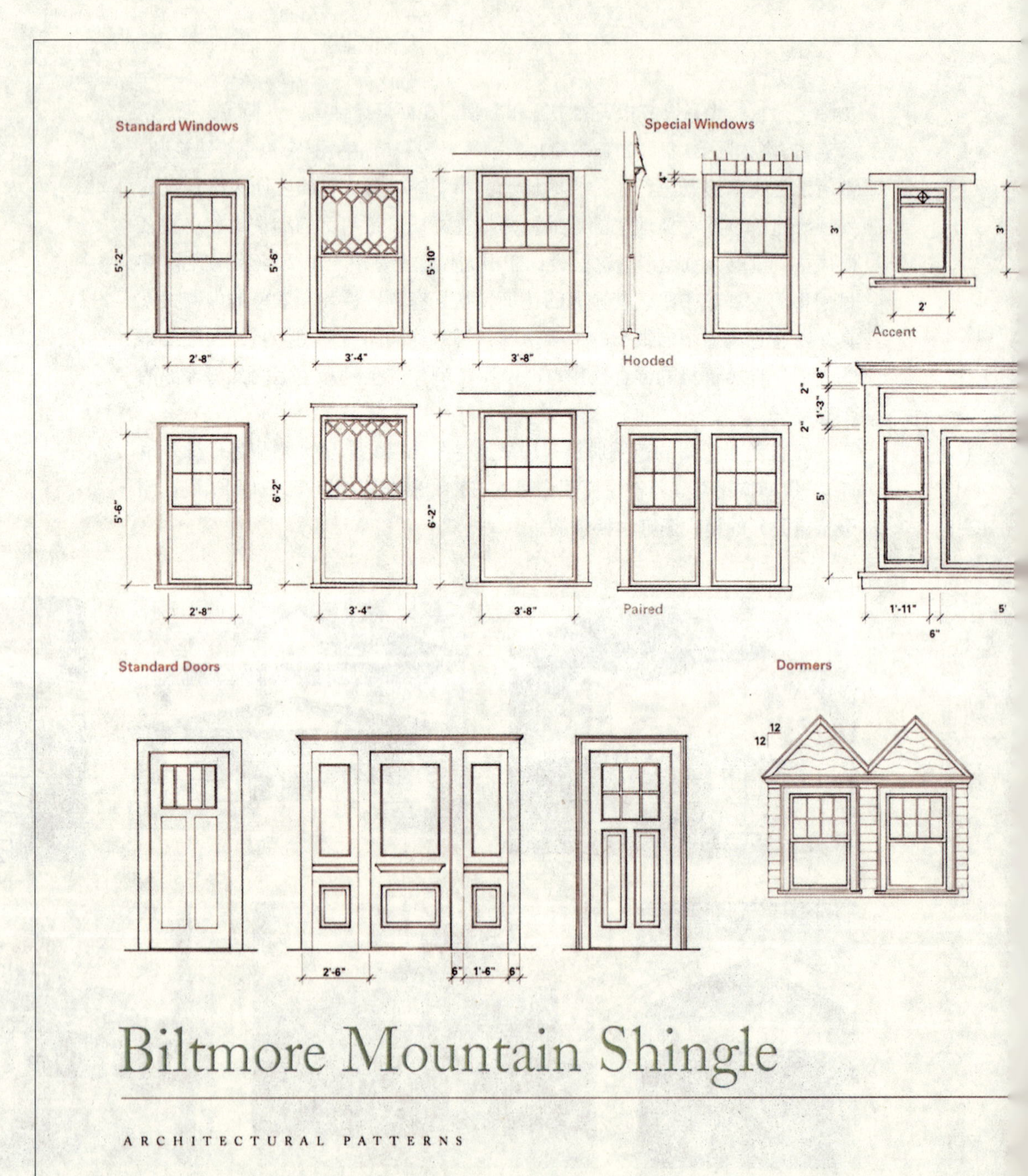
Standard Windows
Special Windows
5'-2"
2'-8"
5'-6"
3'-4"
5'-10"
3'-8"
Hooded
3'
2'
Accent
5'-6"
2'-8"
6'-2"
3'-4"
6'-2"
3'-8"
Paired
8"
2"
1'-3"
2"
5'
1'-11"
6"
5'
Standard Doors
Dormers
12
12
2'-6"
6"
1'-6"
6"
Biltmore Mountain Shingle
ARCHITECTURAL PATTERNS

达克尔山（比尔特摩）

Windows and Doors 窗和门

4"–6"

Detail

Standard Windows

Windows on the first floor are usually arranged in combinations of single openings, pairs and/or strips of three or more, sometimes including large picture windows. Windows on the second floor may be single, paired or triples. Often special accent windows are incorporated into the composition. Window pane patterns include 6 over 1, 12 over 1 and diamond patterned top sash. Dormer windows are commonly ganged together.

Special Windows

Special windows include angled bay windows, picture windows and small, square and rectangular accent windows. Picture windows are typically paired with sidelights and transoms with a special pane pattern or stained glass upper sash.

Standard Doors

Biltmore Mountain Shingle doors are often stained wood with either wood plank design or panel doors with integrated transoms. Doors may have decorative, stained glass sidelights and transoms in Arts and Crafts patterns.

Trim

Trim may either be flat board, typically 5¼ inches wide with a head that extends beyond the jamb trim to the sides, or a more formal casing with a backband.

比尔特摩山地木瓦风格

21

达克尔山（比尔特摩）

北卡罗来纳州阿什维尔附近

达克尔山的工艺美术风格被定义为山地木瓦风格。当 1910 年代至 1930 年代艺匠风格的住宅在整个北卡罗来纳州建造起来的时候，北卡罗来纳州的西部山区除了正在形成的阿什维尔温泉城镇之外都是相对欠发达的地区。由 H·H· 理查森设计的比尔特摩住宅区的建设，需要一些当地建筑师来完成施工图并在村庄和工地上监督工作。建筑师们，其中最为突出的是理查德·夏普·史密斯，从纽约来到阿什维尔监督施工。在他停留在阿什维尔的期间，史密斯开始在为其他当地客户建造的房地产项目和私人住房项目中设计各种附属结构。很多当地的建筑师也设计了这种风格的住宅，并且将其与这一地区众多的温泉疗养旅馆的发展结合起来，阿什维尔很快就拥有了大量经过精心设计和良好施工的木瓦风格或工艺美术风格的住宅，形成了北卡罗来纳州山区的特色。这种风格的主要特征包括对地方石材和砖的运用、经过油漆的木瓦、深深的挑檐、成型木材立柱和屋檐处的切割托架。达克尔的山地木瓦风格住宅发展形成的基础，是工艺美术运动的英国变体和纽约州与康涅狄格州建筑师所设计的乡村住宅。在蒙特福德等历史性邻里中的美妙住宅和诸如格罗夫公园乡村酒店等具有异国情调的旅馆，都是这种地方性风格的实例代表。

Mountain Shingle house

Essential Elements of the Biltmore Mountain Shingle

1. Continuity of roof and wall surfaces.
2. Deep, broad porch elements with expressive structural components.
3. Strong horizontal lines such as eaves, water tables and window heads.
4. A mixture of materials such as stone, shingles and siding in horizontal bands.
5. Asymmetrical window and door compositions.

History and Character

BILTMORE MOUNTAIN SHINGLE HOUSES are derived from the uniquely American expression of architectural design for country houses termed *Shingle Style*, that originated in the northeastern region – Massachusetts, Rhode Island, Connecticut, New York, and Maine – from about 1878-1916. Architects began to look to early American Colonial houses that were simple, wood-shingled forms added onto in an organic way over time. The intention was to develop a uniquely American style for country houses and cottages that reinforced the notion of a more informal, leisure use. Notable practioners included McKim, Mead and White, Henry Hobson Richardson, and the Boston firms of Peabody & Stearns, Arthur Little, and Bruce Price. Early cottage settlements, such as Tuxedo Park in New York, serve as a good example of *Shingle Style* architecture.

In the Asheville region, we find many interpretations of this style in houses by architects such as Richard Sharp Smith and William Henry Lord who built in neighborhoods such as Montford. The use of mountain stone, pebble dash stucco and shingle cladding add a unique regional flavor to this style.

Biltmore Mountain Shingle 比尔特摩山地木瓦风格

ARCHITECTURAL PATTERNS

18

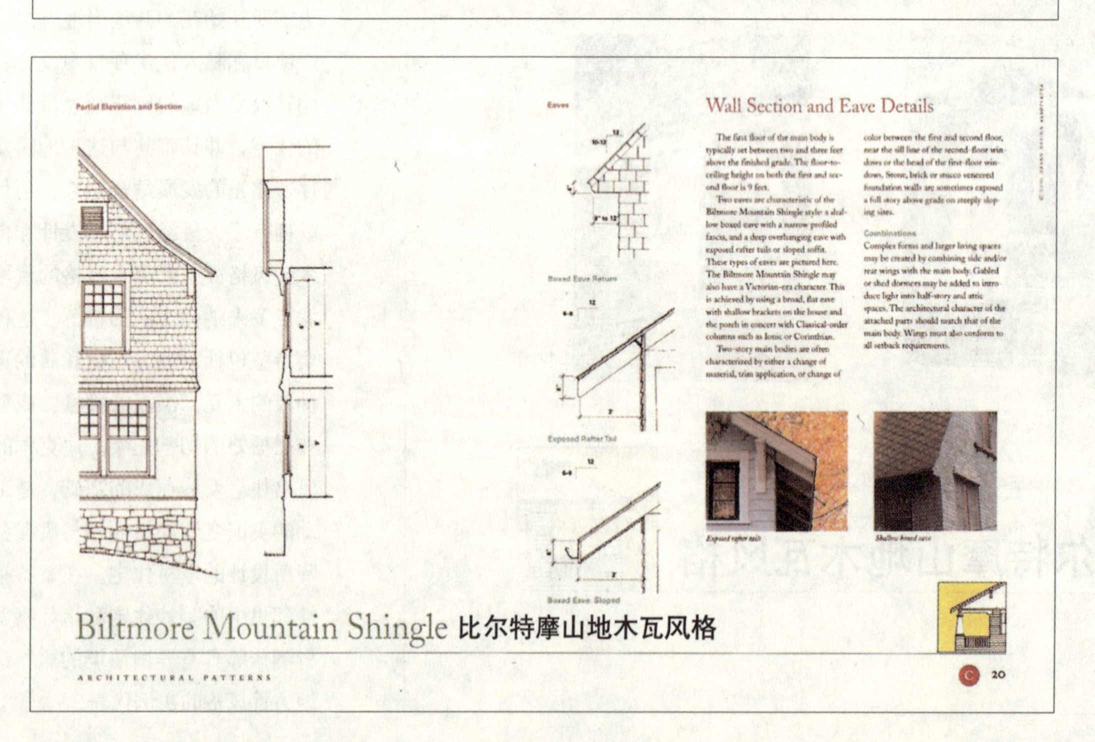

Partial Elevation and Section

Eaves

Boxed Eave Return

Exposed Rafter Tail

Boxed Eave Sloped

Wall Section and Eave Details

The first floor of the main body is typically set between two and three feet above the finished grade. The floor-to-ceiling height on both the first and second floor is 9 feet.

Two eaves are characteristic of the Biltmore Mountain Shingle style: a shallow boxed eave with a narrow profiled fascia, and a deep overhanging eave with exposed rafter tails or sloped soffit. These types of eaves are pictured here. The Biltmore Mountain Shingle may also have a Victorian-era character. This is achieved by using a broad, flat eave with shallow brackets on the house and the porch in concert with Classical-order columns such as Ionic or Corinthian.

Two-story main bodies are often characterized by either a change of material, trim application, or change of color between the first and second floor, near the sill line of the second-floor windows or the head of the first-floor windows. Stone, brick or stucco veneered foundation walls are sometimes exposed a full story above grade on steeply sloping sites.

Combinations

Complex forms and larger living spaces may be created by combining side and/or rear wings with the main body. Gabled or shed dormers may be added to introduce light into half-story and attic spaces. The architectural character of the attached parts should match that of the main body. Wings must also conform to all setback requirements.

Exposed rafter tails

Shallow boxed eave

Biltmore Mountain Shingle 比尔特摩山地木瓦风格

ARCHITECTURAL PATTERNS

20

达克尔山（比尔特摩）

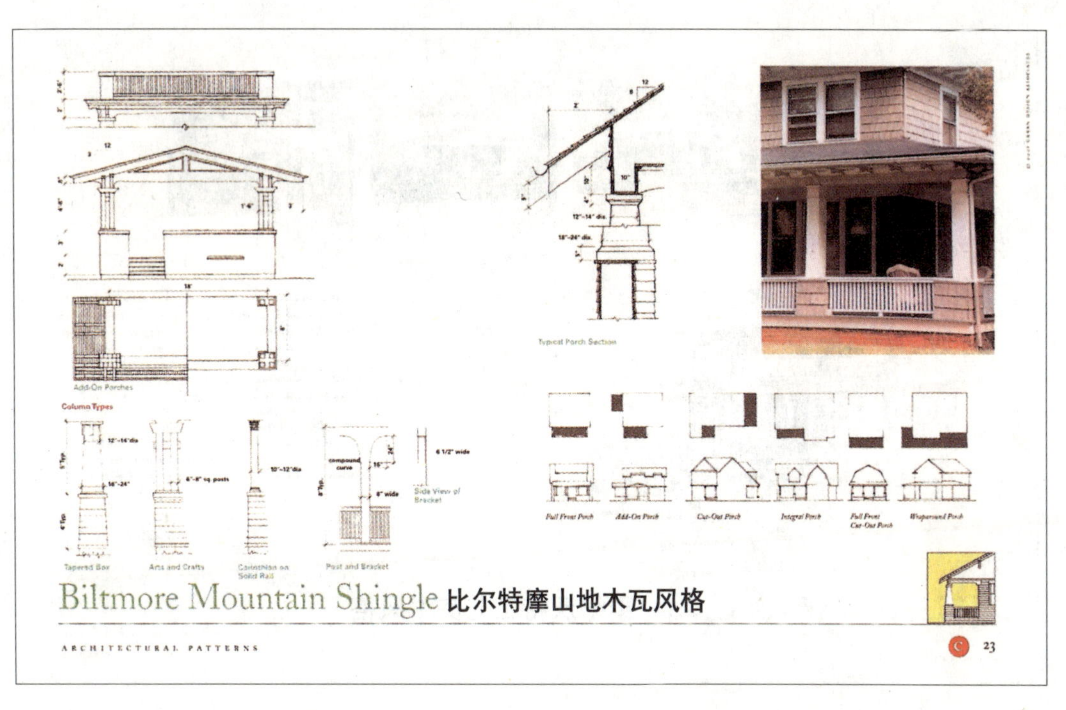
Add-On Porches
Typical Porch Section
Column Types
Tapered Box
Arts and Crafts
Corinthian on Solid Rail
Post and Bracket
Side View of Bracket
Full Front Porch
Add-On Porch
Cut-Out Porch
Integral Porch
Full Front Cut-Out Porch
Wraparound Porch
Biltmore Mountain Shingle 比尔特摩山地木瓦风格
ARCHITECTURAL PATTERNS
23

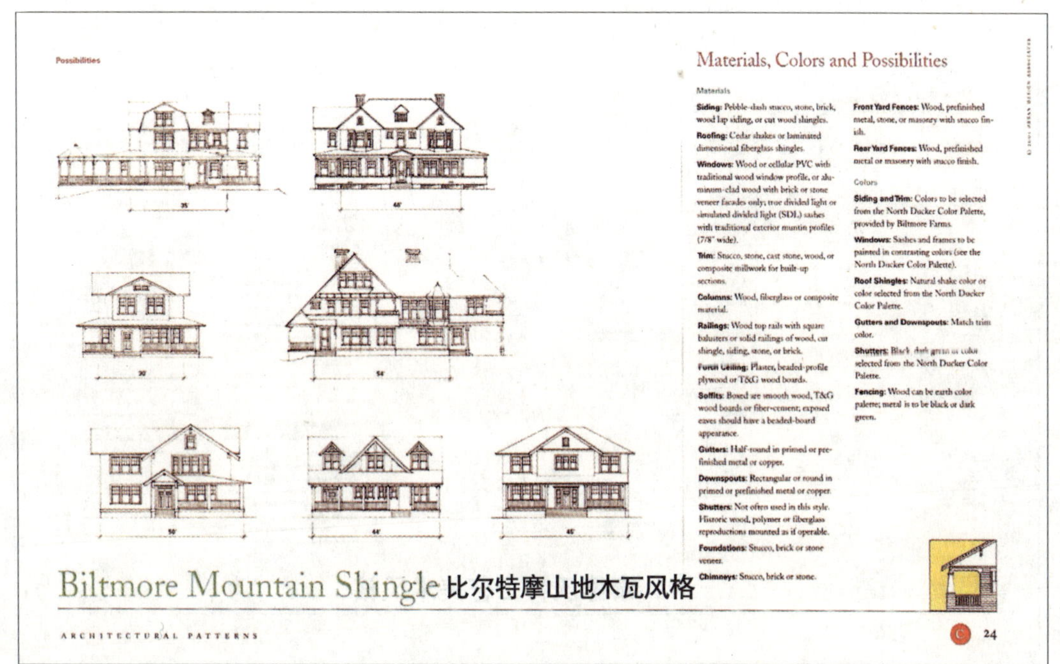
Possibilities
Materials, Colors and Possibilities
Materials
Siding: Pebble-dash stucco, stone, brick, wood lap siding, or cut wood shingles.
Roofing: Cedar shakes or laminated dimensional fiberglass shingles.
Windows: Wood or cellular PVC with traditional wood window profile, or aluminum-clad wood with brick or stone veneer facades only; true divided light or simulated divided light (SDL) sashes with traditional exterior muntin profiles (7/8" wide).
Trim: Stucco, stone, cast stone, wood, or composite millwork for built-up sections.
Columns: Wood, fiberglass or composite material.
Railings: Wood top rails with square balusters or solid railings of wood, cut shingle, siding, stone, or brick.
Porch Ceiling: Plaster, beaded-profile plywood or T&G wood boards.
Soffits: Boxed are smooth wood, T&G wood boards or fiber-cement; exposed eaves should have a beaded-board appearance.
Gutters: Half-round in primed or pre-finished metal or copper.
Downspouts: Rectangular or round in primed or prefinished metal or copper.
Shutters: Not often used in this style. Historic wood, polymer or fiberglass reproductions mounted as if operable.
Foundations: Stucco, brick or stone veneer.
Chimneys: Stucco, brick or stone.
Front Yard Fences: Wood, prefinished metal, stone, or masonry with stucco finish.
Rear Yard Fences: Wood, prefinished metal or masonry with stucco finish.
Colors
Siding and Trim: Colors to be selected from the North Ducker Color Palette, provided by Biltmore Farms.
Windows: Sashes and frames to be painted in contrasting colors (see the North Ducker Color Palette).
Roof Shingles: Natural shake color or color selected from the North Ducker Color Palette.
Gutters and Downspouts: Match trim color.
Shutters: Black, dark green or color selected from the North Ducker Color Palette.
Fencing: Wood can be earth color palette; metal is to be black or dark green.
Biltmore Mountain Shingle 比尔特摩山地木瓦风格
ARCHITECTURAL PATTERNS
24

Cape Charles, Virginia

Elizabeth City, North Carolina

Cape Charles, Virginia

Elizabeth City, North Carolina

Essential Elements of the Tidewater Arts & Crafts

1 Shallow-pitched roofs with deep overhangs.
2 Deep, broad porch elements with expressive structural components.
3 Asymmetrical, but balanced window and door compositions.
4 Grouped windows.

History and Character

TIDEWATER ARTS & CRAFTS HOUSES are derived from the traditions of Bungalow design, which was popular in beach cottages. Characterized by an eclectic mix of architectural elements and a response to coastal environments, this enduring style flourished in the early twentieth century both as modest cottages and large houses. Builders used pattern books and mass-marketed house plans and packages to attract a broad spectrum of homebuyers. These comfortable, eclectic homes were often lighter in color and less ornamented than high style Arts & Crafts houses. It is this more eclectic style that serves as the basis for the Tidewater Arts & Crafts.

The Tidewater Arts & Crafts is characterized by broad open porches; low sloping roofs with deep overhangs; multiple gables; asymmetric compositions; oversized first-floor windows; expressive trim; exposed rafters; and bracketed porches.

Tidewater Arts & Crafts 泰德沃特工艺美术风格

ARCHITECTURAL PATTERNS

C 22

Massing Diagrams

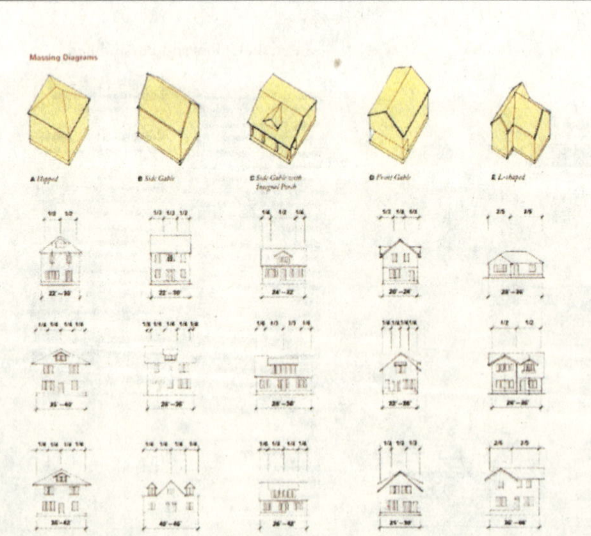

Massing and Composition

Massing

A Hipped
Rectangular or square volume with a 6 in 12 to 8 in 12 roof pitch; the ridge line, if any, runs parallel with the front of the house. Front gabled and/or shed roofed porches with a 3 in 12 to 5 in 12 pitch are placed symmetrically or asymmetrically on the front façade or as full-façade elements. Porches are typically one story and may wrap one or both corners.

B Side Gable
Rectangular volume with a 6 in 12 to 8 in 12 roof pitch. Asymmetrically placed gabled and/or shed roofed porches are common. Porches are typically one story.

C Side Gable with Integral Porch
Rectangular one-and-one half story volume with a 6 in 12 to 8 in 12 roof pitch. The integral porch is set under occupiable interior space, made possible by a dormer and high knee wall on the second floor. Integral front porch ranges from half to the full length of the front façade. Symmetrically placed gabled or shed dormers have a 3 in 12 roof pitch.

D Front Gable
Rectangular volume with a 6 in 12 to 8 in 12 roof pitch and gable facing the street. Symmetrically or asymmetrically placed front and/or shed roofed porches are common and either one- or two-story. An inset, one-story porch may also run the full width of the house.

E L-Shaped
Cross-gabled volume with a 6 in 12 to 8 in 12 gable facing the street. The width of the gable facing the street is typically two-fifths, or less commonly, half that of the main body. Often an in-line front gabled porch or wing is added to the front leg of the **L**. Shed porches may also fill the space between the wings of the **L**.

Massing Combinations

Complex forms and larger living spaces may be created by combining side and/or rear wings with the main body. Gabled or shed dormers may be added to introduce light into half-story and attic spaces. The architectural character of the attached parts should match that of the main body.

Facade Composition

Arts & Crafts façade composition is characterized by an asymmetrical yet balanced placement of doors and windows. Typically, windows occur in pairs and multiples, or as sidelights for oversized ground floor windows. Entrance doors are most often under porches and off center.

Tidewater Arts & Crafts 泰德沃特工艺美术风格

ARCHITECTURAL PATTERNS

C 23

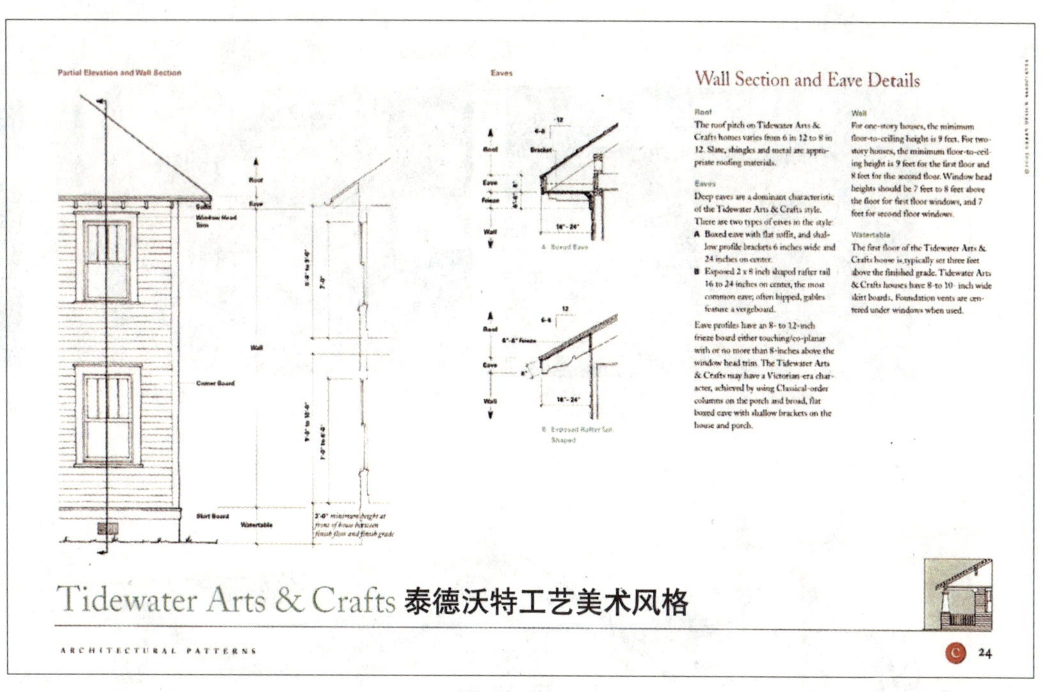
Partial Elevation and Wall Section
Eaves
Wall Section and Eave Details
Roof
The roof pitch on Tidewater Arts & Crafts homes varies from 6 in 12 to 8 in 12. Slate, shingles and metal are appropriate roofing materials.
Eaves
Deep eaves are a dominant characteristic of the Tidewater Arts & Crafts style. There are two types of eaves in the style:
A Boxed eave with flat soffit, and shallow profile brackets 6 inches wide and 24 inches on center.
B Exposed 2 x 8 inch shaped rafter tail 16 to 24 inches on center, the most common eave; often hipped, gables feature a vergeboard.
Eave profiles have an 8- to 12-inch frieze board either touching/co-planar with or no more than 8-inches above the window head trim. The Tidewater Arts & Crafts may have a Victorian-era character, achieved by using Classical-order columns on the porch and broad, flat boxed eave with shallow brackets on the house and porch.
Wall
For one-story houses, the minimum floor-to-ceiling height is 9 feet. For two-story houses, the minimum floor-to-ceiling height is 9 feet for the first floor and 8 feet for the second floor. Window head heights should be 7 feet to 8 feet above the floor for first floor windows, and 7 feet for second floor windows.
Watertable
The first floor of the Tidewater Arts & Crafts house is typically set three feet above the finished grade. Tidewater Arts & Crafts houses have 8-to 10- inch wide skirt boards. Foundation vents are centered under windows when used.
Tidewater Arts & Crafts 泰德沃特工艺美术风格
ARCHITECTURAL PATTERNS
C 24

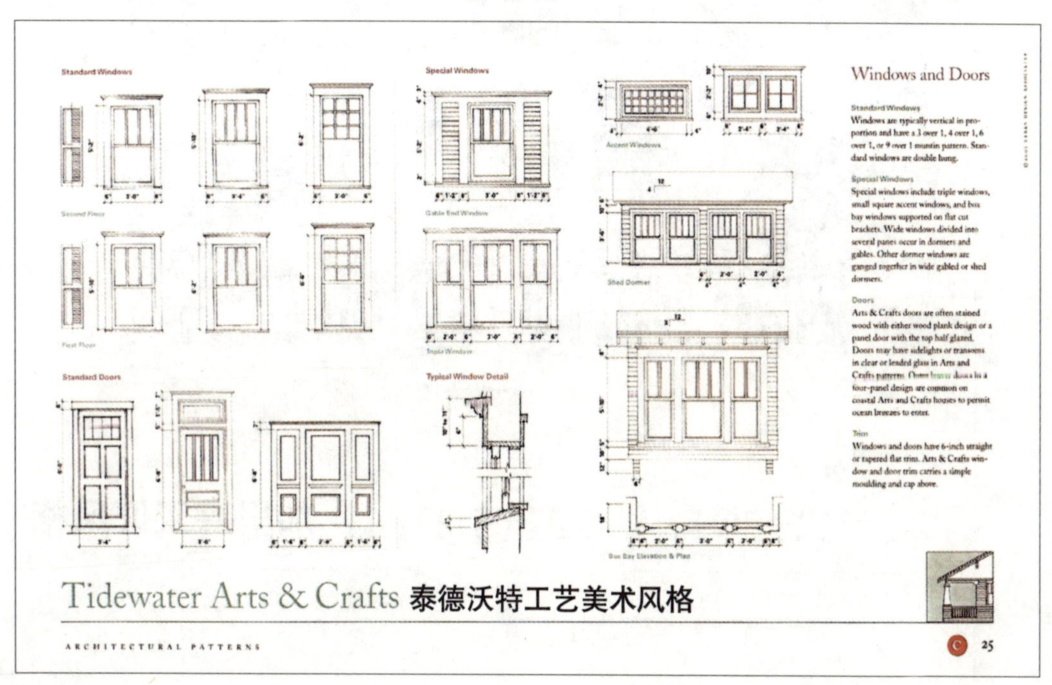
Standard Windows
Special Windows
Standard Doors
Typical Window Detail
Windows and Doors
Standard Windows
Windows are typically vertical in proportion and have a 3 over 1, 4 over 1, 6 over 1, or 9 over 1 muntin pattern. Standard windows are double hung.
Special Windows
Special windows include triple windows, small square accent windows, and box bay windows supported on flat cut brackets. Wide windows divided into several panes occur in dormers and gables. Other dormer windows are ganged together in wide gabled or shed dormers.
Doors
Arts & Crafts doors are often stained wood with either wood plank design or a panel door with the top half glazed. Doors may have sidelights or transoms in clear or leaded glass in Arts and Crafts patterns. Outer louver doors in a four-panel design are common on coastal Arts and Crafts houses to permit ocean breezes to enter.
Trim
Windows and doors have 6-inch straight or tapered flat trim. Arts & Crafts window and door trim carries a simple moulding and cap above.
Tidewater Arts & Crafts 泰德沃特工艺美术风格
ARCHITECTURAL PATTERNS
C 25

Asymmetrical window and door composition and a mix of siding materials are characteristic of the Garrison Craftsman style.

Simple forms with exposed structural members and deep overhangs are typical.

Tower elements will be an optima the Garrison Craftsman homes.

Essential Elements of the Garrison Craftsman

1 Shallow-pitched roofs with deep overhangs.

2 Deep, broad porch elements wi expressive structural component

3 Exposed structural elements in eaves such as rafters and bracke

4 A mixture of materials such as stucco, shingles, and siding.

5 Asymmetrical window and door compositions.

Garrison Craftsman 加里森工匠风格

ARCHITECTURAL PATTERNS

Illustration of Craftsman house from 500 Small Houses of the Twenties

© 2001 URBAN DESIGN ASSOCIATES

some of

History and Character

GARRISON CRAFTSMAN HOUSES are derived from the unique qualities of the Craftsman tradition found throughout the Northern California and Central Coast region. Many regional builders constructed houses influenced by the Arts & Crafts movement. California versions are characterized by exposed or expressive structural elements such as rafters, columns, beams, lintels, and porch elements. The floor plans were generally open with built-in cabinet work often in natural, stained woods. House exteriors were clapboard or shingle siding mixed with stone and brick or stucco accents and were painted in robust color palettes. The California craftsman house was influenced by the Japanese and English Arts and Crafts movements.

For houses in East Garrison, the emphasis in this style is on simple, structural expression of porch and eave elements using a vocabulary of architectural elements including Prairie style, Japanese, and Swiss, as well as influences from the Arts & Crafts movement. A coastal character is important to this style and should be reflected in the use of high contrast color for body and trim details. Forms are simple and reflect dimensioned lumber elements. Windows in this style tend to be vertical in proportion and are typically ganged or paired. Exposed eave brackets on roofs and porches contribute to this image and detail.

Horizontal siding, square and shaped shingle siding patterns, and a mix of stucco and siding materials are key cladding elements. This style also may include unpainted metal roofing and shingled roofs.

C 3

伊斯特加里森

加利福尼亚州蒙特雷县

这种风格是通过对历史性的艺匠风格和正在出现的乡村别墅和住宅建设中的工艺美术风格趋势所进行的广泛调研基础上演变出来的。这种风格的特征是对住宅结构的简洁而直接的表达，它借助暴露在外的椽子、方形截面的柱子与梁、平屋顶和边缘为直角的装饰元素来强化住宅的框架和结构。门和窗则是矩形的简单形状。

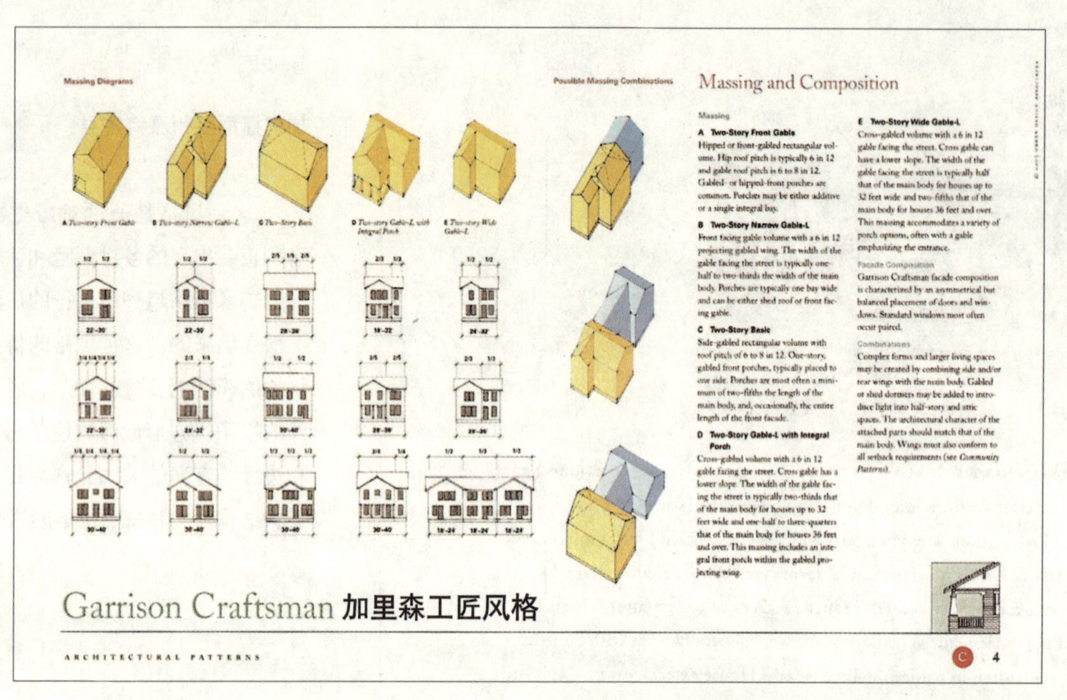

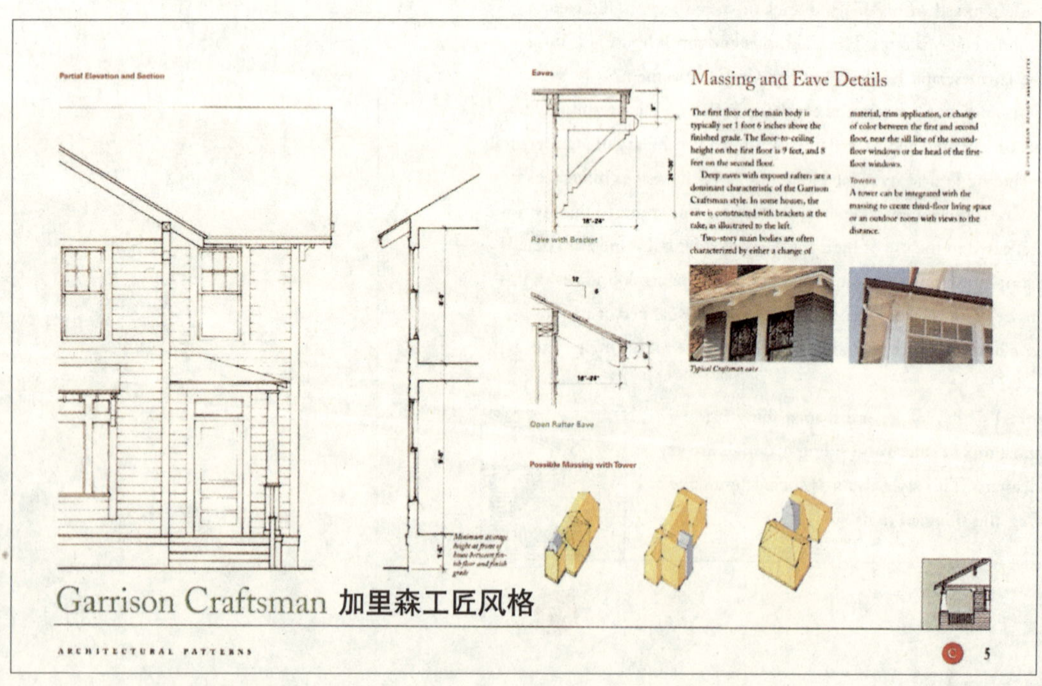

伊斯特加里森

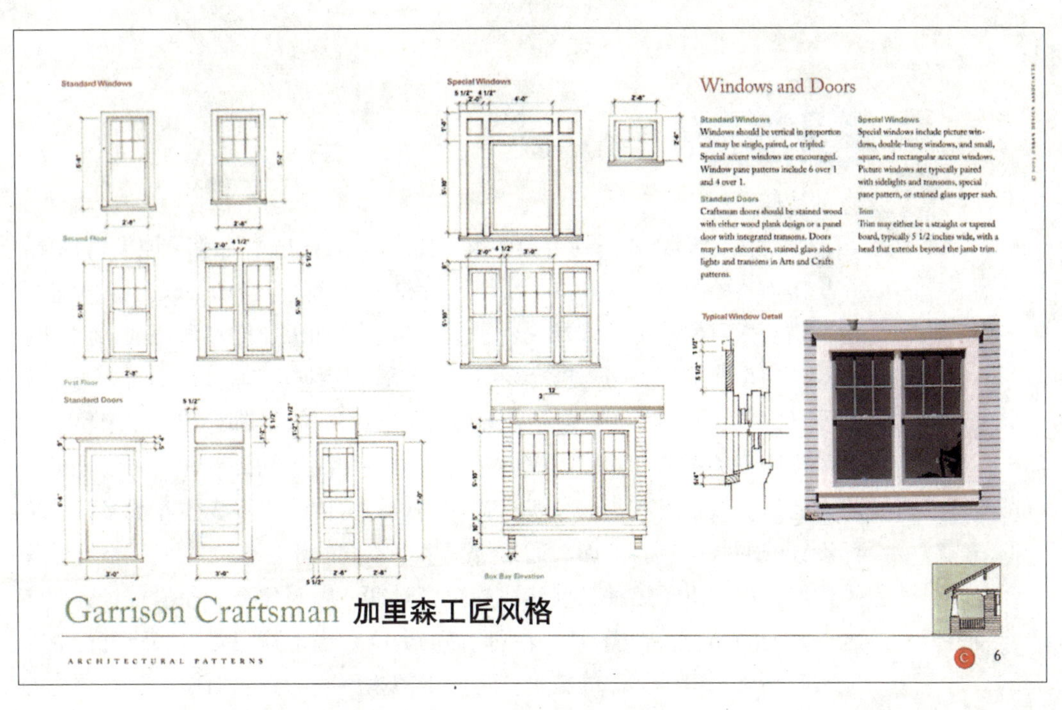
Standard Windows
Second Floor
First Floor
Standard Doors
Special Windows
Box Bay Elevation
Windows and Doors
Standard Windows
Windows should be vertical in proportion and may be single, paired, or tripled. Special accent windows are encouraged. Window pane patterns include 6 over 1 and 4 over 1.
Standard Doors
Craftsman doors should be stained wood with either wood plank design or a panel door with integrated transoms. Doors may have decorative, stained glass side-lights and transoms in Arts and Crafts patterns.
Special Windows
Special windows include picture win-dows, double-hung windows, and small, square, and rectangular accent windows. Picture windows are typically paired with sidelights and transoms, special pane pattern, or stained glass upper sash.
Trim
Trim may either be a straight or tapered board, typically 5 1/2 inches wide, with a head that extends beyond the jamb trim.
Typical Window Detail
Garrison Craftsman 加里森工匠风格
ARCHITECTURAL PATTERNS
6

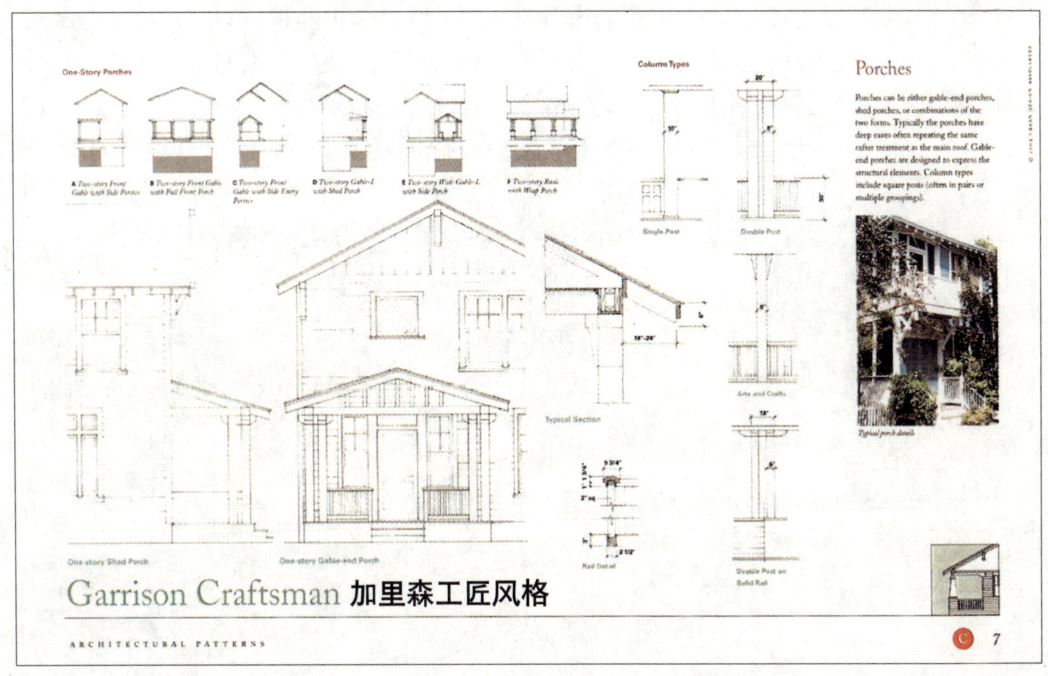
One-Story Porches
Column Types
Single Post
Double Post
Arts and Crafts
Typical Section
Rail Detail
Double Post on Solid Rail
One-story Shed Porch
One-story Gable-end Porch
Porches
Porches can be either gable-end porches, shed porches, or combinations of the two forms. Typically the porches have deep eaves often repeating the same rafter treatment as the main roof. Gable-end porches are designed to express the structural elements. Column types include square posts (often in pairs or multiple groupings).
Typical porch details
Garrison Craftsman 加里森工匠风格
ARCHITECTURAL PATTERNS
7

古典风格

比较分析

在美国大部分地区的古典主义住宅都是来自早期的殖民地风格住区，并与作为其根源的乔治风格、联邦风格和亚当风格等密切关联。UDA 通常将希腊复兴风格作为不同的一种独立风格。早期工匠的参考图则，例如阿舍·本杰明所著的《美国建造者手册》、巴蒂·兰利的《城乡建造者》与《工匠设计宝典》或威廉·哈夫彭尼所著的《实用建筑》，都介绍了很多符合古典风格比例关系的门、窗、室内装饰线脚、饰面等细部设计。对当地或区域内的建筑模式所进行的分析，揭示出这一主体的很多变体和很多奇特细部设计的混合实例。然而，通常情况下，不同的地区会对古典风格有特殊的偏好，这取决于该种风格建成的全盛时期。因为建造者和消费者会追随这一时期的品味，人们可以发现风格的聚类和当地建造者对其所进行的处理实际上显示出了共同的特征。在南卡罗来纳州的查尔斯顿市及其周边的地区性影响，会产生与密苏里州的圣路易斯市非常不同的古典风格建筑。在当地进行调查能确定对这种风格的地区性变体的说明重点所在。然而，模式图则所进行的说明并不是全面的，它更为关注的是看起来与项目相适合的地方模式和在当地环境中存在的邻里的基本特征。

古典风格的范例册页，是从巴克斯特（南卡罗来纳州福特米尔市）、达克尔山（北卡罗来纳州阿什维尔附近）和弗吉尼亚州诺福克市的模式图则中摘录的。

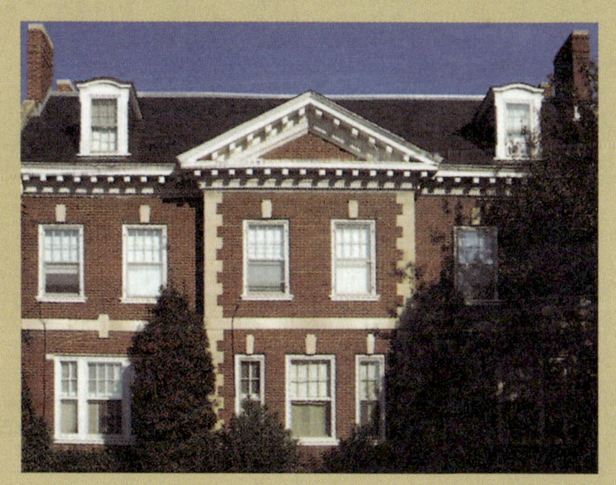

范例图片

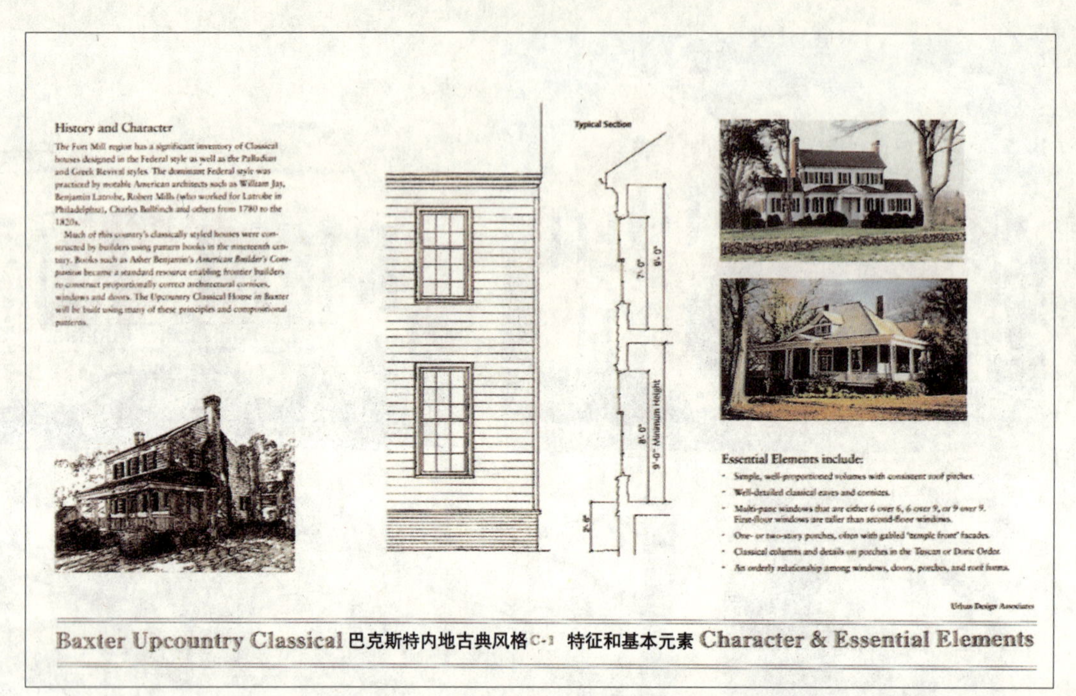
History and Character
The Fort Mill region has a significant inventory of Classical houses designed in the Federal style as well as the Palladian and Greek Revival styles. The dominant Federal style was practiced by notable American architects such as William Jay, Benjamin Latrobe, Robert Mills (who worked for Latrobe in Philadelphia), Charles Bulfinch and others from 1780 to the 1820s.
Much of this country's classically styled houses were constructed by builders using pattern books in the nineteenth century. Books such as Asher Benjamin's American Builder's Companion became a standard resource enabling frontier builders to construct proportionally correct architectural cornices, windows and doors. The Upcountry Classical House in Baxter will be built using many of these principles and compositional patterns.
Typical Section
7'-0"
9'-0"
8'-0"
9'-0" Minimum Height
2'-6"
Essential Elements include:
• Simple, well-proportioned volumes with consistent roof pitches.
• Well-detailed classical eaves and cornices.
• Multi-pane windows that are either 6 over 6, 6 over 9, or 9 over 9. First-floor windows are taller than second-floor windows.
• One- or two-story porches, often with gabled "temple front" facades.
• Classical columns and details on porches in the Tuscan or Doric Order.
• An orderly relationship among windows, doors, porches, and roof forms.
Urban Design Associates
Baxter Upcountry Classical 巴克斯特内地古典风格 C-1　特征和基本元素 Character & Essential Elements

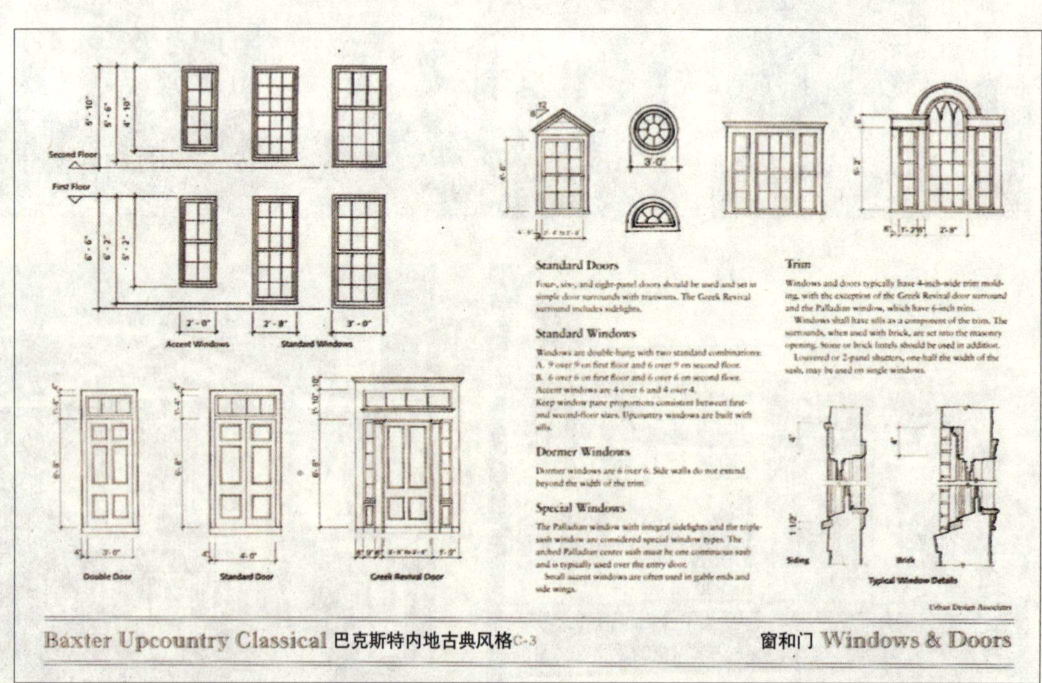
Second Floor
First Floor
5'-10"
5'-6"
4'-10"
6'-6"
6'-2"
5'-2"
2'-0"
2'-8"
3'-0"
Accent Windows
Standard Windows
3'-0"
Double Door
Standard Door
Greek Revival Door
Standard Doors
Four-, six-, and eight-panel doors should be used and set in simple door surrounds with transoms. The Greek Revival surround includes sidelights.
Standard Windows
Windows are double-hung with two standard combinations:
A. 9 over 9 on first floor and 6 over 9 on second floor.
B. 6 over 6 on first floor and 6 over 6 on second floor.
Accent windows are 4 over 6 and 4 over 4.
Keep window pane proportions consistent between first- and second-floor sizes. Upcountry windows are built with sills.
Dormer Windows
Dormer windows are 6 over 6. Side walls do not extend beyond the width of the trim.
Special Windows
The Palladian window with integral sidelights and the triple-sash window are considered special window types. The arched Palladian center sash must be one continuous sash and is typically used over the entry door.
Small accent windows are often used in gable ends and side wings.
Trim
Windows and doors typically have 4-inch-wide trim molding, with the exception of the Greek Revival door surround and the Palladian window, which have 6-inch trim.
Windows shall have sills as a component of the trim. The surrounds, when used with brick, are set into the masonry opening. Stone or brick lintels should be used in addition.
Louvered or 2-panel shutters, one-half the width of the sash, may be used on single windows.
Siding
Brick
Typical Window Details
Urban Design Associates
Baxter Upcountry Classical 巴克斯特内地古典风格 C-3　窗和门 Windows & Doors

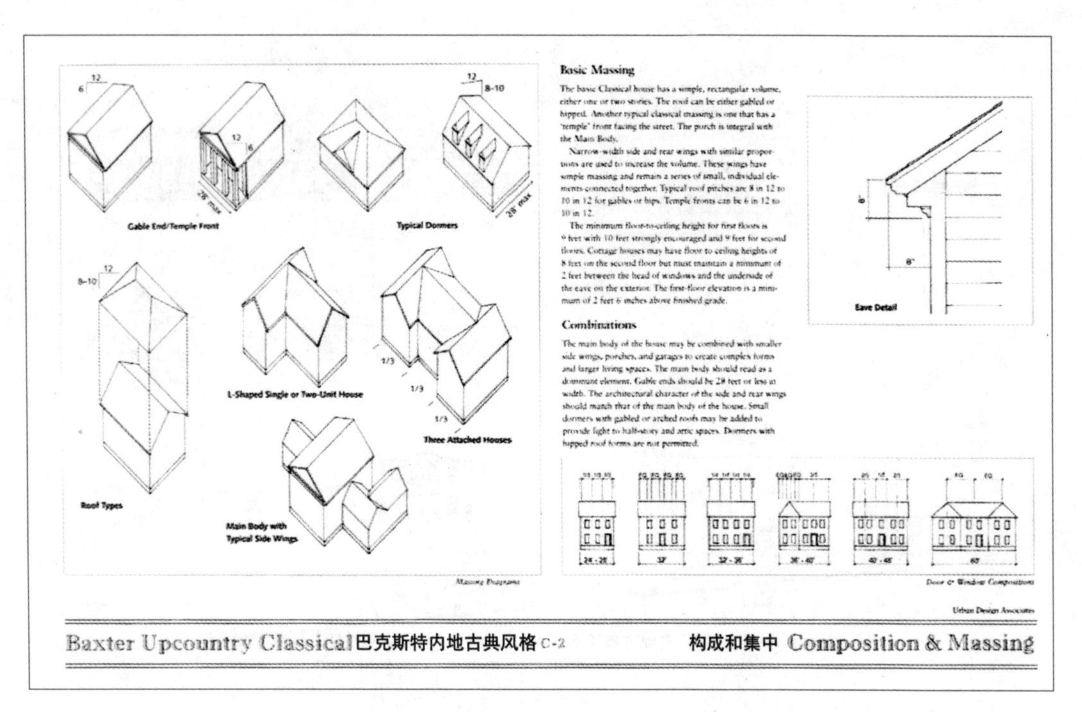

Gable End/Temple Front

Typical Dormers

Roof Types

L-Shaped Single or Two-Unit House

Three Attached Houses

Main Body with Typical Side Wings

Massing Diagrams

Basic Massing

The basic Classical house has a simple, rectangular volume, either one or two stories. The roof can be either gabled or hipped. Another typical classical massing is one that has a "temple" front facing the street. The porch is integral with the Main Body.

Narrow-width side and rear wings with similar proportions are used to increase the volume. These wings have simple massing and remain a series of small, individual elements connected together. Typical roof pitches are 8 in 12 to 10 in 12 for gables or hips. Temple fronts can be 6 in 12 to 10 in 12.

The minimum floor-to-ceiling height for first floors is 9 feet with 10 feet strongly encouraged and 9 feet for second floors. Cottage houses may have floor to ceiling heights of 8 feet on the second floor but must maintain a minimum of 2 feet between the head of windows and the underside of the eave on the exterior. The first-floor elevation is a minimum of 2 feet 6 inches above finished grade.

Eave Detail

Combinations

The main body of the house may be combined with smaller side wings, porches, and garages to create complex forms and larger living spaces. The main body should read as a dominant element. Gable ends should be 28 feet or less in width. The architectural character of the side and rear wings should match that of the main body of the house. Small dormers with gabled or arched roofs may be added to provide light to half-story and attic spaces. Dormers with hipped roof forms are not permitted.

Door & Window Compositions

Urban Design Associates

Baxter Upcountry Classical 巴克斯特内地古典风格 C-2　　构成和集中 Composition & Massing

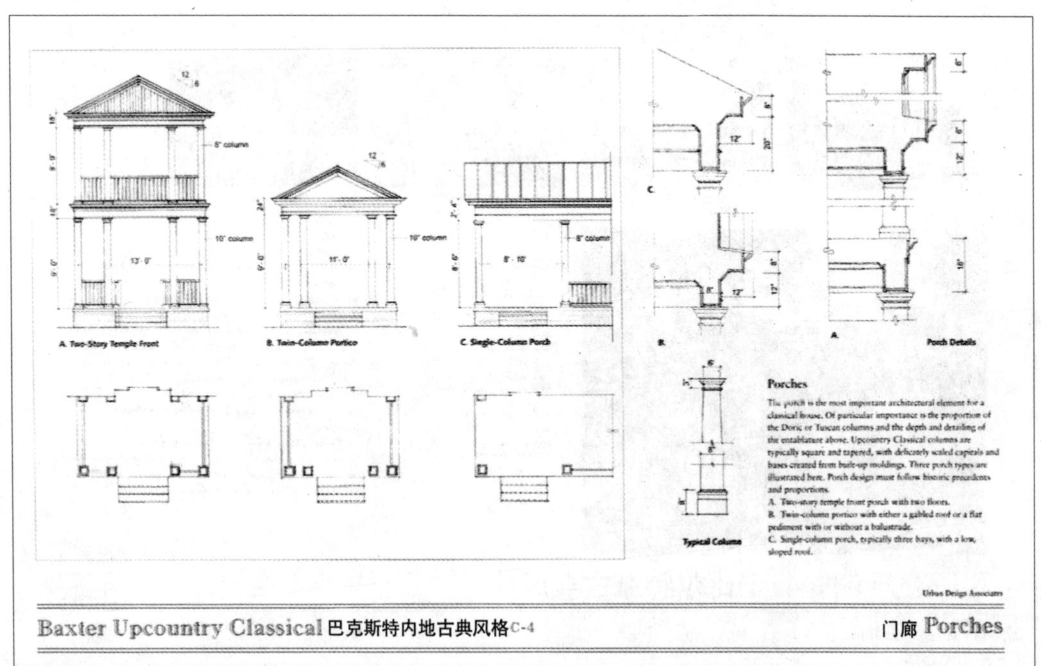

A. Two-Story Temple Front

B. Twin-Column Portico

C. Single-Column Porch

Porch Details

Typical Column

Porches

The porch is the most important architectural element for a classical house. Of particular importance is the proportion of the Doric or Tuscan columns and the depth and detailing of the entablature above. Upcountry Classical columns are typically square and tapered, with delicately scaled capitals and bases created from built-up moldings. Three porch types are illustrated here. Porch design must follow historic precedents and proportions.

A. Two-story temple front porch with two floors.

B. Twin-column portico with either a gabled roof or a flat pediment with or without a balustrade.

C. Single-column porch, typically three bays, with a low, sloped roof.

Urban Design Associates

Baxter Upcountry Classical 巴克斯特内地古典风格 C-4　　门廊 Porches

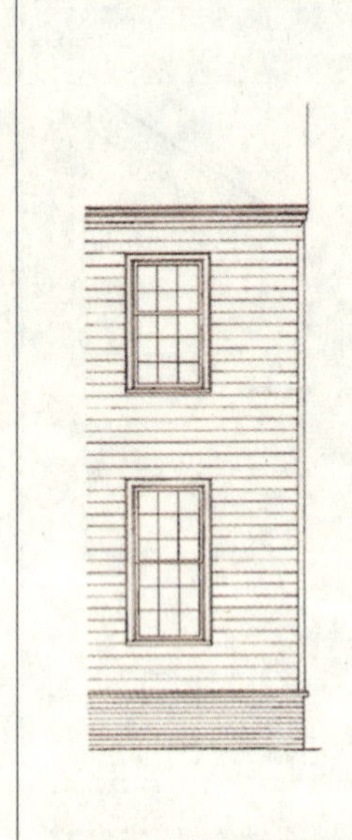

Materials

Shutters: Panel or louver, painted; encourage mounting as if operable. Width of shutter equals 50% of sash width.
Siding: Wood, composition board, fiber-cement board, or vinyl with Beveled, Shiplap or Beaded Lap profile; common brick in Flemish, Common, or English bond patterns and stucco with smooth/sand finish.
Roofing: Standing seam metal, slate, asphalt or fiberglass shingles.
Windows: Divided light sash with sill or GBG
Lintels: Brick jack arch, or stone/cast stone.
Trim: Wood, composition board, fiber-cement board and molded millwork for built-up sections; Corner boards 5/4 x 4.
Soffits and Porch Ceilings: GWB, plaster, T&G wood or composite. (Continuous perforated soffit materials not permitted.)
Gutters: Metal, half-round, or ogee.
Downspouts: Metal, round, or rectilinear.
Foundations: Common brick, stone/cast stone on front facade and side street facade when on corner lot. Stucco permitted on others.
Columns: Wood, fiberglass, or composite.
Railings: Straight or turned balusters.
Chimneys: Common brick, stone, siding or stucco.
Front Yard Fences: Painted white picket, or prefinished composite material such as PVC or vinyl and metal.
Rear Yard Fences: Painted wood, white composite material such as PVC or vinyl; prefinished metal.

Colors

Siding: White. Other colors to be selected from approved list. (Refer to Baxter's extensive color palette.)
Roof Shingles: Roof shingles to be typically black or dark grey. Standing seam may be black, dark green, or dark red.
Accent: White or a rich color to match trim.
Brick: All brick to be selected from Baxter palette.
Windows: Sashes and frames to be white.
Trim: All window, door, corner and eave trim to be selected from Baxter color palette. Columns should be white.
Gutters: White; terne metal or copper.
Downspouts: White; terne metal or copper.
Shutters: Black or dark green.
Fencing: Wood or PVC/vinyl is to be white; metal is to be black or dark green.

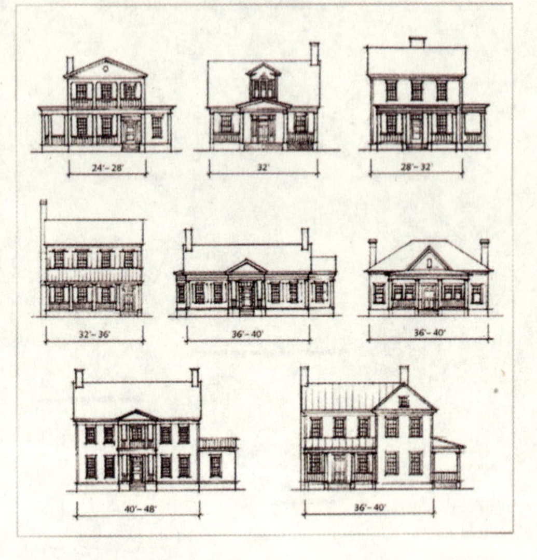

Urban Design Associates

Baxter Upcountry Classical 巴克斯特内地古典风格 C-5　　材料和性能 Materials & Possibilities

Two-story brick Classical house

Two-story Classical house with siding

One- and-one-half story Classical house with a court entry

Early twentieth century Classical house

Essential Elements of the Biltmore Classical

1. Simple volumes with one-story side wings and porches added to make more complex shapes.
2. Symmetrical composition of doors and windows.
3. Simplified versions of Classical details and columns, often with robust and exotic Classical orders such as Ionic and Corinthian used in the porch element.
4. Multi-pane windows that are wide in proportion usually with 6-over-6 or 9-over-9 pane patterns.

History and Character

THE BILTMORE CLASSICAL HOUSE is based on Federal and Classical Revival houses from the mid-19th century. The western North Carolina region has significant examples of houses from this period. The dominant Federal style was practiced by notable architects such as Robert Mills and Benjamin Latrobe; however many houses from this period were constructed using Pattern Books such as Asher Benjamin's *American Builder's Companion*. These houses are typically developed as simple, additive massing types with a dominant center pavilion, or Main Body, which can be one or two stories, and additive side wings, rear wings and pavilions. Palladian compositions became a principal organizing and proportioning reference for many houses from this period.

Biltmore Classical houses will be designed using these elements and massing principles. On steeply sloping sites found within Ducker Mountain, these houses will use raised terraces to create formal entry relationships particularly important on lots sloping upward from the street.

Biltmore Classical 比尔特摩古典风格

ARCHITECTURAL PATTERNS　　C　I

巴克斯特和达克尔山（比尔特摩）

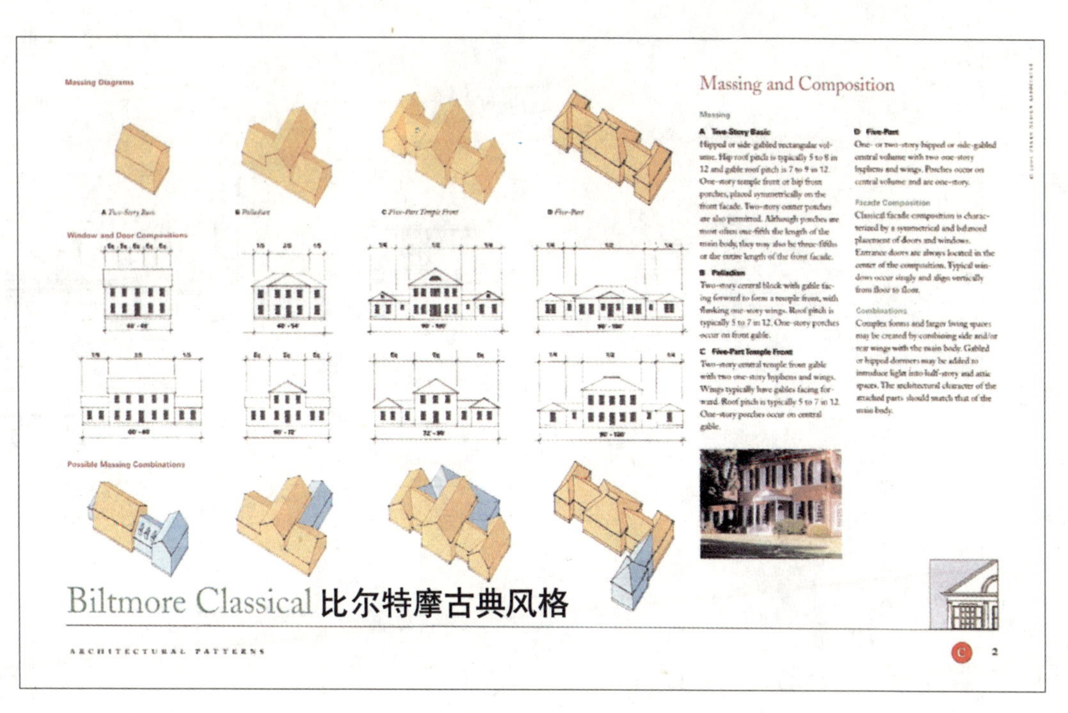

Massing Diagrams

A *Two-Story Basic*　B *Palladian*　C *Five-Part Temple Front*　D *Five-Part*

Window and Door Compositions

Possible Massing Combinations

Massing and Composition

Massing

A Two-Story Basic
Hipped or side-gabled rectangular volume. Hip roof pitch is typically 5 to 8 in 12 and gable roof pitch is 7 to 9 in 12. One-story temple front or hip front porches, placed symmetrically on the front facade. Two-story center porches are also permitted. Although porches are most often one-fifth the length of the main body, they may also be three-fifths or the entire length of the front facade.

B Palladian
Two-story central block with gable facing forward to form a temple front, with flanking one-story wings. Roof pitch is typically 5 to 7 in 12. One-story porches occur on front gable.

C Five-Part Temple Front
Two-story central temple front gable with two one-story hyphens and wings. Wings typically have gables facing forward. Roof pitch is typically 5 to 7 in 12. One-story porches occur on central gable.

D Five-Part
One- or two-story hipped or side-gabled central volume with two one-story hyphens and wings. Porches occur on central volume and are one-story.

Facade Composition

Classical facade composition is characterized by a symmetrical and balanced placement of doors and windows. Entrance doors are always located in the center of the composition. Typical windows occur singly and align vertically from floor to floor.

Combinations

Complex forms and larger living spaces may be created by combining side and/or rear wings with the main body. Gabled or hipped dormers may be added to introduce light into half-story and attic spaces. The architectural character of the attached parts should match that of the main body.

Biltmore Classical 比尔特摩古典风格

ARCHITECTURAL PATTERNS　C 2

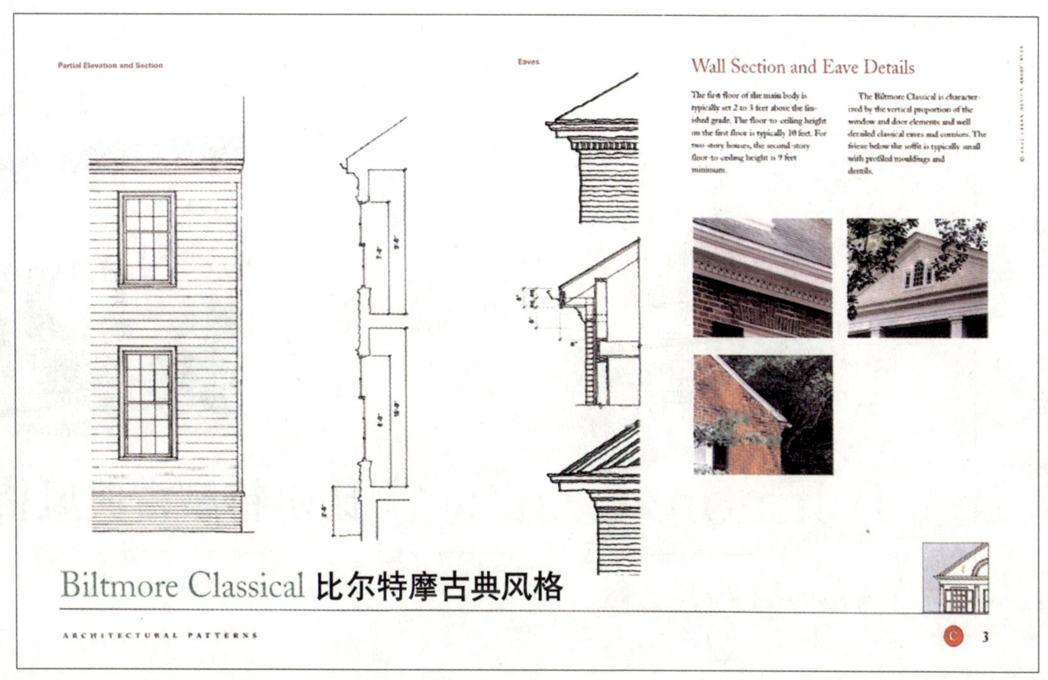

Partial Elevation and Section

Eaves

Wall Section and Eave Details

The first floor of the main body is typically set 2 to 3 feet above the finished grade. The floor-to-ceiling height on the first floor is typically 10 feet. For two-story houses, the second-story floor-to-ceiling height is 9 feet minimum.

The Biltmore Classical is characterized by the vertical proportion of the window and door elements and well detailed classical eaves and cornices. The frieze below the soffit is typically small with profiled mouldings and dentils.

Biltmore Classical 比尔特摩古典风格

ARCHITECTURAL PATTERNS　C 3

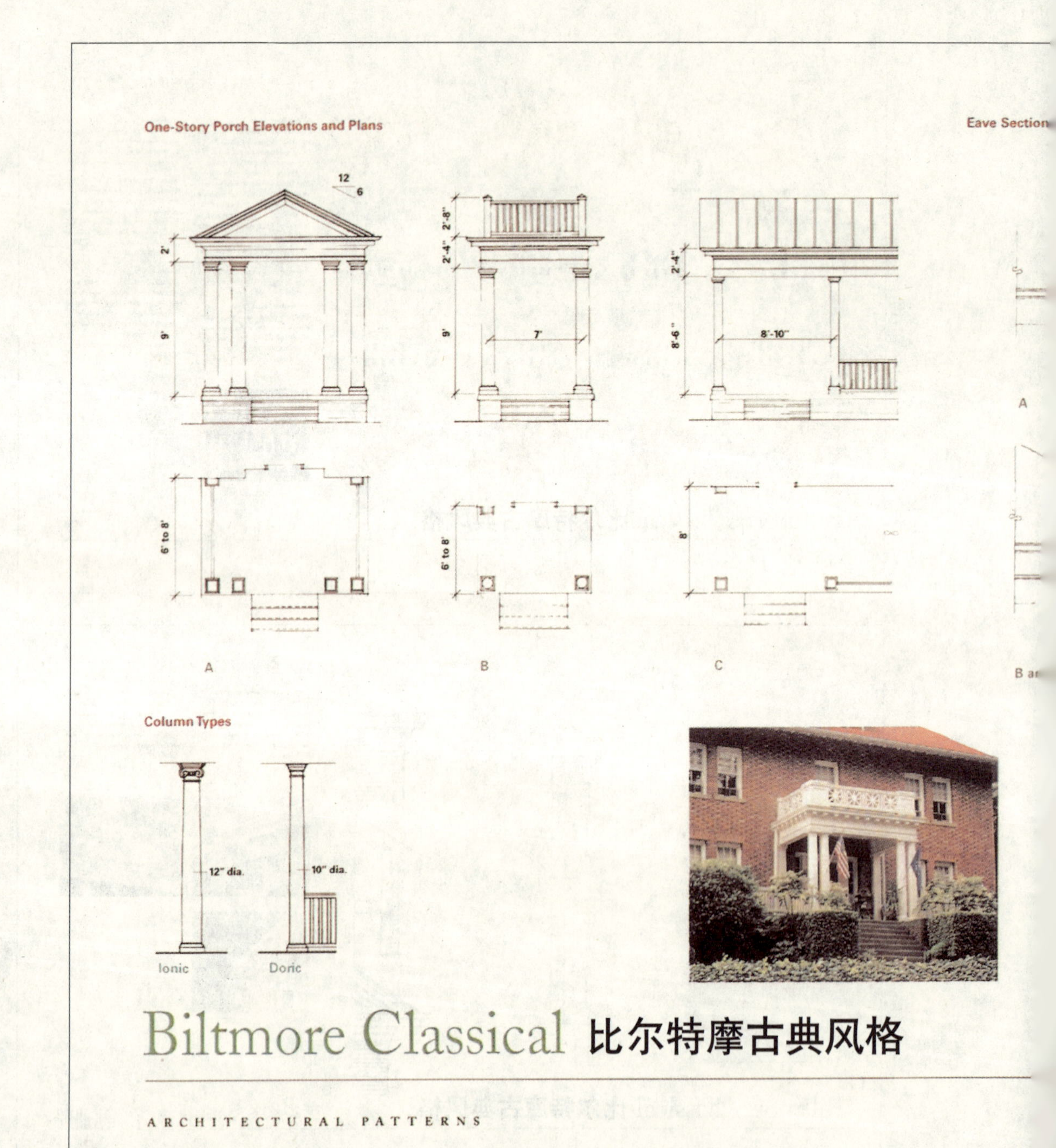
One-Story Porch Elevations and Plans
Eave Section
12
6
2'
9'
6' to 8'
2'-8"
2'-4"
7'
2'-4"
8'-6"
8'-10"
8'
A
B
C
Column Types
12" dia.
10" dia.
Ionic
Doric
Biltmore Classical
比尔特摩古典风格
ARCHITECTURAL PATTERNS

达克尔山（比尔特摩）

Typical Porch Locations

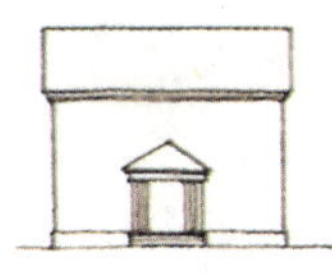
Two-story basic with portico

Two-story basic with full front porch

Two-story basic with three-fifths front porch

6

达克尔山（比尔特摩）

北卡罗来纳州阿什维尔附近

比尔特摩古典风格住宅的基础是从19世纪中期以来的联邦风格和古典复兴风格的住宅。北卡罗来纳州的西部地区拥有这一时期以来的重要的住宅实例。占主导地位的联邦风格建筑是由像罗伯特·米尔斯和本杰明·拉特罗布那样的著名建筑师设计建造的，然而，这一时期以来的很多住宅都是使用阿舍·本杰明的《美国建造者手册》等模式图册建造起来的。这些住宅通常发展成简洁的附加体块类型，带有处于支配地位的可以是一层或两层的中心空间或主体，和以及附加的侧翼部分、后翼部分与亭阁。对这一时期的很多住宅来说，帕拉第奥式构图成为进行组织和确定比例关系的主要参考。

比尔特摩古典风格住宅的设计将应用这些元素和集成原则。在北达克尔地区的陡峭坡地上，这些住宅将使用加高的台地来营造整齐的入口关系，这对于从街道向后升高的坡地场地来说是很重要的。

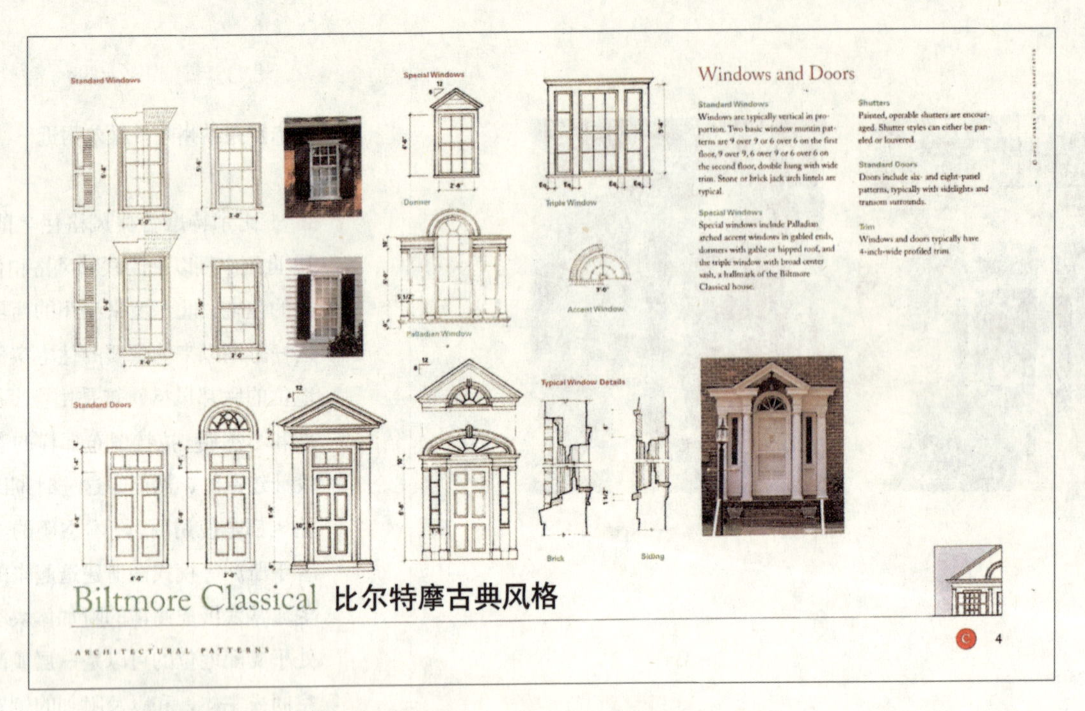

Standard Windows

Special Windows

Dormer

Triple Window

Palladian Window

Accent Window

Standard Doors

Typical Window Details

Brick

Siding

Windows and Doors

Standard Windows
Windows are typically vertical in proportion. Two basic window muntin patterns are 9 over 9 or 6 over 6 on the first floor, 9 over 9, 6 over 9 or 6 over 6 on the second floor, double hung with wide trim. Stone or brick jack arch lintels are typical.

Special Windows
Special windows include Palladian arched accent windows in gabled ends, dormers with gable or hipped roof, and the triple window with broad center sash, a hallmark of the Biltmore Classical house.

Shutters
Painted, operable shutters are encouraged. Shutter styles can either be paneled or louvered.

Standard Doors
Doors include six- and eight-panel patterns, typically with sidelights and transom surrounds.

Trim
Windows and doors typically have 4-inch-wide profiled trim.

Biltmore Classical 比尔特摩古典风格

ARCHITECTURAL PATTERNS

4

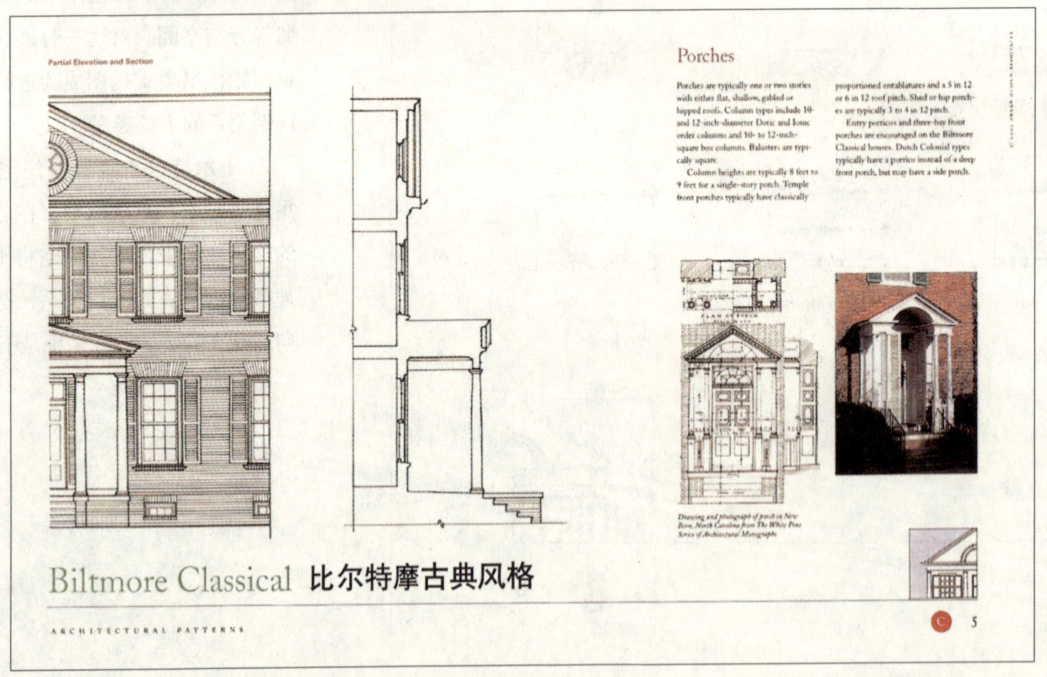

Partial Elevation and Section

Porches

Porches are typically one or two stories with either flat, shallow, gabled or hipped roofs. Column types include 10- and 12-inch-diameter Doric and Ionic order columns and 10- to 12-inch-square box columns. Balusters are typically square.

Column heights are typically 8 feet to 9 feet for a single-story porch. Temple front porches typically have classically proportioned entablatures and a 5 in 12 or 6 in 12 roof pitch. Shed or hip porches are typically 3 to 4 in 12 pitch.

Entry porticos and three-bay front porches are encouraged on the Biltmore Classical houses. Dutch Colonial types typically have a portico instead of a deep front porch, but may have a side porch.

Drawing and photograph of porch in New Bern, North Carolina from The White Pine Series of Architectural Monographs

Biltmore Classical 比尔特摩古典风格

ARCHITECTURAL PATTERNS

5

莱杰斯、亨茨维尔、阿拉巴马，古典建筑模式建成效果

Essential Elements of the Norfolk Classical Revival Style

1 Simple volumes with one-story side wings and porches added to make more complex shapes.

2 Symmetrical composition of doors and windows.

3 Simplified versions of Classical details and columns, often with robust and exotic Classical orders such as Ionic and Corinthian used in the porch element.

4 Multi-pane windows that are wide in proportion, usually with 6 over 6 or 9 over 9 pane patterns.

NORFOLK CLASSICAL REVIVAL

古典复兴风格

The Norfolk Classical Revival style is based on Federal and Classical Revival houses from the mid-nineteenth century. Norfolk and the Mid-Atlantic Region have significant examples of houses from this period.

The dominant Federal style was practiced by notable architects such as Robert Mills and Benjamin Latrobe, however many houses from this period were constructed using Pattern Books such as Asher Benjamin's American Builder's Companion.

Classical Revival houses are typically developed as simple, additive massing types with a dominant center pavilion, or Main Body, which can be one or two stories, and additive side wings, rear wings and pavilions. Palladian compositions became a principle organizing and proportioning reference for many houses from this period.

Partial Elevation and Wall Section

Massing & Composition 集中和构成

MASSING DIAGRAMS

A Broad Front

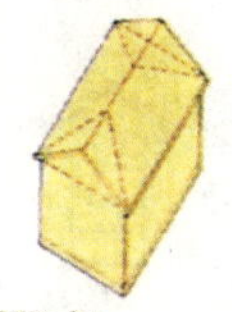

B Narrow Front

MASSING COMBINATIONS

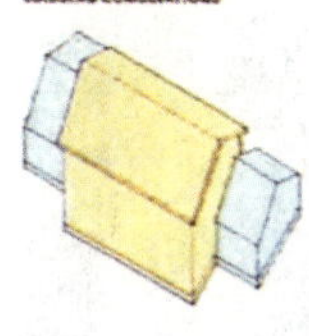

FACADE COMPOSITION DIAGRAMS

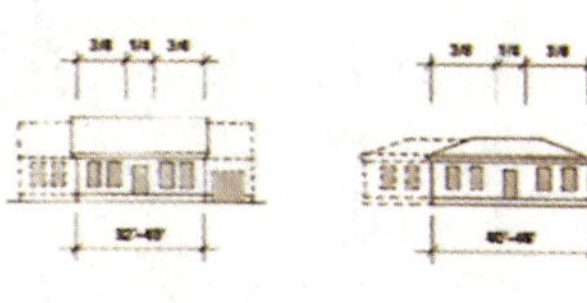

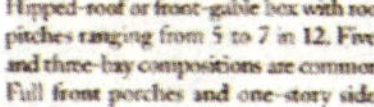

Massing

A BROAD FRONT

Hipped-roof or side-gable rectangular volume with roof pitches ranging from 5 to 7 in 12. One-story shed or hip roofed porches are often placed symmetrically on the front facade. One-story side wings often occur. Although porches are most often one-third or one-fifth the length of the main body, they may also be three-fifths or the entire length of the front facade.

B NARROW FRONT

Hipped-roof or front-gable box with roof pitches ranging from 5 to 7 in 12. Five- and three-bay compositions are common. Full front porches and one-story side-wings are common to this massing type.

Facade Composition

The Norfolk Classical Revival facade composition is characterized by a symmetrical and balanced placement of doors and windows. Entrance doors are always located in the center of the composition. Typical windows occur singly and align vertically from floor to floor.

Combinations

Complex forms and larger living spaces may be created by combining side and/or rear wings with the main body. Gabled or hipped dormers may be added to introduce light into half-story and attic spaces. The architectural character of the attached parts should match that of the main body.

Wall Section & Eave Details

The first floor of the main body is typically set 2 to 3 feet above the finished grade. The floor-to-ceiling height on the first floor is typically 10 feet. For two-story houses, the second-story floor-to-ceiling height is 9 feet minimum.

The Norfolk Classical Revival style is characterized by the vertical proportion of the window and door elements and well-detailed Classical eaves and cornices. The frieze below the soffit is typically small with profiled moldings and dentils.

TYPICAL EAVE DETAILS

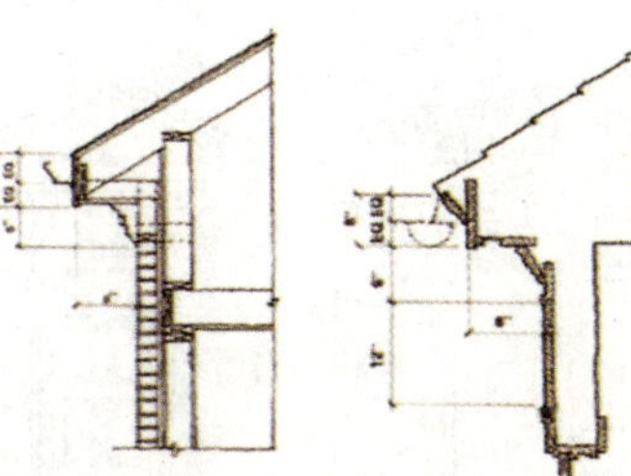

Windows & Doors 窗和门

STANDARD WINDOWS

Second floor

First floor

SPECIAL WINDOWS

Palladian window

Dormer

Accent

Triple window

Standard Windows

Windows are typically vertical in proportion. Two basic window muntin patterns are 9 over 9 or 6 over 6 on the first floor, 9 over 9, 6 over 9 or 6 over 6 on the second floor, double hung with wide trim. Stone or brick jack arch lintels are typical.

Special Windows

Special windows include Palladian arched accent windows in gabled ends, dormers with gable or hipped roof, and the triple window with broad center sash, a hallmark of the Norfolk Classical Revival style house.

Shutters

Shutters should be sized and mounted to appear operable. Shutter styles can either be paneled or louvered.

Doors

Doors include six- and eight-panel patterns, typically with sidelights and transom surrounds.

Trim

Windows and doors typically have 4-inch-wide profiled trim.

TYPICAL WINDOW DETAILS

Brick

Siding

DOORS

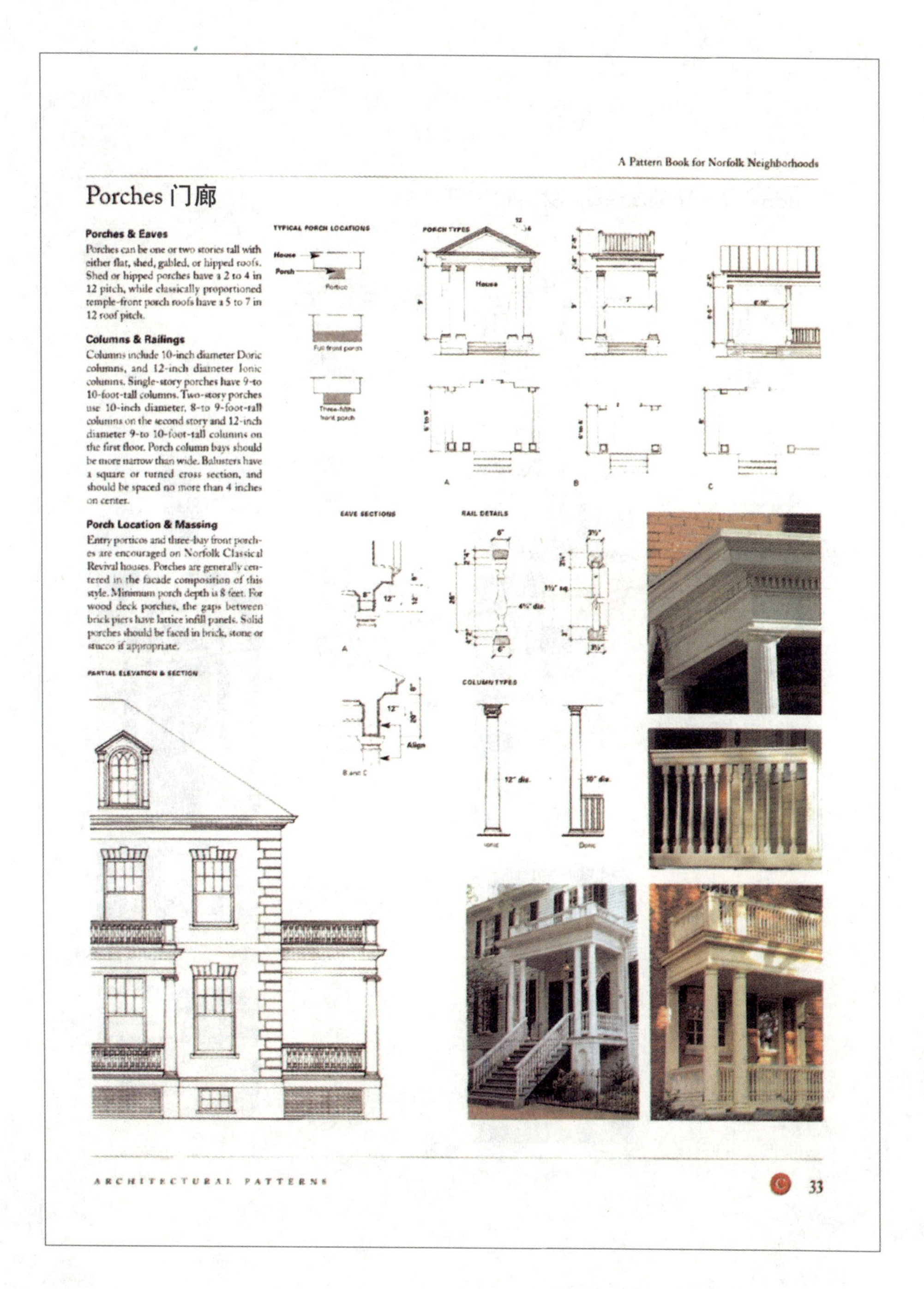

Porches 门廊

Porches & Eaves

Porches can be one or two stories tall with either flat, shed, gabled, or hipped roofs. Shed or hipped porches have a 2 to 4 in 12 pitch, while classically proportioned temple-front porch roofs have a 5 to 7 in 12 roof pitch.

Columns & Railings

Columns include 10-inch diameter Doric columns, and 12-inch diameter Ionic columns. Single-story porches have 9-to 10-foot-tall columns. Two-story porches use 10-inch diameter, 8-to 9-foot-tall columns on the second story and 12-inch diameter 9-to 10-foot-tall columns on the first floor. Porch column bays should be more narrow than wide. Balusters have a square or turned cross section, and should be spaced no more than 4 inches on center.

Porch Location & Massing

Entry porticos and three-bay front porches are encouraged on Norfolk Classical Revival houses. Porches are generally centered in the facade composition of this style. Minimum porch depth is 8 feet. For wood deck porches, the gaps between brick piers have lattice infill panels. Solid porches should be faced in brick, stone or stucco if appropriate.

Materials & Applications 材料和应用

Roofing

- Slate (including manufactured slate products), laminated asphalt or composition shingles with a slate pattern, flat clay tile, or painted metal standing seam or 5-V crimp panels

Soffits

- Smooth finish composition board, tongue-and-groove wood boards, or fiber-cement panels

Gutters & Downspouts

- Half-round or ogee profile gutters with round or rectangular downspouts in copper, painted or prefinished metal

Cladding

- Sand-molded or smooth-finish brick in Common, English or Flemish bond patterns
- Smooth-finish wood or fiber-cement lap siding, 6 to 8 inches wide
- Light sand-finish stucco

Foundations & Chimneys

- Brick, stucco or stone veneer

Columns

- Architecturally correct Classical proportions and details in wood, fiberglass, cast stone, or composite material

Railings

- Milled wood top and bottom rails with square or turned balusters; square balusters in Chippendale patterns
- Wrought iron or solid bar stock decorative metal

Porch Ceilings

- Plaster, tongue-and-groove wood or composite boards, or beaded-profile plywood

Windows

- Painted wood or solid cellular PVC, or clad wood or vinyl with brick veneer only; true divided light or simulated divided light (SDL) sash with traditional exterior muntin profile (7/8 inch wide)

Trim

- Wood, composite, cellular PVC or polyurethane millwork; stucco, stone or cast stone

Doors

- Wood, fiberglass or steel with traditional stile-and-rail proportions and raised panel profiles, painted or stained

Shutters

- Wood or composite, sized to match window sash and mounted with hardware to appear operable

Front Yard Fences

- Wood picket or wood, wrought iron or solid bar stock metal picket with brick or stucco finish masonry piers

Lighting

- Porch pendant or wall-mounted carriage lantern

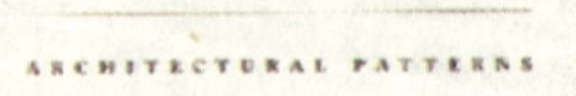

Gallery of Examples 楼座范例

欧洲浪漫主义风格

比较分析

在 19 世纪后半叶直到 1940 年代，欧洲浪漫主义风格住宅在美国变得非常流行。这是另一种折衷的类别，并具有非常广泛的地区性处理方式。这种类型的很多住宅都是根据商品目录为中产阶级建造的，很多更为精致的住宅则是由建筑师为富裕客户设计建造的。欧洲浪漫主义风格被认为是这一时期的适用庄园形象。在美国，在较大的庄园住宅和一些城市邻里中可以看到作为一种具有异国情调的建造者住宅，法国的影响是很突出的。运用英国语汇的乡村别墅和都铎风格的住宅类型，比任何其他的欧洲运动都更多地影响了美国住宅设计的这一趋势。风格变体，例如意大利文艺复兴风格，出现在同一时期，但更为拘泥形式，并且与英国影响的欧洲浪漫主义风格相比更接近新古典主义风格。在 1920 年代后期和 1930 年代的短暂期间在商业建造者市场中具有显著的影响。很多邻里的特征表现为小型的砖砌乡村别墅，带有正面三角山墙、不对称的入口形式、半木构和以使用灰泥为特征的细部设计。这种风格被表达为框架和窗子、木瓦或构成壁板。通常情况下，门廊是较为次要的元素，而带有简单托架的沉重木柱，则被用来补充住宅体块并表现有顶的入口或作为主要屋顶形式外延的小型游廊。坡度很大的正面三角山墙和体块、装饰性烟囱是这一时期很多住宅实例的特点。

欧洲浪漫主义风格的范例册页，是从伊格尔帕克（北卡罗来纳州贝尔蒙特市）、弗吉尼亚州诺福克市和达克尔山（北卡罗来纳州阿什维尔附近）的模式图则中摘录的。

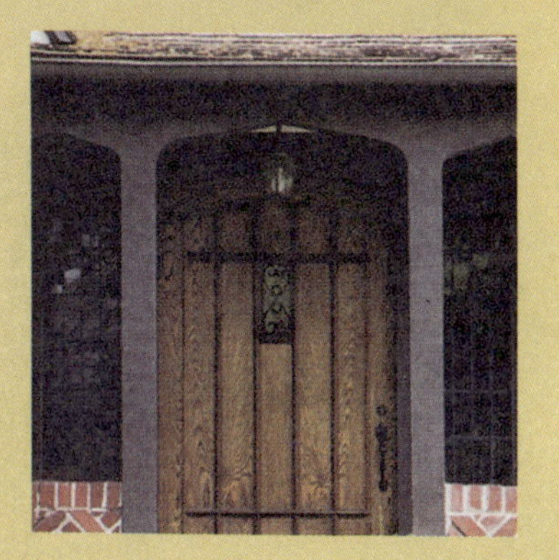

范例图片

An English house in Charlotte

A small Romantic house in Belmont

Half-timbering on an English

Essential Elements of Belmont European Romantic

1. Steep roof pitches with dormers.
2. Balanced window and door locations.
3. Vertical windows in groupings.
4. Large, simple roof planes.
5. Porches typically notched out, or extended roof
6. Roof lines extending below windo at second floor, and to top of wind at first floor.
7. Simple detailing.
8. Shallow overhangs.
9. Massive chimneys.

Eagle Park European Romantic

ARCHITECTURAL PATTERNS

伊格尔帕克

European Romantic house

History and Character

THE EAGLE PARK EUROPEAN ROMANTIC STYLE is based on the early-twentieth-century interpretations of English architecture by American architects and builders. The source for the design comes from Medieval English cottages, manor houses, and rural village vernacular houses. The American version is normally a house with simple volumes, and often front facing gables. Gables have steeply pitched roofs between 8 in 12 and 20 in 12. Half-timbering, shingles, and horizontal siding are often used as infill in gables. The decorative half-timbering may occur at the entire second story. Gable, hip, and shed dormers are dominant features of the style.

Windows include single and paired double-hung types mixed with vertically proportioned casement windows arranged in groups of two to five. There are relatively few windows in the facade; the general impression is a solid mass with small openings.

Chimneys are often the dominant element in the massing of the house. These massive chimneys may be finished in brick or plaster. They feature simple detailing and chimney pots.

伊格尔帕克欧洲浪漫主义风格

13

伊格尔帕克

北卡罗来纳州贝尔蒙特市

北卡罗来纳州贝尔蒙特市伊格尔帕克邻里的欧洲浪漫主义风格，吸收了1920年代在美国非常流行的英国乡村别墅的样式。这些设计既出现在平面图则上，也出现在像西尔斯和阿拉丁这样的公司所提供的住宅销售目录中。很多这样的住宅建成了，形成了邻里街道的完整肌理——其中大多数是砖砌的单层乡村别墅，带有坡度很陡的正面三角山墙，而结实的、浑然一体的烟囱常常附加于简洁的侧面山墙的住宅形式上。尽管这种借鉴了带有经提炼的都铎式细部设计的考斯特伍德乡村别墅的风格流行了8–10年，但这一朴素的住宅类型并没有形成广受喜爱的独特形象、成功和乡村特色。为这一风格制定的伊格尔帕克模式图则，所关注的是在整个南部非常常见的结构和木瓦或壁板的样式。贝尔蒙特拥有这种风格和这一时期的一些很好的实例。模式图则关于欧洲浪漫主义风格的部分，调整并列出了一系列体块类型，其中所包括的前置门廊成为住宅的基本元素。这种对通常并不将前置门廊作为标准住宅组成部分的传统住宅类型所进行的改变，尝试将该风格置于一套将门廊作为标准部件的图卡之中。

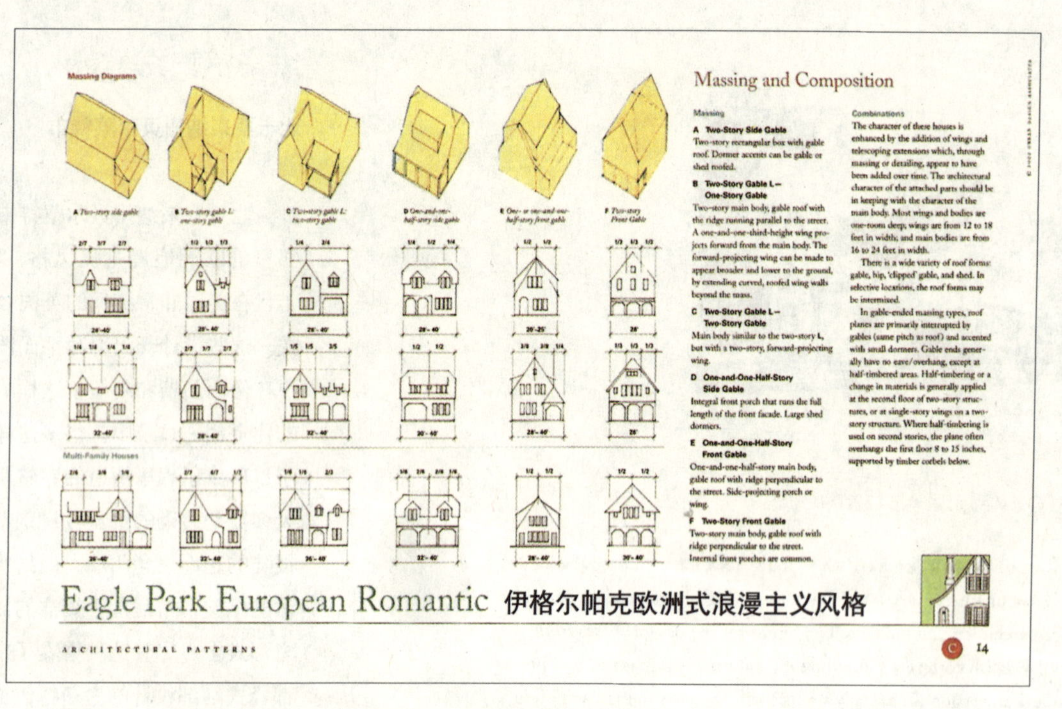

Massing Diagrams

Multi-Family Houses

Massing and Composition

Massing

A　Two-Story Side Gable
Two-story rectangular box with gable roof. Dormer accents can be gable or shed roofed.

B　Two-Story Gable L – One-Story Gable
Two-story main body, gable roof with the ridge running parallel to the street. A one-and-one-third-height wing projects forward from the main body. The forward-projecting wing can be made to appear broader and lower to the ground, by extending curved, roofed wing walls beyond the mass.

C　Two-Story Gable L – Two-Story Gable
Main body similar to the two-story L, but with a two-story, forward-projecting wing.

D　One-and-One-Half-Story Side Gable
Integral front porch that runs the full length of the front facade. Large shed dormers.

E　One-and-One-Half-Story Front Gable
One-and-one-half-story main body, gable roof with ridge perpendicular to the street. Side-projecting porch or wing.

F　Two-Story Front Gable
Two-story main body, gable roof with ridge perpendicular to the street. Internal front porches are common.

Combinations

The character of these houses is enhanced by the addition of wings and telescoping extensions which, through massing or detailing, appear to have been added over time. The architectural character of the attached parts should be in keeping with the character of the main body. Most wings and bodies are one-room deep; wings are from 12 to 18 feet in width; and main bodies are from 16 to 24 feet in width.

There is a wide variety of roof forms: gable, hip, 'clipped' gable, and shed. In selective locations, the roof forms may be intermixed.

In gable-ended massing types, roof planes are primarily interrupted by gables (same pitch as roof) and accented with small dormers. Gable ends generally have no eave/overhang, except at half-timbered areas. Half-timbering or a change in materials is generally applied at the second floor of two-story structures, or at single-story wings on a two-story structure. Where half-timbering is used on second stories, the plane often overhangs the first floor 8 to 15 inches, supported by timber corbels below.

Eagle Park European Romantic　伊格尔帕克欧洲式浪漫主义风格

ARCHITECTURAL PATTERNS　C 14

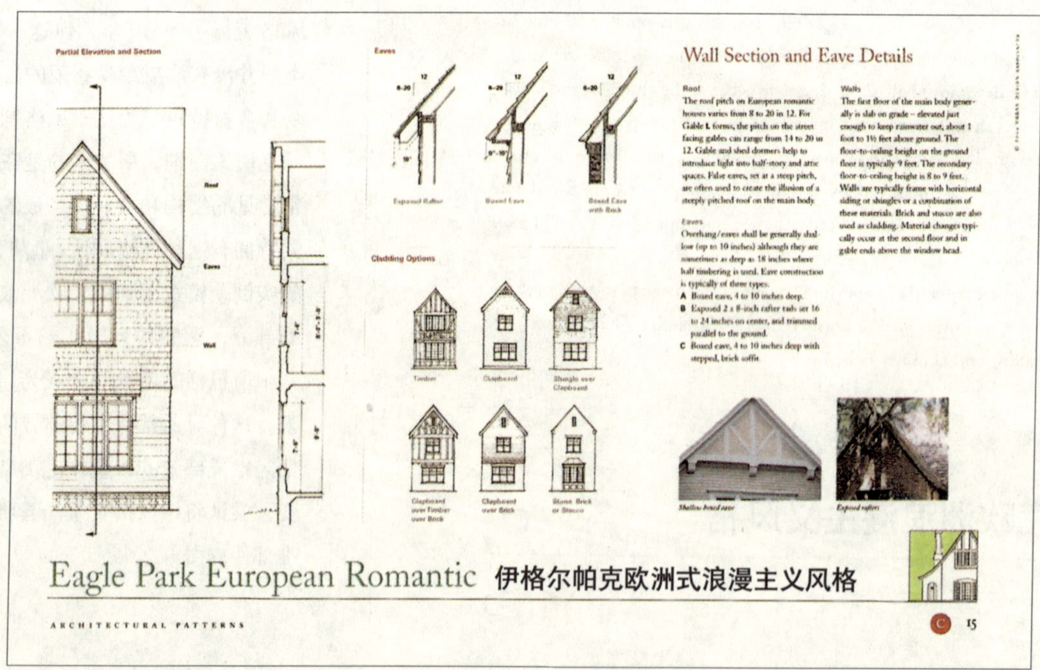

Partial Elevation and Section

Eaves

Cladding Options

Wall Section and Eave Details

Roof

The roof pitch on European romantic houses varies from 8 to 20 in 12. For Gable L forms, the pitch on the street facing gables can range from 14 to 20 in 12. Gable and shed dormers help to introduce light into half-story and attic spaces. False eaves, set at a steep pitch, are often used to create the illusion of a steeply pitched roof on the main body.

Eaves

Overhang/eaves shall be generally shallow (up to 10 inches) although they are sometimes as deep as 18 inches where half timbering is used. Eave construction is typically of three types.

A Board eave, 4 to 10 inches deep.
B Exposed 2 x 8-inch rafter tails set 16 to 24 inches on center, and trimmed parallel to the ground.
C Boxed eave, 4 to 10 inches deep with stepped, brick soffit.

Walls

The first floor of the main body generally is slab on grade – elevated just enough to keep rainwater out, about 1 foot to 1½ feet above ground. The floor-to-ceiling height on the ground floor is typically 9 feet. The secondary floor-to-ceiling height is 8 to 9 feet. Walls are typically frame with horizontal siding or shingles or a combination of these materials. Brick and stucco are also used as cladding. Material changes typically occur at the second floor and in gable ends above the window head.

Eagle Park European Romantic　伊格尔帕克欧洲式浪漫主义风格

ARCHITECTURAL PATTERNS　C 15

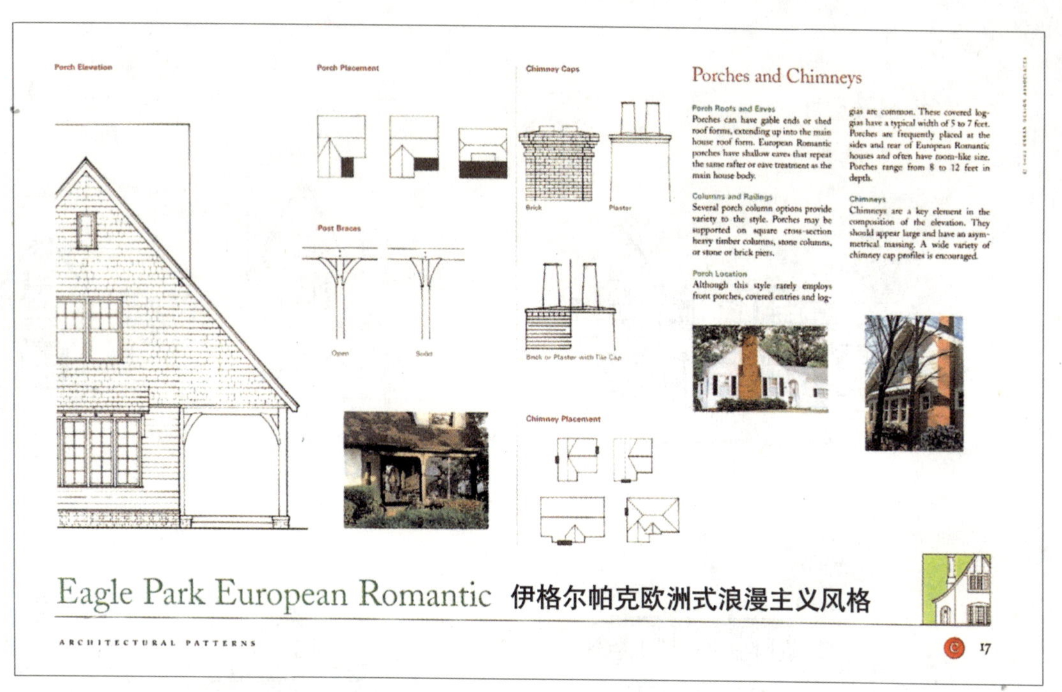

Porches and Chimneys

Porch Roofs and Eaves

Porches can have gable ends or shed roof forms, extending up into the main house roof form. European Romantic porches have shallow eaves that repeat the same rafter or eave treatment as the main house body.

Columns and Railings

Several porch column options provide variety to the style. Porches may be supported on square cross-section heavy timber columns, stone columns, or stone or brick piers.

Porch Location

Although this style rarely employs front porches, covered entries and loggias are common. These covered loggias have a typical width of 5 to 7 feet. Porches are frequently placed at the sides and rear of European Romantic houses and often have room-like size. Porches range from 8 to 12 feet in depth.

Chimneys

Chimneys are a key element in the composition of the elevation. They should appear large and have an asymmetrical massing. A wide variety of chimney cap profiles is encouraged.

Eagle Park European Romantic 伊格尔帕克欧洲式浪漫主义风格

ARCHITECTURAL PATTERNS

17

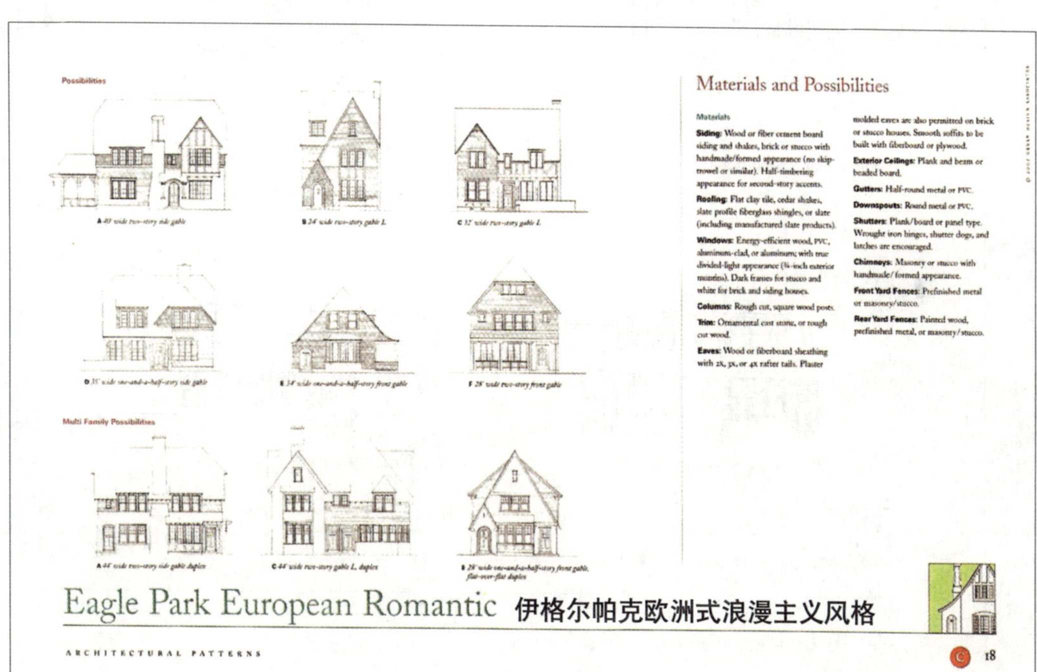

Materials and Possibilities

Materials

Siding: Wood or fiber cement board siding and shakes, brick or stucco with handmade/formed appearance (no skip-trowel or similar). Half-timbering appearance for second-story accents.

Roofing: Flat clay tile, cedar shakes, slate profile fiberglass shingles, or slate (including manufactured slate products).

Windows: Energy-efficient wood, PVC, aluminum-clad, or aluminum; with true divided-light appearance (⅞-inch exterior muntins). Dark frames for stucco and white for brick and siding houses.

Columns: Rough cut, square wood posts.

Trim: Ornamental cast stone, or rough cut wood.

Eaves: Wood or fiberboard sheathing with 2x, 3x, or 4x rafter tails. Plaster molded eaves are also permitted on brick or stucco houses. Smooth soffits to be built with fiberboard or plywood.

Exterior Ceilings: Plank and beam or beaded board.

Gutters: Half-round metal or PVC.

Downspouts: Round metal or PVC.

Shutters: Plank/board or panel type. Wrought iron hinges, shutter dogs, and latches are encouraged.

Chimneys: Masonry or stucco with handmade/ formed appearance.

Front Yard Fences: Prefinished metal or masonry/stucco.

Rear Yard Fences: Painted wood, prefinished metal, or masonry/stucco.

Eagle Park European Romantic 伊格尔帕克欧洲式浪漫主义风格

ARCHITECTURAL PATTERNS

18

Possibilities

A 40' wide two-story side gable

B 24' wide two-story gable L

C 32' wide two-story gable L

D 35' wide one-and-a-half-story side gable

E 34' wide one-and-a-half-story front gable

F 28' wide two-story front gable

Multi Family Possibilities

A 44' wide two-story side gable duplex

C 44' wide two-story gable L, duplex

E 28' wide one-and-a-half-story front ga
flat-over-flat duplex

Eagle Park European Romantic

ARCHITECTURAL PATTERNS

伊格尔帕克

Materials and Possibilities 材料和性能

© 2002 URBAN DESIGN ASSOCIATES

Materials

Siding: Wood or fiber cement board siding and shakes, brick or stucco with handmade/formed appearance (no skip-trowel or similar). Half-timbering appearance for second-story accents.

Roofing: Flat clay tile, cedar shakes, slate profile fiberglass shingles, or slate (including manufactured slate products).

Windows: Energy-efficient wood, PVC, aluminum-clad, or aluminum; with true divided-light appearance (⅞-inch exterior muntins). Dark frames for stucco and white for brick and siding houses.

Columns: Rough cut, square wood posts.

Trim: Ornamental cast stone, or rough cut wood.

Eaves: Wood or fiberboard sheathing with 2x, 3x, or 4x rafter tails. Plaster molded eaves are also permitted on brick or stucco houses. Smooth soffits to be built with fiberboard or plywood.

Exterior Ceilings: Plank and beam or beaded board.

Gutters: Half-round metal or PVC.

Downspouts: Round metal or PVC.

Shutters: Plank/board or panel type. Wrought iron hinges, shutter dogs, and latches are encouraged.

Chimneys: Masonry or stucco with handmade/formed appearance.

Front Yard Fences: Prefinished metal or masonry/stucco.

Rear Yard Fences: Painted wood, prefinished metal, or masonry/stucco.

伊格尔帕克欧洲式浪漫主义风格

C 18

伊格尔帕克

北卡罗来纳州贝尔蒙特市

伊格尔帕克可能建筑形式的册页，用图片说明了体块类型的设计范例和在模式图则中所展示的构图选择。可能建筑形式通过剖面图展示了窗子、门、门廊和屋檐要素的细部设计。与1920年代晚期以来的多数欧洲浪漫主义住宅不同，这些新住宅的特征表现为前置的和侧面布置的门廊——在以前的欧洲浪漫主义设计的复兴中常常被遗忘的一种要素。此外，这些住宅主要采用壁板或壁板与砖材结合使用。这反映了这种风格的地区适用性，在大多数地区它主要是用砖建造的。这一册页图解说明了分离式和联排式的住宅类型。

Essential Elements of the Norfolk European Romantic Style

1. Large, steeply-pitched roof planes with dormers and shallow overhangs.
2. Roof lines extend below windows at second floor, and top of window at first floor.
3. Broad expanses of wall with a limited number of deep-set openings.
4. Asymmetrical window and door locations.
5. Vertically proportioned windows in groups.

NORFOLK EUROPEAN ROMANTIC

欧洲浪漫主义风格

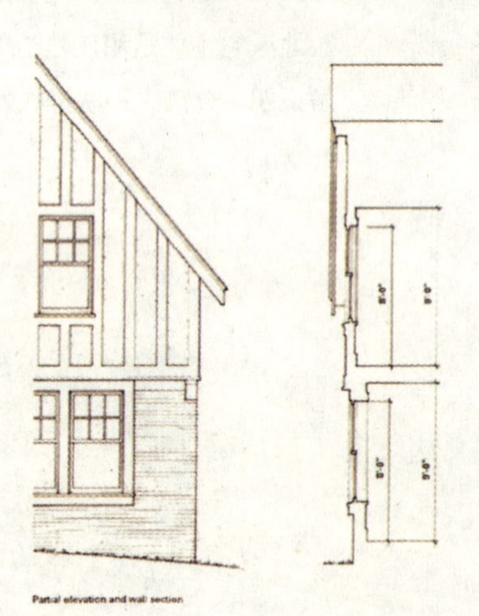

Partial elevation and wall section

The European Romantic style is based on the early twentieth century interpretations of English architecture by American architects and builders. The source for design comes from medieval English cottages, manor houses, and rural village vernacular houses. The American interpretations include houses with simple volumes often with front-facing gables that have steeply pitched roofs between 12 in 12 and 16 in 12. Gable, hip, and shed dormers are a dominant feature of the style. There is often a mix of exterior materials including stone, plaster, or brick. Half-timbering and horizontal siding are often used as infill in gables.

Chimneys act as principal forms for the massing of the house. These are usually very massive, with simple detailing and chimney pots. Decorative half-timbering in the gables is common and can occur on the entire second story or in the upper gables. Windows are typically casements, vertical in proportion and arranged in groups.

Massing & Composition 集中和构成

MASSING DIAGRAMS

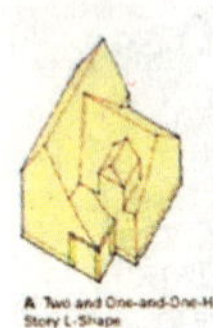

A Two and One-and-One-Half-Story L-Shape

B Two and One-Story L-Shape

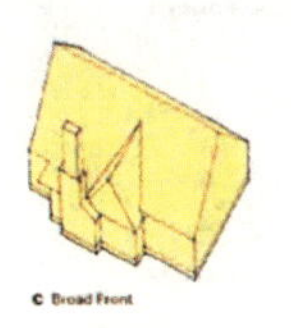

C Broad Front

D One-Story Gable L

FACADE COMPOSITION DIAGRAMS

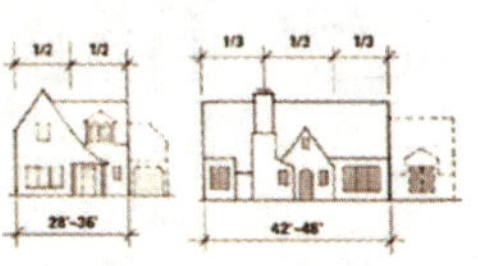

MASSING COMBINATIONS

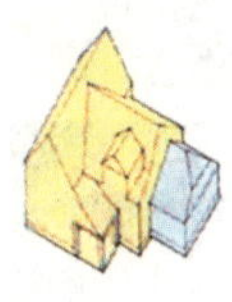

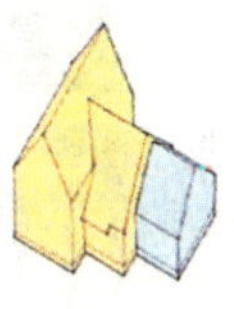

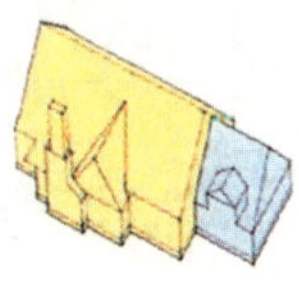

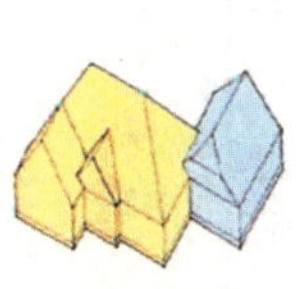

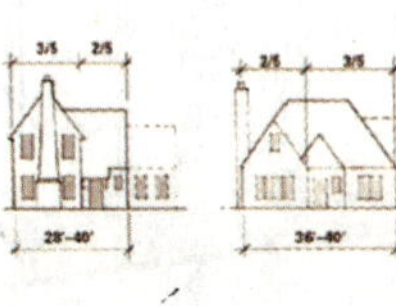

Massing

A TWO- & ONE-AND-ONE-HALF-STORY L-SHAPE

L-shaped plan with a two-story front-facing gable paired with a one-and-one-half story roof expression parallel to the street. The roof of the front-facing gable slides down to provide a covered entry. Dormers can have gable or shed roofs.

B TWO- & ONE-STORY L-SHAPE

L-shaped plan with a two-story front facing gable paired with a one-story roof expression parallel to the street. The one-story roof may curve out to provide a covered shelter over the door.

C BROAD FRONT

Rectangular shaped plan with a one, one-and-one-half, or two-story expression. A small gable or two may project to provide visual relief and to provide balance to large chimneys and other architectural elements.

D ONE-STORY GABLE L

Rectangular volume with hipped roof with a front facing gabled wing. Mass may have a one- or one-and-one-half-story expression. A series of nested gables may provide balance to chimneys and other architectural elements.

Facade Composition

European Romantic facade composition is characterized by an asymmetrical and balanced placement of doors and windows. Grouped double-hung windows are common. Front doors are generally located at the center of the composition, especially in wide houses. There is typically a material change from the first to the second floor.

Roof

The roof pitch on European Romantic houses varies from 12 to 20 in 12. For Gable L forms, the pitch on the street-facing gables ranges from 14 to 20 in 12. Gable and shed dormers help to introduce light into half-story and attic spaces. False eaves, set at a steep pitch, are often used to create the illusion of a steeply pitched roof on the main body.

Eaves

Overhangs tend to be generally shallow (up to 10 inches) although they are sometimes as deep as 18 inches where half timbering is used. Eave construction is typically of three types:

A Boxed eave, 4 to 10 inches deep.

B Exposed 2 x 8-inch rafter tails set 16 to 24 inches on center, and trimmed parallel to the ground.

C Bricked eave, 4 to 10 inches deep with stepped, brick soffit.

Wall Section & Eave Details

The first floor is typically set 12 to 18 inches above finished grade. The floor-to-ceiling height on the ground floor is typically 9 feet. The secondary floor-to-ceiling height is 8 to 9 feet. Walls are typically framed with horizontal siding or shingles or a combination of these materials. Brick and stucco are also used as cladding. Material changes typically occur at the second floor and in gable ends above the window head. Clapboard or shake cladding materials should never come within 8 inches of finished grade; only durable materials like brick, stone, and stucco may come into direct contact with the soil.

TYPICAL EAVE DETAILS

Boxed eave

Exposed rafter

Bricked eave

Windows & Doors 窗和门

Standard Windows

European Romantic is characterized by tall, relatively narrow windows typically in groups of two, three, four, or five windows separated by posts or mullions. The height of windows should diminish with each succeeding story. All windows have a divided-light appearance. Windows surrounded by stucco should be deeply recessed from the facade to create the illusion of thick walls.

Special Windows

Special windows include bay windows, oriel windows, dormer windows, and small accent windows. Bays should be constructed of a light material, such as wood, and should be composed of casement windows with a divided-light appearance. Dormers should be shed or gabled type, typically with paired casement windows. Small decorative accent windows are encouraged, especially in powder rooms, closets and halls.

Shutters

Plank/board or panel-style shutters are encouraged as an accent. Wrought iron lift-off hinges, shutter dogs and latches are also encouraged. Shutters are not permitted on half-timbered areas.

Doors

Plank/board or panel-style single doors with a round or arched top and wrought iron accents are preferred. As with windows, doors should be recessed as deeply as possible. Detailing around doors is encouraged.

Trim

Four-to six-inch wood trim is typical for windows and doors when siding is used. Ornamental cast stone or wood lintels and sills are encouraged for masonry window and door openings. Stucco should typically return to the window – eliminating the need for most trim. Half-timbering functions as trim in these areas.

Porches & Chimneys 门廊和烟囱

Porches

Although porches are less common on European Romantic houses than other styles, porches and carriage porches were common on larger houses. They should feature post-and-beam construction, shed roofs and rough-sawn clapboard siding. Arched braces between posts and beams are encouraged. The covered patios and loggias may be constructed of either post-and-beam or masonry.

Porch Roofs & Eaves

Porches can have gable ends or shed roof forms, extending up into the main house roof form. European Romantic porches have shallow eaves that repeat the same rafter or eave treatment as the main house body.

Columns & Railings

Several porch column options provide variety to the style. Porches may be supported on square cross-section heavy timber columns, stone columns, or stone or brick piers.

Porch Location & Massing

Although this style rarely employs front porches, covered entries and loggias are common. These covered loggias have a typical width of 5 to 7 feet. Porches are frequently placed at the sides and rear of European Romantic houses and often have room-like size. Porches range from 8 to 12 feet in depth.

Chimneys

Chimneys are a key element in the composition of the elevation. They should appear large and have an asymmetrical massing. A wide variety of chimney cap profiles is encouraged.

CHIMNEY CAPS

Plaster

Brick

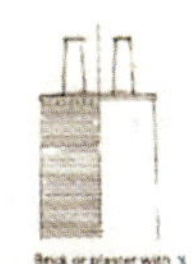

Brick or plaster with tile cap

PORCH ELEVATION

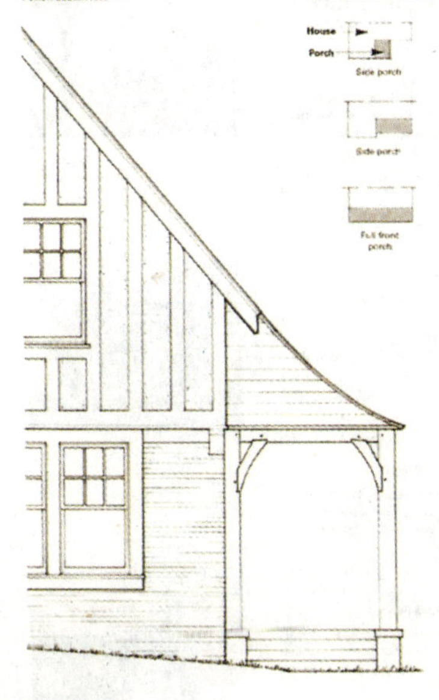

TYPICAL PORCH LOCATIONS

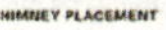

CHIMNEY PLACEMENT

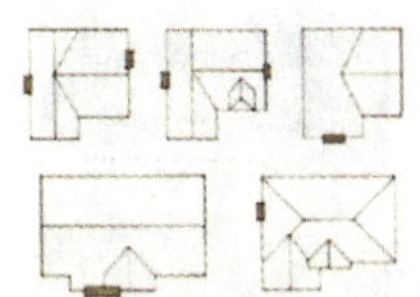

POST BRACES

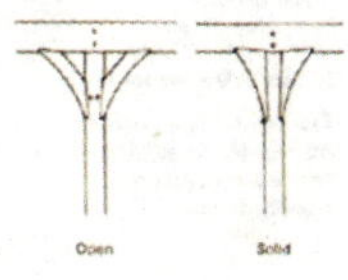

A Pattern Book for Norfolk Neighborhoods

Materials & Applications 材料和应用

Roofing

- Slate (including manufactured slate products), laminated asphalt or composition shingles with a slate pattern, or clay tile with flat or barrel profile

Soffits

- Smooth-finish composition board, tongue-and-groove wood boards, or fiber-cement panels

Gutters & Downspouts

- Half-round or ogee profile gutters with round or rectangular downspouts in copper, painted or prefinished metal

Windows

- Painted wood or solid cellular PVC, or clad wood or vinyl with brick veneer only; true divided light or simulated divided light (SDL) sash with traditional exterior muntin profile (7/8 inch wide)

Trim

- Wood, composite, cellular PVC or polyurethane millwork; stucco, stone or cast stone

Doors

- Wood, fiberglass or steel with traditional stile-and-rail proportions and panel profiles, painted or stained

Shutters

- Wood or composite, sized to match window sash and mounted with hardware to appear operable

Cladding

- Smooth-finish brick in Common bond pattern
- Stucco with handmade/formed appearance (no slop-trowel or similar); half-timbering for second story accents
- Random-width cut wood or fiber-cement shingles with mitered corners
- Smooth-finish wood or fiber-cement lap siding, 6 to 8 inches exposure, with mitered corners

Foundations, Chimneys & Piers

- Brick or stucco with handmade/formed appearance

Columns

- Wood posts and brackets

Railings

- Wood top and bottom rails with square balusters
- Wrought iron or solid bar stock square metal picket
- Brick or masonry with stucco finish

Porch Ceilings

- Plank-and-beam or flat plaster, tongue-and-groove wood or composite boards, or beaded-profile plywood

Front Yard Fences

- Wood picket, masonry with stucco, brick or stone finish, or combination

Lighting

- Porch pendant or wall-mounted lantern

C 46　ARCHITECTURAL PATTERNS

Gallery of Examples 楼座范例

Essential Elements of the Biltmore English Romantic

1. Steep roof pitches with dormers.
2. Apparently random window and door locations.
3. Vertical windows in groupings.
4. Thick walls with deep-set doors and windows.
5. Asymmetrical massing with large, simple roof planes.
6. Broad expanses of wall with few door and window penetrations.
7. Roof lines extending below windows at second floor and to top of window at first floor.
8. Simple detailing.

History and Character

THE BILTMORE ENGLISH ROMANTIC STYLE is based on the early-twentieth-century interpretations of English architecture by American architects and builders. The source for design comes from Medieval English cottages, manor houses, and rural village vernacular houses. The American interpretations are houses with simple volumes, often with front-facing gables that have steeply pitched roofs between 12 in 12 and 16 in 12. Dormers – gable, hip, and shed – are a dominant feature of the style. There is often a mix of exterior materials, including stone, plaster, and brick. Pebble-dash plaster is characteristic of the Asheville area, and was used on many of the Biltmore Estate's buildings. Half-timbering, pebble-dash, and horizontal siding are often used as infill in gables.

Chimneys typically act as principal forms for the massing of the house. These are usually very massive, often with plaster finish, simple detailing, and chimney pots. Decorative half-timbering in the gables is common and can occur on the entire second story or in the upper gables. Often there is a mix of timber patterns on the same house. Windows are typically casements, vertical in proportion and arranged in groups of two to five.

There are relatively few windows in the facade, the dominant form is one of a solid mass with small openings.

Biltmore English Romantic 比尔特摩英式浪漫主义风格

ARCHITECTURAL PATTERNS

C 26

Partial Elevation and Section

Eaves

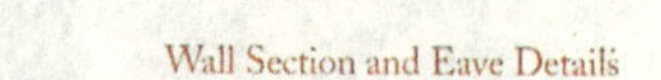

Wall Section and Eave Details

Vertical Section

The first floor of the main body is generally on grade – elevated just enough to keep rainwater out. The floor-to-ceiling height on the principal floor is typically 9 feet. The secondary floor-to-ceiling height is 8 to 9 feet.

Eave Details

Overhang/eaves are generally shallow (0 to 10 inches) although they are sometimes as deep as 18 inches where half-timbering is used. Eaves may be constructed of either building wall material (plaster, brick) or wood.

Biltmore English Romantic 比尔特摩英式浪漫主义风格

ARCHITECTURAL PATTERNS

C 28

达克尔山（比尔特摩）

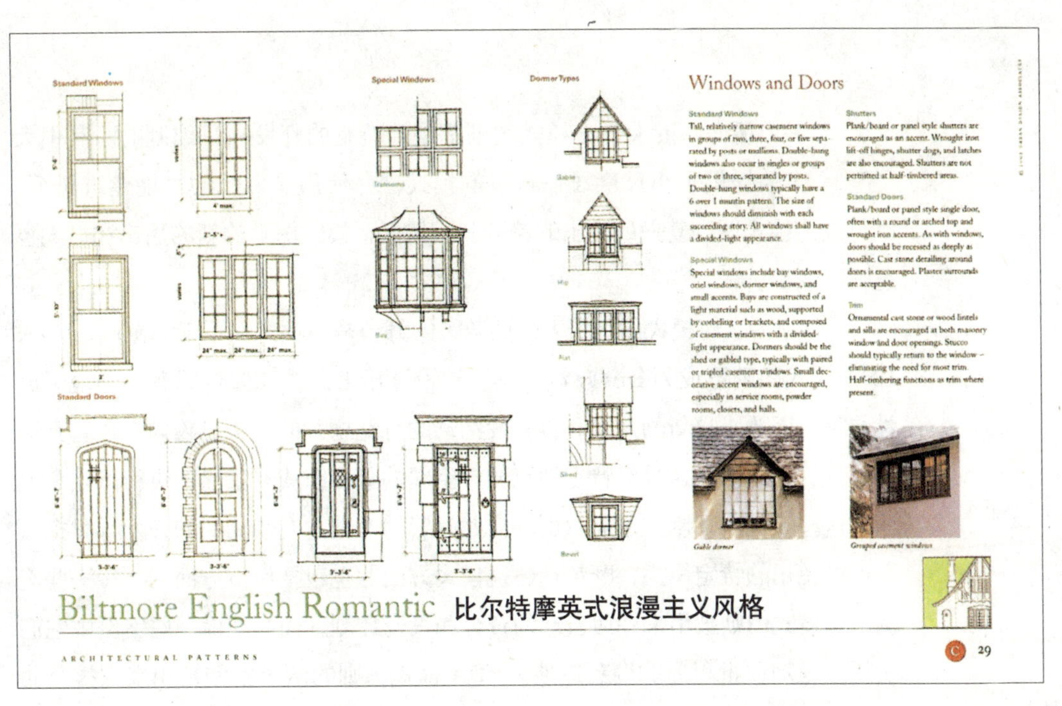

Standard Windows

Special Windows

Dormer Types

Standard Doors

Windows and Doors

Standard Windows

Tall, relatively narrow casement windows in groups of two, three, four, or five separated by posts or mullions. Double-hung windows also occur in singles or groups of two or three, separated by posts. Double-hung windows typically have a 6 over 1 muntin pattern. The size of windows should diminish with each succeeding story. All windows shall have a divided-light appearance.

Special Windows

Special windows include bay windows, oriel windows, dormer windows, and small accents. Bays are constructed of a light material such as wood, supported by corbeling or brackets, and composed of casement windows with a divided-light appearance. Dormers should be the shed or gabled type, typically with paired or tripled casement windows. Small decorative accent windows are encouraged, especially in service rooms, powder rooms, closets, and halls.

Shutters

Plank/board or panel style shutters are encouraged as an accent. Wrought iron lift-off hinges, shutter dogs, and latches are also encouraged. Shutters are not permitted at half-timbered areas.

Standard Doors

Plank/board or panel style single door, often with a round or arched top and wrought iron accents. As with windows, doors should be recessed as deeply as possible. Cast stone detailing around doors is encouraged. Plaster surrounds are acceptable.

Trim

Ornamental cast stone or wood lintels and sills are encouraged at both masonry window and door openings. Stucco should typically return to the window – eliminating the need for most trim. Half-timbering functions as trim where present.

Gable dormer

Grouped casement windows

Biltmore English Romantic 比尔特摩英式浪漫主义风格

ARCHITECTURAL PATTERNS

C 29

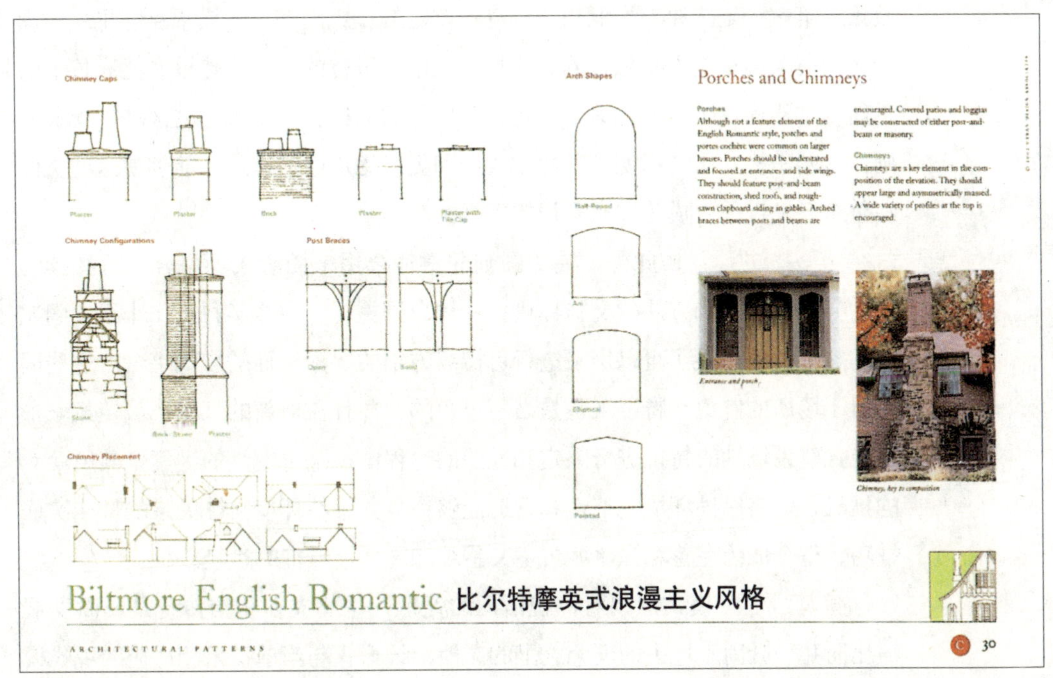

Chimney Caps

Arch Shapes

Chimney Configurations

Post Braces

Chimney Placement

Porches and Chimneys

Porches

Although not a feature element of the English Romantic style, porches and portes cochère were common on larger houses. Porches should be understated and focused at entrances and side wings. They should feature post-and-beam construction, shed roofs, and rough-sawn clapboard siding in gables. Arched braces between posts and beams are encouraged. Covered patios and loggias may be constructed of either post-and-beam or masonry.

Chimneys

Chimneys are a key element in the composition of the elevation. They should appear large and asymmetrically massed. A wide variety of profiles at the top is encouraged.

Entrance and porch

Chimney, key to composition

Biltmore English Romantic 比尔特摩英式浪漫主义风格

ARCHITECTURAL PATTERNS

C 30

后记 模式图则的复兴

本书中所介绍的大部分模式图则都是为特定的开发项目编制的。其中大多数提供了小尺度居住建筑的模式，并使用了传统的当地建筑语汇。这已成为模式图则复兴所迈出的第一步。然而，模式图则的新的用途和形式也已呼之欲出。

*公共政策。*将模式图则引入公共政策的舞台是一项意义重大的范例。近来，UDA 模式图则已成为公共政策工具，被设计用来保护和强化固有的当地区域性特征，因为来自新的开发和再开发活动的日益增加的压力已趋向于危害到这种特征的完整性。通过对特定地点城市空间的众多组成要素逐一进行限定和说明，这些模式图则制定了能够引导公共政策的准则。它们为新的开发项目提供了可供使用的特定模式，既包括建筑模式方面的也包括城市方面的。在一些案例中，模式图则被用来评估备选的资金和投资计划，而在其他一些案例中它们被用来推动审批程序的顺利进展。UDA 模式图则的两个案例被用来支撑公共政策。第一个是为弗吉尼亚州诺福克市制定的模式图则——诺福克邻里模式图则——它为诺福克很多独特的居住性邻里中所进行的新住宅建设和已有住宅的更新与加建提供了导则。第二个是为英格兰的西约克郡所制定的区域性模式图则。这两份模式图则指明了发挥模式图则更广泛用途的道路，能够使这些文件在整个开发过程中成为非常有用的工具。

*可持续的设计。*此外，模式图则非常适合用来描绘成功整合了可持续的设计标准、技术和方法的文脉环境。从理念到施工，由多学科工作小组共同完成的整个过程提供了可以用来进行可持续方案的要素设计的广阔平台。这些因素对于场所的意象和特征来说是必不可少的，并且在对新的邻里、城镇或城市填充式建设场地的特征进行限定和说明的过程中也是关键性的因素。其所面对的挑战，是将保护环境完整性和营造能够恢复并维持环境文脉关系的最佳实践技巧，整合成为完善和强化城市主义感观和区域性特征的途径。

*建筑语汇。*模式图则对传统语汇的处理，是建立在相信传统是不断发展演化而并非凝固于过去的理念之上的。使传统形式与当前技术相适应的现代模

式图则，是在建造和开发产业与这些传统之间重新建立联系的过程中所迈出的第一步。随着工业技术的发展，我们希望能够利用这一理解建筑形式的系统方法来发明和发展新的语汇。例如，为伊斯特加里森所制定的模式图则，包括了在对传统艺术家工作室所进行的现代诠释基础上所确定的一处艺术街区的模式。我们希望从现在开始的100年中，传统模式图则能够获得更为丰富的建筑语汇。

*建筑类型。*模式图则能够在新的建筑类型和市场以及建设所处的文脉环境之间建立极重要的联系，其中不仅包括商业性的混合用途建筑的模式，也包括商业街区特征和设计的模式。这些模式图则还包括了处理标志牌位置和寻路系统的册页。右图所介绍的是为巴克斯特城镇中心所制定的标志牌设计导则。

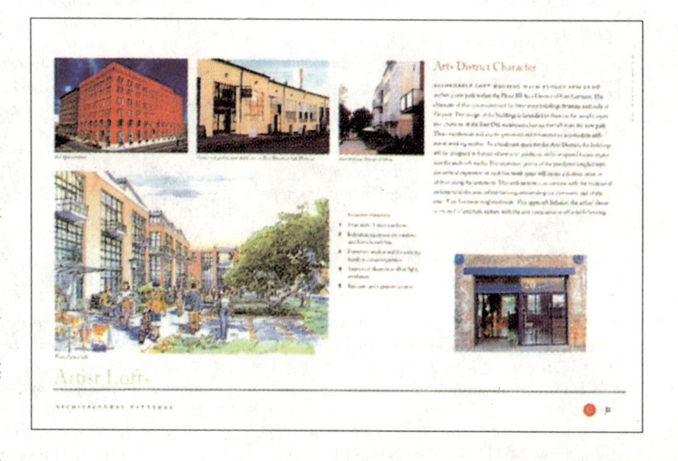

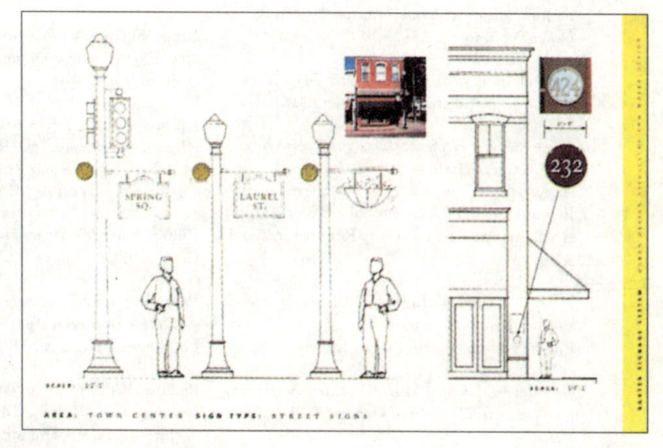

模式图则方法——它的模式和建筑语汇——现在也正在应用于市政建筑。在一些案例中，UDA已使用模式图则来设计开发项目中的一栋或两栋关键性的建筑。当其与不同建筑师所设计的建筑融为一体的时候取得了良好的效果，模式图则所进行的所有工作都是为了符合当地的文脉环境。

鼓舞人心的是，在很多案例中，我们也许正在设计和建造那些能够强化场所、景观和文化基本属性的城市场所的适当途径方面达成共识。

参考文献

Alberti, Leon Battista. *On the Art of Building in Ten Books.* 1442-1452. Reprint, translation by Joseph Rykwert, et al. Cambridge, Mass.: The MIT Press, 1988.

Benjamin, Asher. *The American Builder's Companion.* 6th ed. Reprint, introduction by William Morgan. New York: Dover Publications, 1969.

———. *The Architect, or Practical House Carpenter.* 1830. Reprint. New York: Dover Publications, 1988.

———. *The Country Builder's Assistant.* 1797. Reprint. Bedford, Mass.: Applewood Books, n.d.

Biddle, Owen. *The Young Carpenter's Assistant.* Philadelphia: B. Johnson, 1805.

Brunner, A. W. *Cottages.* New York: William T. Comstock, 1890 (5th edition).

Campbell, Colen. *Vitruvius Britannicus, (3 vols.).* London, 1715-1725. Reprint. New York: Benjamin Blom, 1967.

Comstock, William T., *Modern Architectural Design and Details.* New York. 1926. Reprinted as: *Victorian Domestic Architectural Plans and Details.* New York: Dover Publications, 1987.

Davis, Alexander Jackson. *Rural Residences.* 1837. Reprint, introduction by Jane B. Davies, New York: Da Capo Press, 1980.

Downing, Andrew Jackson. *Cottage Residences.* 4th ed. 1842. Reprint, unabridged replication. *Victorian Cottage Residences.* New York: Dover Publications, 1981.

———. *The Architecture of Country Houses.* New York: D. Appleton & Company, 1850. Reprint, unabridged and corrected. New York: Dover Publications, 1969.

Durand, Jean Nicolas Louis. *Précis Des Leçons D'Architecture.* Paris: Firmin Didot, 1823.

———. *Précis of the Lectures on Architecture with Graphic Portion of the Lectures on Architecture.* Paris, 1802-05 and 1821 respectively. Reprint, text and documents. Julia Bloomfield, et al, ed. Los Angeles: The Getty Research Institute, 2000.

Fréart de Chambray, Roland. *A Parallel of the Ancient Architecture with the Modern.* 1650. 1st English edition by John Evelyn. London, 1664.

Gibbs, James. *A Book of Architecture, Containing Designs of Buildings and Ornaments.* London, 1728.

———. *Rules for Drawing the Several Parts of Architecture.* 1732. Reprint, reduced facsimile. London: Hodder & Stoughton Limited, 1947.

Hafertepe, Kenneth and James F. O'Gorman. *American Architects and Their Books to 1848.* Amherst, Mass.: University of Massachusetts Press, 2001.

Halfpenny, William. *Practical Architecture.* 1736. Reprint, facsimile reduction. Privately printed for Richard Sidwell, n.d.

Halfpenny, William, J. Halfpenny, R. Morris and T. Lightoler. *The Modern Builder's Assistant: or, A Concise Epitome of the Whole System of Architecture.* 1742. Reprint. Westmead, England: Gregg International, 1971.

Kent, William. *The Designs of Inigo Jones, Consisting of Plans and Elevations for Public and Private Buildings.* London: 1727 (2 vols.)

Lafever, Minard. *The Modern Builder's Guide.* 1833. Reprint, introduction by Jacob Landy. New York: Dover Publications, 1969.

Langley, Batty. *The City and Country Builder's and Workman's Treasury of Designs.* 1740 original. Reprint, New York: Benjamin Blom, 1967.

Maynard, W. Barksdale. *Architecture in the United States, 1800-1850.* New Haven, Conn.: Yale University Press, 2002.

Mitrović, Branko. *Learning from Palladio.* New York: W. W. Norton & Company, 2004.

Morris, Robert. *Select Architecture: Being Regular Designs of Plans and Elevations Well suited to Both Town and Country.* 1755. Reprint. New York: Da Capo Press, 1973.

Pain, William. *The Practical House Carpenter.* 1789. First American edition. Boston: William Norman, 1796.

Palladio, Andrea. *The Four Books on Architecture.* Venice, 1570. Reprint, translation by Robert Tavernor and Richard Schofield. Cambridge, Mass.:The MIT Press, 1997

Palliser. *Palliser's Model Homes.* Bridgeport, Conn.: Palliser, Palliser & Co., 1878.

Park, Helen. *A List of Architectural Books Available in America Before the Revolution.* Los Angeles: Henessy & Ingalls, 1961.

Radford, William A. *Radford's Details of Building Construction.* Chicago: The Radford Architectural Company, 1911.

———. *Radford's Portfolio of Plans.* Chicago: The Radford Architectural Company, 1909.

Reiff, Daniel D. *Houses From Books.* University Park, Penn.: The Pennsylvania University Press, 2000.

Salmon, William. *Palladio Londinensis or the London Art of Building.* 1734. Reprint, Westmead, Eng.: Gregg International, 1969.

Schimmelman, Janice G. *Architectural Books in Early America.* New Castle, Del.: Oak Knoll Press, 1999.

Sears, Roebuck and Co., *Honor Bilt Modern Homes.* Chicago. 1926. Reprinted as: *Sears, Roebuck Catalog of Houses 1926.* New York: Dover Publications, 1991.

Serlio, Sebastiano. *Sebastiano Serlio on Architecture, Vol. 1. 1584.* Reprint, translation by Vaughan Hart and Peter Hicks. New Haven, Conn.: Yale University Press, 1996.

———. *The Five Books of Architecture.* 1584. Dutch ed. 1606, First English ed. 1611. Reprint. New York: Dover Publications, 1982.

Shoppell, R. W. et al. *Turn-of-the-Century Houses, Cottages and Villas.* 1890-1900. Reprint, selected sections, New York: Dover Publications, 1984.

Sloan, Samuel. *The Model Architect* (2 vols.). Philadelphia: E. S. Jones & Co. 1852. Reprinted as: *Sloan's Victorian Buildings.* New York: Dover Publications, 1980.

Swan, Abraham. *The British Architect.* 1745. Reprint, introduction by Adolf Placzek. New York: Da Capo Press, 1967.

Vaux, Calvert. *Villas and Cottages.* 1857. Reprint facsimile. New York: Da Capo Press, 1968.

Vignola, Giacomo Barozzi da. *Canon of the Five Orders of Architecture.* 1562. Reprint, translation by Branko Mitrović. New York: Acanthus Press, 1999.

Vitruvius (Marcus Vitruvius Pollio). *Ten Books on Architecture.* trans Morris Hickey Morgan. (Cambridge: Harvard University Press, 1914). Reprint, New York: Dover Publications, 1960.

Ware, William R. *The American Vignola.* 1903. Reprint of 1977, W.W. Norton & Company. New York: Dover Publications, 1994.

Weyerhaeuser Forest Products, *A Dozen Modern Small Houses* (portfolio), 1926.

White Pine Series of Architectural Monographs. 1914-1940. Reprint. *Architectural Treasures of Early America.* Robert G. Miner, ed. New York: Arno Press Inc., 1977.

索 引

斜体印刷的页码参见范例图解。

A

Adam, James　詹姆斯・亚当，27
Adam, Robert　罗伯特・亚当，27，46
addresses　场所，52，113
Alabama, Huntsville　亚拉巴马州，亨茨维尔，149，201
Alberti, Leon Battista　莱昂・巴蒂斯塔・阿尔伯蒂，16–17
American Builder's Companion　《美国建造者手册》，28，192，199
American Vignola: A Guide to the Making of Classical Architecture　《美国的维尼奥拉：古典建筑建造指南》，25
architectural patterns　建筑模式，49，53–56，147–148
　Arts & Crafts examples　工艺美术风格范例，180–181，*181–191*
　Classical style examples　古典主义风格范例，192，*193–207*
　Colonial Revival examples　殖民复兴风格范例，150，*151–165*
　European Romantic examples　欧洲浪漫主义风格范例，208，*209–223*，211
　historical consideration　历史考虑因素，54
　pattern book descriptions　模式图则描述，53–56
　pattern book development　模式图则制定，76，85
　selection　选择，53
　urban assembly kit approach　城市集成工具包方法，49
　use of local design tradition　地方设计传统的应用，67–68，225
　Victorian examples　维多利亚风格范例，166，167–179，也可参见“住宅设计”
Architecture of Country Homes, The　《乡村住宅建筑》，166
Arts & Crafts style　工艺美术风格，148
　examples　范例，180–181，*181–191*

B

Barnett, Jonathan　乔纳森・巴奈特，41
Baron, Richard　理查德・巴伦，43
Benjamin, Asher　阿舍・本杰明，28，192，299
Bicknell, Amos J.　阿莫斯・J・比克内尔，31
Biddle, Owen　欧文・比德尔，28
Book of Architecture　《建筑一书》，26
Bramante, Donato　多纳托・伯拉孟特，17
British architect, The: Or, The Builder's Treasury　《英国建筑师：又名，建造者宝典》，27
Builder's Assistant　《建造者辅助参考》，28
Bulfinch, Charles　查尔斯・布尔芬奇，27

C

California　加利福尼亚州
　Fort Ord　福特雷德，82，166，*176–179*，181，*188–191*
　Lake Elsinore　艾尔希诺湖，81，*106–108*，150，*156–159*，166
Campbell, Colen　科伦・坎贝尔，22
Carpenter Gothic style　哥特木构风格，166
catalogs　目录图册，14，25，31
Chicago House Wrecking Company　芝加哥住宅拆除公司，31
City and Country Builder's, The　《城乡建造者》，192
civic buildings　市政建筑，225
Classical patterns　古典主义模式，148
　examples　范例，192，*193–207*
College of William an Mary　威廉和玛丽学院，26
Colonial Revival patterns　殖民复兴模式，148，150，*151–165*
color, pattern book description　色彩，模式图则描述，56
commercial buildings　商业建筑，225
community patterns　社区模式，52，67，69，77–78，113
　samples　范例，114，115，*116–143*
Comstock, William　威廉・康斯托克，29
construction manuals　建造指南，14
construction techniques　建造技术
　evolution in early America　美国早期的发展，25–26
　standardization in 20th century　20世纪的标准化，35
　Vitruvius's recommendations　维特鲁威的建议，15
consultants, in pattern book development　顾问，模式图则的制定中，78
Co-operative Building Association　合作建筑协会，31
Cooper Robertson　库珀・罗伯逊，70
Cottage Residences　《农舍住宅》，28，166
Country Builder's Assistant, The　《乡村建造者辅助参考》，28
Craftsman style　工匠风格，181，183，189
Cummings, M.F　卡明斯，M.F，31

D

Davis, Alexander Jackson　亚历山大・杰克逊・戴维斯，28，166

design guidelines 设计导则，48
Designs of Inigo Jones 《伊尼戈·琼斯的设计》，27
development process 开发过程，22
architects and design professionals in 建筑师和设计专业人员，36
benefits of pattern book method 模式图则方法的益处，64–70
current practice 当前实践，59–64
environmental considerations 环境考虑因素，63，64，69
failures of 20th century urban design 20 世纪城市设计的失败，35–36
flexibility of master plan 总体设计的灵活性，60–61，66–67
implementation 实施，63–64，69–70
information gathering 信息收集，64–65
market considerations 市场考虑因素，59–60，65，70
master plan 总体设计，59，60–61，64–65，78
pressures for standardization 标准化的压力，61–62
role of pattern books in 模式图则的作用，25，45，47–48
use of local design tradition in 地方设计传统的应用，67–68
zoning and planning approvals 区划和规划的批准，62，68–69
diversity 多样性
current development practice and 当前开发实践，60
mixed–income neighborhoods 混合收入的社区，43–44
diversity *continued* 多样性
mixed–use communities 混合用途的社区，60
pressures for standardization 标准化的压力，61–62
traditional community characteristics 传统社区特征，60
Downing, Andrew Jackson 安德鲁·杰克逊·唐宁，28，166
Drayton Hall 德雷顿市政厅，26
Durand, Jean Nicolas Louis 让·尼古拉斯·路易·迪朗，23–24

E

eaves and soffits 屋檐和拱腹结构，55
European Romantic patterns 欧洲浪漫主义模式，148，211
examples 范例，208，209–223

F

facades and entranceways 立面和入口
evolution of UDA pattern book UDA 模式图则的发展，39–40
Georgian England townhouses 乔治时期英格兰城市住宅，22
pattern book descriptions 模式图则描述，55
relationship to street in Alberti's design 在阿尔伯蒂设计中与街道的关系，17
Vitruvius's design 维特鲁威的设计，15
Federal style 联邦风格，199
Florida 佛罗里达州
Celebration 塞里布瑞恩，44
Seaside 锡赛德，37
WaterColor 沃特卡勒，44，70，81，*92–94*，112，124–127
Four Books on Architecture 《建筑四书》，20–21

G

garages 车库，36
Gibbs, James 詹姆斯·吉布斯，26，27
graphic design 美术设计
pattern book cover 模式图则封面，57
pattern book page layout 模式图则册页版式，79
Greek Revival style 希腊复兴风格，28，192

H

Halfpenny, John 约翰·哈夫彭尼，27
Halfpenny, William 威廉·哈夫彭尼，22，23，27，192
Hammer, Philip 菲利普·海默，41
harmonic relationships 协调关系，16–17
Haviland, John 约翰·哈维兰，28
house design 住宅设计
colors 色彩，56
early American 早期美国，25–26，27，29
eaves and soffits 屋檐和拱腹结构，55
facade composition 立面构图，55
Georgian England townhouses 乔治时期英格兰城市住宅，22–23
local design tradition 地方设计传统，67–68
massing 体块，55，148
materials of construction 建造材料，56
pattern book descriptions 模式图则描述，53–56，148
pattern book development 模式图则制定，73–76，85
porches 门廊，56
pressures for standardization 标准化的压力，61–62
twentieth century American 20 世纪的美国，32，36
Vitruvius's recommendations 维特鲁威的建议，15
windows and doors 窗和门，55
see also architectural patterns 也可参

见“建筑模式”

I

individual expression 个性的表达，23，33
Italianate style 意大利风格，66
Italian Renaissance style 意大利文艺复兴风格，208

J

Jefferson, Thomas 托马斯·杰斐逊，27
Jones, Inigo 伊尼戈·琼斯，20–21，27

K

Kent, William 威廉·肯特，26–27
Kentucky, Louisville 肯塔基州，路易斯维尔，44，49，70，81，*98–100*，*132–137*

L

Lafever, Minard 米纳尔·拉费沃尔，28，29
landscape design 景观设计，53，63，64
Langley, Batty 巴蒂·兰利，27，192
Latrobe, Benjamin Henry 本杰明·亨利·拉特罗布，27，199
legal controls on development 对开发的合法控制，48
lot description 地块描述，53

M

marketing 营销
in development process 在开发过程中，59–60
use of pattern books in 模式图则的使用，29–31，65，70
Maryland, Kentlands 马里兰州，肯特兰，37
mass production 批量生产，23
materials 材料
architectural pattern descriptions in pattern book 模式图则中的建筑模式描述，56
evolution of early America building 早期美国建筑的发展，25–26
Vitruvius's recommendations 维特鲁威的建议，15
McComb, John 约翰·麦考康伯，27
Michigan, Monroe 密歇根州，门罗，44，82，*138–141*
Mills, Robert 罗伯特·米尔斯，199
mixed–income neighborhoods 混合收入的邻里，43–44
Model Architect, The 《典范建筑师》，29
model charrette 模型研讨会，76，77–78
Modern Builder's assistant 《现代建造者辅助参考》，27
Modern Builder's Guide, The 《现代建造者指南》，28
Modern Movement 现代主义运动，36
Morris, Robert 罗伯特·莫里斯，27
Morris, William 威廉·莫里斯，180

N

New Urbanism 新城市主义，37
North Carolina 北卡罗来纳州
Asheville 阿什维尔，81，*102–106*，181，*182–185*，192，*196–200*，208，*222–223*
Belmont 贝尔蒙特，81，*101–102*，208，*210–215*

O

On the Art of Building in Ten Books 《建筑十书里的建筑艺术》，16–17
orders 柱式
Durand's review 迪朗的评论，24
Palladio's 帕拉第奥的评论，20
Serlio's 塞利奥的评论，18
Vitruvius's 维特鲁威的评论，14–15
Overview section 总则部分，45，50–51，83
sample pages 范例册页，84–111

P

Pain, William 威廉·佩恩，27
Palazzo Rucellai 鲁切拉伊宫，17
Palladio, Andrea 安德烈亚·帕拉第奥，14，20–21，26，27
Palladio Londiensis 《伦敦的帕拉第奥风格》，27
Palliser, Palliser & Co. 帕利泽，帕利泽及其公司，31
pattern books 模式图则
Architectural Patterns section 建筑模式部分，49，53–56，147–148，*149–223*
benefits of 益处，64–70
Community Patterns section 社区模式部分，52，67，69，77–78，113，*114*，115，*116–143*
cover design 封面设计，57
defining patterns in development of 限定开发模式，77–78
documenting neighborhood characteristics for 记录邻里特征，75–76
early American 早期美国，25–31
examples 范例，81
final layout 最终排版，79
flexibility 灵活性，66
format 版式，20，25，31，50
Georgian England 乔治时期的英格兰，23
historical origins 历史起源，14–21
historical use 历史用途，13
illustrations 图解说明，79
legal status 合法地位，48
marketing applications 市场应用，29–31，65，70

model charrette in development of 开发中的模型研讨会，77–78
in Napoleonic France 在法国拿破仑时期，23–24
new town development 新城镇开发，44
Overview section 总则部分，45，50–51，83–111
phases in development of 开发的阶段，73
in promoting design diversity 促进设计的多样性，33
as public policy tool 作为公共政策工具，224
recent trends 最近趋势，46
revival in 20th century America 在20世纪美国的复兴，37
site conditions and 场地条件，69
UDA approach UDA 方法，37，38–40，42，44，45，46
use of local design tradition in 对地方设计传统的应用，67–68，73–76，83，147–148，225
zoning and planning approval process 区划和规划审批程序，68–69
see also urban assembly kits 也可参见“城市集成工具包”
Pennsylvania 宾夕法尼亚州
Pittsburgh 匹兹堡，38，43–44
York 约克，38
Piazzo Pio 皮奥广场，17
Pienza 皮恩扎，17
plan books 平面图则，14，25，31，32
political environment 政治环境，62
porches 门廊，36
pattern book descriptions 模式图则描述，56
Practical Architecture 《实用建筑》，22，192
Practical House Carpenter 《实用住宅木工技术》，27
precedent books 范例图则，14
Précis of the Lectures on Art 《艺术讲义概要》，24
preservation movement 保护运动，37，38
public spaces 公共空间
importance of 重要性，43，67
pattern book descriptions 模式图则描述，52

Q

Queen Anne style 安妮女王风格，166

R

Radford, William 威廉·拉德福德，31
regional design 区域设计，45
Classical style variations 古典风格变体，192
pattern book development 模式图则制定，73–76
relationship of parts 部件关系
Alberti's design 阿尔伯蒂的设计，16，17
Community Patterns description 社区模式描述，113
Durand's approach 迪朗的方法，24
effects of early American pattern books 早期美国模式图则的效果，33
evolution of UDA pattern book UDA模式图则的发展，38，39–40
Georgian England design 乔治时期英格兰设计，23
harmonic systems 和谐体系，16–17
role of pattern book in urban design 模式图则在城市设计中的作用，47
Serlio's design 塞利奥的设计，18
Vitruvius's design 维特鲁威的设计，16
Renaissance design 文艺复兴风格的设计，16–17
Rice, David 戴维·赖斯，42
Richardsonian Romanesque 理查森罗马风格，166
Roman building practices 罗马建筑实践，14
Rossellino 罗塞利诺，17
Rules for Drawing the Several Parts of Architecture 《若干建筑部件的制图规则》，26
Rural Residences, Etc. 《乡村住宅及其他》，28，167

S

Salmon, William 威廉·萨尔蒙，27
Sanborn Maps 卫星图片，74
scale models 比例模型，77
Scully, Vincent 文森特·斯卡利，180
Sears, Roebuck and Company 西尔斯、罗巴克及其公司，31
Second Empire style 第二帝国风格，166
Select Architecture: Being Regular Designs of Plans and Elevations Well Suited to Both Towns and Country 《建筑精选：城乡皆宜的平、立面的设计规则》，27
Serlio, Sebastiano 塞巴斯蒂亚诺·塞利奥，18，26
Shaw, Richard Norman 理查德·诺曼·肖，180
Shingle style 木瓦风格，180–181
Shoppell, R.W. R·W·肖佩尔，31
signage 标志牌，225
site selection and evaluation for cities 场地选择和城市发展演化，15
development process 开发过程，63，64
pattern book approach 模式图则方法，69
Sloan, Samuel 塞缪尔·斯隆，29

Smith, Richard Sharpe　理查德·夏普·史密斯，183
social interaction　社会交往，36
South Carolina, Fort Mill　南卡罗来纳州，福特米尔，44，70，81，*95–97*，114，*115*，*116–123*，192，*194–196*
street layout and design　街道布局和设计
Alberti's design　阿尔伯蒂的设计，17
development planning　发展计划，63，64
failures of 20th century design　20 世纪设计的失败，36
Georgian England　乔治时期的英格兰，22，23
relationship to facades　与建筑立面的关系，17
Vitruvius's recommendations　维特鲁威的建议，15
sustainable design　可持续的设计，224
Swan, Abraham　亚伯拉罕·斯沃恩，27

T

Ten Books on Architecture　《建筑十书》，14–16
theory of design　设计理论，14
townhouse design　城市住宅设计，22
Treasury of Designs　《设计宝典》，27
treatises　论文丛集，14

U

United States 美国
early pattern books　早期模式图则，25–31
expression of individuality in design　在设计中个性的表达，33
twentieth century design　20 世纪设计，32，35–36
urban assembly kits 城市集成工具包
conceptual basis　概念基础，49
development of UDA pattern book　UDA 模式图则的发展，38–40
flexibility of　灵活性，66–67
historical origins　历史起源，24
role of　角色，49–50
see also pattern books　也可参见“模式图则”
urban design　城市设计
current practice　当前实践，59–64
failures of 20th century　20 世纪的失败，35–36
in Georgian England　在乔治时期的英格兰，22–23
new towns　新城，44–45
New Urbanist movement　新城市主义运动，37
public policy development　公共政策发展，224
public spaces　公共空间，43，52
role of pattern books　模式图则的作用，47–48
scales of design　设计的尺度，47
Vitruvius's approach　维特鲁威的方法，15
Urban Design Associates　城市设计事务所，37
Baxter (Fort Mill, SC) project　巴克斯特（福特米尔，南卡罗来纳州）项目，44，70，81，*95–97*，*114*，115，*116–123*，192，*194–196*
Crawford Square (Pittsburgh, PA) project　克劳福德广场（匹兹堡，宾夕法尼亚州）项目，43–44
Diggs Town (Norfolk, VA) project　迪格斯镇(诺福克，弗吉尼亚州）项目，42–43
Ducker Mountain (Asheville, NC) project　达克尔山（阿什维尔，北卡罗来纳州市）项目，18，*102–106*，181，*182–185*，192，*196–200*，199，208，*222–223*
Eagle Park (Belmont, NC) project　伊格尔帕克（贝尔蒙特，北卡罗来纳州）项目，81，*101–102*，208，*210–215*，211，215
East Beach (Norfolk NA) project　伊斯特比奇（诺福克，弗吉尼亚州）项目，75–76，81，*85–91*，*128–131*，150，*152–155*，153，166，*173–175*，181，*186–187*
East Garrison (Fort Ord, CA) project　伊斯特加里森（福特雷德，加利福尼亚州）项目，82，166，*176–179*，177，181，*188–191*
Génitoy East (Bussy St. Georges, France) Project，热尼特伊斯特（比西圣乔治，法国）项目，82，*142–143*
Ledges (Huntsville, AL) project　莱奇斯（亨茨维尔，亚拉巴马州）项目，70，*149*，201
Liberty (Lake Elsinore, CA) project　利伯蒂（艾尔希诺湖，加利福尼亚州）项目，81，*106–108*，150，*156–159*，166，*168–172*
Mason Run (Monroe, MI) project　梅森兰（门罗，密歇根州）项目，82，*138–141*
Middletown Arch (Norfolk, VA) project　米德尔敦阿尔赫（诺福克，弗吉尼亚州）项目，40
Norfolk, VA pattern book　诺福克，弗吉尼亚州，模式图则，82，*109–111*，150，*160–165*，192，*202–207*，208，*216–221*
Ocean View (Norfolk, VA) project　欧申维尤（诺福克，弗吉尼亚州）项目，41
Park DuValle (Louisville, KY) project　帕克杜瓦拉（路易斯维尔，肯塔基州）项目，70，81，

98–100, *132–137*
pattern book development 模式图则制定, 37, 38–40, 42, 44, 45, 46
Randolph (Richmond, VA) project 伦道夫（里士满，弗吉尼亚州）项目, 39–40
regional design 区域设计, 45
WaterColor (Florida) project 沃特卡勒（佛罗里达州）项目, 44, 70, 81, *92–94*, *112*, *124–127*

V

Vaux, Calvert 沃克斯，卡尔弗特, 29, 31
Victorian architectural Patterns 维多利亚风格建筑模式, 166, *167–179*
Victorian patterns 维多利亚风格模式, 148
Villas and Cottages 《乡村别墅和村舍》, 29
Virginia 弗吉尼亚州
Norfolk 诺福克, 40–43, 45, 75–76, 81, 83, *85–91*, *109–111*, *128–131*, 150, *152–155*, *160–165*, 166, *173–175*, 181, *186–187*, 192, *202–207*, 208, *216–221*
Richmond 里士满, 39
Vitruvius 维特鲁威, 14–16
Vitruvius Britannicus 《英国的维特鲁威》, 22
Voysey, C.F.A. C·F·A·沃伊齐, 180

W

Ware, William R. 威廉·R·威尔, 25
Weyerhaeuser Company 惠好公司, 32
White Pine plan books 白松平面图则, 32
Williamsburg style 威廉斯堡式风格, 40
Wills, Royal Barry 罗伊尔·巴里·威尔斯, 150
windows and doors 窗和门
Colonial Revival 殖民复兴风格, 150
Georgian England pattern books 乔治时期英格兰模式图则, 23
pattern book descriptions 模式图则描述, 55
Workman's Treasury of Designs 《工匠设计宝典》, 192

Y

Young Carpenter's Assistant, The 《青年木工辅助参考》, 28

Z

zoning 区划, 36, 48
advantages of pattern book approach 模式图则方法的优势, 68–69
development process 开发过程, 62